5000090638

SCIENCE AND PARTIAL TRUTH

OXFORD STUDIES IN THE PHILOSOPHY OF SCIENCE

The Book of Evidence
Peter Achinstein

Science, Truth, and Democracy
Philip Kitcher

The Devil in the Details: Asymptotic Reasoning in Explanation, Reduction, and Emergence
Robert W. Batterman

Science and Partial Truth: A Unitary Approach to Models and Scientific Reasoning
Newton C. A. da Costa and Steven French

SCIENCE AND PARTIAL TRUTH

A Unitary Approach to Models and Scientific Reasoning

NEWTON C. A. DA COSTA
STEVEN FRENCH

2003

OXFORD
UNIVERSITY PRESS

Oxford New York
Auckland Bangkok Buenos Aires Cape Town Chennai
Dar es Salaam Delhi Hong Kong Istanbul Karachi Kolkata
Kuala Lumpur Madrid Melbourne Mexico City Mumbai Nairobi
São Paulo Shanghai Taipei Tokyo Toronto

Published by Oxford University Press, Inc.
198 Madison Avenue, New York, New York 10016

www.oup.com

Library of Congress Cataloging-in-Publication Data
Costa, Newton C. A. da
Science and partial truth : a unitary approach to models and scientific reasoning /
by Newton C. A. da Costa and Steven French.
p. cm. — (Oxford studies in philosophy of science)
Includes bibliographical references and index.
ISBN 0-19-515651-X
1. Reasoning. 2. Science—Philosophy. I. French, Steven.
II. Title III. Series.
Q175.32.R45C67 2003
501—dc21 2002070449

9 8 7 6 5 4 3 2 1

Printed in the United States of America
on acid-free paper

To Neusa and Dena, with profound gratitude

Acknowledgments

In developing the ideas presented here, we bear an enormous debt to certain colleagues, students, and friends. In particular, we would like to thank Otávio Bueno, Décio Krause, and James Ladyman for their stimulating comments, criticisms, and overall support. Although they are not, of course, to be held responsible for any of the views expressed in this book, it would have been very much diminished if not for them. The bulk of this work was completed while Steven French was on sabbatical, funded by the University of Leeds and the School of Philosophy. He would like to express his deep gratitude to the university for supporting this work but most especially to all his colleagues in the School and the Division of History and Philosophy of Science.

Our account is based on a number of papers published over the past ten years or so. The technical counterpart to chapter 1 can be found in Mikenberg, da Costa, and Chuaqui (1986), and chapter 2 is based in part upon da Costa and French (1990a and 2000). Much of chapter 3 is new, but the central idea of "partial isomorphism" is from French and Ladyman (1999). The basis of chapter 4 can be found in da Costa and French (1993a), whereas aspects of chapter 5 can be found scattered through a variety of publications, including da Costa and French (1991a) and French (1990). Elements of chapter 6 are mostly based on French (1997), and chapter 7 draws on earlier work presented in da Costa (1987b) and da Costa and French (1988, 1989a, and 1991b). Chapter 8 is almost entirely new, but all the chapters in this book represent substantial additions to and extensions of our earlier published works. We would like to thank *The American Philosophical Quarterly*, *Boletim da Sociedade Paranaense de Matemática*, Elsevier, Erkenntnis, International Studies in the Philosophy of Science, Kluwer Academic Publishers, Oxford University Press, and Rodopi Publishers for permission to reproduce elements of our published papers here.

Contents

SCIENCE AND PARTIAL TRUTH

Introduction

Aims and Overview

One of the enduring consequences of the naturalist turn in the philosophy of science has been a reappraisal of the nature and importance of scientific *practice*. A less palatable consequence, at least for many, has been the conclusion that no unitary account of reasoning can accommodate the vagaries and complexities of such practice. And so in both the philosophy and sociology of science, we find attempts to construct accounts of practice which can accommodate the supposedly "local" character of knowledge production and the apparently fragmented nature of representations and experimental techniques. This tendency can be seen in the recent focus on model building in science, where it is claimed that a significant portion of this particular practice is concerned with low-level, "phenomenological" models, rather than the high-level theoretical forms typically examined by philosophers of science. Reprising similar claims made in the 1960s, it is then argued that the nature and role of these low-level models is such that they cannot be captured by any kind of unitary framework, either because they are supposedly "autonomous" from theory or because they are mutually inconsistent.

What these claims demonstrate is the continued unfolding of the post-positivist gulf between our actual knowledge-gathering activities and the philosophical characterization of these activities. The suggestion that in the face of this breach it is our activities, our practice, which must be brought back in line rings increasingly hollow as the counterexamples proliferate. It is our contention, however, that a conceptually unitary and broadly rationalist account *can* be constructed which pays close attention to and incorporates two fundamental aspects of epistemic practice in general. The first, which will be the primary focus of this work, concerns the nature of the representations employed in scientific reasoning and the epistemic attitudes adopted toward them. These representations are, crucially, conceptually incomplete and open-ended, and the overall attitude adopted is a fallibilistic one. Doxastically, the representations employed in scientific practice are regarded, not as true in the correspondence sense, but as partially true or approximately true, or as containing an element of truth. However, attempts to capture such notions within an analytic framework have typically foundered on the lack of their formal expression in a manner similar to that which Tarski provided for correspondence truth.

We believe that such a barrier has now been overcome with the development of a formal notion of "pragmatic," "partial," or "quasi" truth. Having such a notion to hand then allows one to claim that the high-level theories of physics, for example, while clearly not true in the correspondence sense, are nevertheless true in this partial respect and in a way that removes the ill-motivated and ultimately implausible doxastic separation between such theories and the phenomenological models. Even more important, perhaps, these formal developments offer a new perspective on the nature of the representations themselves. Tarski's innovation not only formally secured truth-as-correspondence, at least to some, but it also significantly furthered, if not initiated, the model-theoretic account of these representations. This account has had an enormous influence in the philosophy of science, where it offers the only unitary perspective in these philosophically fractured post-Kuhnian times. By introducing the notion of "partial structures" into the model-theoretic or "semantic" approach, the formalism of "quasi" truth offers a way of accommodating the conceptual incompleteness inherent in scientific representations and thus ties together both "theories" and "models" in all their various manifestations.

The second aspect of practice concerns methodology rather than representation and truth. Although methodological issues per se are not our primary concern, we claim that some light can be shed on them from the perspective developed here. In particular, if models and theories in their various guises are regarded as partially true only, then the classic discovery-justification distinction crumbles through doxastic parity. Our unitary epistemology then drives a unitary account of theory development in which a variety of heuristic procedures are employed and guidelines followed, the results of which can be represented in terms of structural interconnections. The impact of epistemic factors—such as good old-fashioned *evidence*—can also be captured, by suitably modifying Bayesian inductive logic in order to accommodate quasi truth.

In chapter 1 we shall review the formalism of "pragmatic" or "quasi" truth, which was originally presented by Mikenberg, da Costa, and Chuaqui (1986), highlighting those aspects that we shall draw upon in our further discussion. Just as Tarski's formalization attempted to capture what he called the "intentions" of the correspondence view of truth, so the formalism outlined here attempts to represent the "intentions" of the pragmatists, notably Peirce and James. Of these, perhaps the most significant is a concern with those representations of the world that are not perfect copies but are, in certain respects, incomplete and partial. The nature of the agreement between such representations and the world is then spelled out in pragmatic terms, which emphasize the set of empirical consequences of a particular idea or "concept." The fundamental formal device behind this formalism is that of a "partial structure," and it is upon this that our epistemic construction rests.

Tarski's innovation performed a double duty in both mathematics and philosophy. In formalizing an intuitive notion of truth-as-correspondence in terms of "sentence s is true in a structure $\mathcal{A}$" and suggesting that other semantic notions could be defined similarly, he brought model theory into mathematics and was therefore largely responsible for the explosion of developments that resulted. Motivated by Tarski's work, Beth, Suppes, Suppe, van Fraassen, and others argued that the semantic counterpart of scientific theories is of fundamental importance, some of them even suggesting that scientific theories should be viewed not as axiomatized sets of sentences in some suitable formal language but as classes of *models*. This, of course, forms the basis of the "semantic" or "model-theoretic" approach in the philosophy of science, presented in chapter 2. Likewise, the introduction of partial structures performs a similar double duty within our overall approach. It allows for the definition of a notion of "pragmatic" or "partial" truth, which casts a wide variety of issues in a new light. And it provides a more sophisticated concept of "model" or, more generally, "structure," which accommodates the essential incompleteness and partial nature of scientific representations. In this way it further advances the model-theoretic approach by allowing it to accommodate some of the more complex aspects of scientific practice indicated here.

Thus in chapter 3, we explore the consequences of introducing the idea of a "partial" structure into this approach to theories. A fundamental issue within the model-theoretic approach concerns the relationship between theories, viewed as families of mathematical models, and the kinds of models which scientists regularly deploy in their practice. Representing the former in terms of partial structures allows for the accommodation of the latter, covering the whole gamut of kinds from "iconic" to "impoverished," together with Hesse's famous account of positive and negative analogies. At the heart of the model-theoretic approach lies the idea that theories should be regarded as "extralinguistic entities," so that a change in logical, metamathematical axiomatization should not be counted as a change in "theory" per se. This focus on the models or, more generally, the structures in which sentences are satisfied encourages a shift to a broadly *representational* and nonpropositional account of the objects of belief. The consequences of such a move have yet to be fully explored in this context (indeed, it has barely come to be acknowledged), and in chapter 4 we set out a possible doxastic framework for understanding this shift in epistemic attention.

In particular, we draw on Sperber's account of the difference between "factual" and "representational" beliefs, where the object of the latter is a "semipropositional" representation of incomplete conceptual content. Our suggestion is that the fundamental idea behind this notion can be conveniently grasped in terms of partial structures: a "representational" belief that p is then understood as a belief that p is partially true only, whereas a factual belief that p is a belief that p is true in the correspondence sense.

With regard to the latter, there is an awareness of that to which *p* corresponds only, but in the case of the former there is an awareness of a commitment to a particular representation. Since it is commitment in this sense that is typically invoked in accounts of "acceptance," this framework suggests a unitary account of belief and acceptance that cuts across recent discussions of these concepts. Although often employed within the philosophy of science, these notions have not been given the attention their epistemic importance warrants. Here we shall consider two important consequences of our account: the treatment of inconsistencies in science and the blurring of the discovery-justification distinction, in chapters 5 and 6, respectively.

One response to the existence of inconsistent theories in science is to abandon the framework of classical logic while maintaining truth-as-correspondence and admit the possibility of true contradictions. However, this fails to fit the very scientific practice that it draws upon, and in particular, it fails to adequately accommodate the partial representational nature of scientific theories. On our view, inconsistent theories are not be dismissed out of hand as irrational, as on the standard view, nor contemplated as representations of some "Hegelian" reality, as suggested by the first response above, but rather they are to be taken at face value as open-ended structures that point the way to consistent successors. It is worth noting that, although classical logic is retained at the level of the representations themselves, although not, perhaps, at that of factual beliefs, the logic of quasi truth is interestingly paraconsistent (see da Costa, Bueno, and French, 1998a).

Consideration of inconsistent theories is typically relegated to the domains of discovery and pursuit, with unanalyzed notions of "entertainment" produced to cover the doxastic attitudes involved. Our approach does away with the need for such notions and, by regarding both consistent and inconsistent theories as partially true only, significantly blurs the discovery-justification distinction. The collapse of this distinction has long been urged by post-positivist philosophers and historians of science, but in chapter 6 we emphasize that the tendency to allow sociological concerns to flood over into justification can be resisted through an articulation of the epistemic structure of heuristics. In particular, we explore the role of heuristic criteria in establishing structural relations between models and suggest that the discovery-justification distinction should be replaced by a unitary notion along the lines of theory "development." In this way the challenge of the contextualists can be met, and while on the one hand partial structures offer a means of representing the "openness" of theories and models, on the other the focus on heuristics in terms of structure offers a plausible topography of the space of such development.

In chapter 7 we turn our attention from "horizontal" relationships between theories to the "vertical" relationship between theory and evidence. In particular, we consider the impact of our conception of "partial" truth on the dynamics of belief change. By regarding belief in the terms outlined here, a new basis for a logic of induction can be constructed, which, we hold, is faithful to the central principles of induction as set down by Russell and which can handle various well-known problems that are taken to beset standard

Bayesian approaches. In particular, we show that the standard, but implausible, assumption of "logical omniscience" can be appropriately weakened through a formal modification of the notion of "quasi truth."

Of course, whenever one has to consider the nature of truth in science, it is hard to avoid the realism-antirealism debate. The relevant issues are touched on throughout the book, but in our penultimate chapter we try to bring them all together and subject them to critical analysis from our overall "pragmatic" perspective. In particular, we examine the three main "attack routes" taken against realism: the underdetermination of theories by evidence, the pessimistic metainduction, and the realists' use of inference to the best explanation. With regard to the first, underdetermination, we indicate how certain attempts to tackle this problem by appealing to "background" theories can be nicely accommodated within the framework of partial structures and quasi truth. As far as the pessimistic metainduction is concerned, we offer some criticisms of a recent attempt to blunt its force and suggest that Ladyman's form of "structural realism" might offer a more fruitful approach. The latter recommendation should come as no surprise, given Ladyman's explicit advocacy of the partial structures line! The final issue is trickier, involving as it does the entwined questions as to the status of realism itself and how it might best be defended. Recent attempts at a "naturalistic" defense appear to fall foul of accusations that such moves beg the question, and we offer a broadly pragmatist approach as one possible way forward. Alternatively, one might decide to abandon realism entirely and turn to Bueno's "structural empiricism," which incorporates both partial structures and quasi truth, understood from an antirealist perspective. Thus our approach may be put into service by both sides in the debate as a useful framework for clarifying a range of issues.

As we hope to have shown, these twin notions of quasi truth and partial structures form the core of a rich and fruitful program with applications across philosophy and philosophy of science. In our final chapter, we briefly indicate some of these further applications. It is our firm, if ambitious, belief that we can accommodate the complexity and variety of *all* our cognitive practices within a unitary framework by explicitly acknowledging the openness and partiality of our representational structures. There is, of course, a great deal more work still to be done—there always is—but we hope you will find this work as stimulating and provocative as we have.

1

Truth, the Whole Truth, and Partial Truth

Tarski and the Correspondence Theory of Truth

"The present article is almost wholly devoted to a single problem—*the definition of truth*." Thus begins one of the most remarkable papers in logic and philosophy published this century, Tarski's "The Concept of Truth in Formalized Languages," first presented to the Warsaw Scientific Society in 1931. The paper continues by spelling out the task to be undertaken, which is

> to construct—with reference to a given language—a materially adequate and formally correct definition of the term "true sentence." This problem, which belongs to the classical questions of philosophy, raises considerable difficulties. For although the meaning of the term "true sentence" in colloquial language seems to be quite clear and intelligible, all attempts to define this meaning more precisely have hitherto been fruitless, and many investigations in which this term has been used and which started with apparently evident premises have often led to paradoxes and antinomies (for which, however, a more or less satisfactory solution has been found). The concept of truth shares in this respect the fate of other analogous concepts in the domain of the semantics of language. (Tarski 1935, p. 152)

There are two fundamental aspects of Tarski's achievement that we would like to emphasize. First and foremost, as Tarski himself explicitly stated in the introduction to this paper, he was exclusively concerned with what he called the "classical" conception of truth, which understands this notion in terms of "corresponding with reality" (Tarski 1935, p. 153).[1] The intuition underlying what we shall interpret as a "correspondence" view of truth, and which a "semantical definition" would have to grasp, in some sense, was expressed as follows: "*a true sentence is one which says that the state of affairs is so and so, and the state of affairs is indeed so and so*" (Tarski 1935, p. 155; Tarski cites Aristotle's famous dictum, "To say of what is that it is not, or of what is not that it is, is false, while to say of what is that it is, or of what is not that it is not, is true"). Thus, statements of the form "the sentence *x* is true if and only if *p*"—a particular example being the famous "'it is snowing' if and only if it is snowing"—were regarded by Tarski as "partial definitions" of the concept of truth or, more precisely, as he put it, as *explanations* of the sense of expressions of the type, the sentence *x* is true.

It is because Tarski's construction is built upon these kinds of partial definitions, or explanations, that it has elicited a variety of reactions, ranging from the view that it finally resolves the problem of truth, to the claim that it is trivial, to the further claim that it is actually irrelevant (see Fernández Moreno 2001). However, as Mates notes, "however uninformative and trivial [such statements] may seem, . . . they say just what needs to be said about the relation between a true sentence and the world, and they say it *neat*" (Mates 1974, p. 396). Unencumbered by the metaphysical baggage of individuals and attributes and concepts that weigh down the alternatives, Tarski's offers, in our view, a spare, trimmed-down account of what it is for a proposition to be true, in the correspondence sense.

His remarkable technical achievement was then to show how this account—and, in particular, the recursive heart of it—could be elaborated for formalized languages containing quantifiers.[2] This brings us to the second aspect we wish to emphasize here. At the end of the first section of "The Concept of Truth in Formalized Languages," Tarski concludes, pessimistically, that "*the very possibility of a consistent use of the expression 'true sentence' which is in harmony with the laws of logic and the spirit of everyday language seems to be very questionable, and consequently the same doubt attaches to the possibility of constructing a correct definition of this expression*" (Tarski 1935, p. 165, emphasis in text). The "spirit" of everyday or "colloquial," language that creates this disharmony is its "universality," which encourages admission into the language, in addition to sentences and other expressions, the names of these sentences and other expressions, together with sentences containing them, and, of course, semantic expressions like "true sentence" (Tarski 1935, p. 164). It is precisely this which leads to the "semantical antinomies," such as the liar paradox. Tarski's famous response to this problem is to simply abandon the attempt to construct a consistent definition of truth for everyday language and instead restrict attention to formalized languages, such as the example he then explores, the language of classes.

In particular, Tarski notes, such languages do not contain any expressions that themselves denote signs and expressions of the language or are "structural-descriptive" in the sense of describing the structural connections between such expressions. Thus, "we must always distinguish clearly between the language *about* which we speak and the language *in* which we speak" (Tarski 1935, p. 167). A formalized language about which we are interested is regarded as the *object* language in the former sense, whereas the language in which the names of the expressions of the object language and the relations between them are introduced is, of course, the *meta*language.

It is into the metalanguage that the term "true" is introduced via a "formally correct" definition, and if this is done in such a way that every statement of the form above can be proved on the basis of the axioms and rules of inference of the metalanguage, then, according to Tarski, this definition of truth is said to be "materially adequate" (1936, p. 404). It is this latter idea that is expressed in the famous "Convention T":

> *A formally correct definition of the symbol "Tr," formulated in the metalanguage, will be called an* adequate definition of truth *if it has the following [consequence]:*
> *. . . all sentences which are obtained from the expression "x ∈ Tr if and only if p" by substituting for the symbol "x" a structural-descriptive name of any sentence of the language in question and for the symbol "p" the expression which forms the translation of this sentence into the metalanguage.* (1935, p. 188; emphasis in original)

Within the constraint expressed by Convention T, a definition of truth for sentences of the object language can then be provided in terms of the metalanguage. More specifically, what the metalanguage does is provide an *interpretation* for, or *model* of, the object language. This interpretation or model consists of two parts: (1) a universe of discourse, *A*, consisting of a nonempty set of elements, and (2) a function for the vocabulary of the object language which assigns to each individual constant of this language an element of *A*—that is, its denotation—and which assigns to each *n*-ary predicate symbol of the object language a set of *n*-tuples of *A*, that is, the predicate symbol's extension.

Tarski expressed the nature of this latter function in terms of the notion of the "satisfaction" of a given sentence by a sequence of objects and, most importantly, showed how this notion could be recursively defined. This then "leads us directly to the concept of truth" (1935, p. 189) since all that needs to be established, for a particular formalized language, is which sequences of elements of *A* satisfy the fundamental sentences, or well-formed formulas, of the object language and how this notion of satisfaction behaves under the application of the basic operations of this language: "As soon as we have succeeded in making precise the sense of this concept of satisfaction, the definition of truth presents no further difficulty: the true sentences may be defined as those sentences which are satisfied by an arbitrary sequence of objects" (Tarski 1935, p. 215). And so Tarski succeeded in reducing "truth" to the more basic notion of "satisfaction" and effectively created model theory on the side.

What is clear from all this is that truth and our semantic concepts in general have a "relative character" in the sense that "they must always be related to a particular language" (Tarski 1936, p. 402). Putting it more explicitly, in order to talk rigorously of truth in this manner, we require not only a language $\mathbb{L}$ but also an interpretation $\mathcal{I}$ of $\mathbb{L}$ in a structure $\mathcal{A}$. This is what the metalanguage provides. A sentence of $\mathbb{L}$ is then true or false only with reference to $\mathcal{I}$; that is, truth and falsity are properties of sentences of a particular language $\mathbb{L}$, in accordance with an interpretation $\mathcal{I}$ of $\mathbb{L}$ in some structure $\mathcal{A}$. Informally, of course, we may assert a sentence of $\mathbb{L}$ as being true, without mentioning the interpretation; rigorously speaking, this is unacceptable. This is the second aspect of Tarski's achievement that we wish to emphasize, and we do so because it is so often overlooked.

There are two further points to note before we move on. The first is that the temptation to dismiss out of hand Tarskian approaches to truth because they do not apply to "everyday" languages must be resisted. As Tarski himself noted, any attempt to apply some kind of "exact method" to the semantics of everyday language will run into the semantic paradoxes, unless the universality of that language is effectively removed by some sort of reforming measure, in which case this "everyday" language would in fact take on the characteristics of a formalized language.[3]

Furthermore, and more importantly for the point of view to be explored in this work, "The results obtained for formalised languages also have a certain validity for colloquial language, and this is owing to its universality: if we translate into colloquial language any definition of a true sentence which has been constructed for some formalised language, we obtain a fragmentary definition of truth which embraces a wider or narrower category of sentences" (1935, p. 165, fn. 2). Thus, Tarski's formulation can be taken to apply to those portions of everyday languages that are adequately represented by an appropriate formalized language (first-order predicate logic, say). If, as many have held, the *language of science* can be adequately captured by such a formalized language, then Tarski would seem to have provided a formal understanding of truth-as-correspondence within science.

Nevertheless, three caveats remain. The first and most general is that, within everyday or colloquial language, we still have only a "fragmentary" definition of truth on this basis, and the question arises whether it might not be possible to appropriately "defragment" to some extent this definition or, equivalently, widen the category of sentences it embraces by some suitable modification. Second, the hope that Tarski's formulation might provide a formal underpinning to the realist conception of scientific theories is undermined by the claim that what the realist actually needs, given the history of theory change in science, is a formal account of *approximate* truth. Indeed, the lack of just such an account is typically pointed out as rendering the realist's invocation of "approximate" truth as so much vague hand waving.

Third, as is well known, the correspondence view of truth is not the only show in town, and it might be asked whether a similar sort of account might be provided for, say, the pragmatic theory of truth, suitably understood. It is precisely the latter kind of account that Mikenberg, da Costa, and Chuaqui claim to have provided, and it is to that which we now turn.

Pragmatic Truth

We suggest that the three caveats to Tarski's formulation, expressed at the end of the previous section, all share a common origin—namely, that this formulation, albeit a major advance in both logic and philosophy, is simply not rich enough to capture all the characteristics of "truth" as it features in both "everyday" and scientific practice. The source of the concern is revealed when we consider what it is that is taken to be "true": sentences expressing propositions such as

"snow is white."

Such propositions have an air of ultimate simplicity about them, precisely because they can be regarded as straightforwardly true or false in the correspondence sense. Perhaps this simplicity, in turn, is a function of the "observational" nature of the terms involved in the proposition, or perhaps it can be related to the straightforward nature of the procedures required to establish whether the relevant well-formed formula is satisfied by the elements of the relevant universe of discourse and thus to establish whether the sentence is "in fact" true or false (we shall return to these considerations in chapter 4). Yet at least some of the more interesting representations, to coin a general term, about which we have beliefs, which supply the filling in the sandwich of "belief that . . ." and ". . . is true," are not so ultimately simple or straightforward. They are conceptually open and partial in a way that a proposition, understood precisely as that which is a candidate for truth and falsehood in this straightforward correspondence sense, is not. We take this difference to be fundamental.

To see what we mean here, and to introduce the notion of a partial structure, let us begin as Mikenberg, da Costa, and Chuaqui (1986) did, with the pragmatic view of truth.

We note, first of all, that just as Tarski began his work with an informal and idealized notion of truth-as-correspondence, so we shall not pretend to give a formal exegesis of the notion of pragmatic truth; the philosophy of pragmatism is too complex and inhomogeneous for that. Indeed, it can quite fairly be argued that there is no one "pragmatic view of truth" common to Peirce, James and Dewey, say, although with Dewey excluded the elements of commonality between Peirce and James can be excavated (see, for example, the comparison put forward by Smith 1978, chapter 2). Nevertheless, we feel that we can extract the fundamental basis of this view and thus, to paraphrase Tarski, shall be concerned exclusively with grasping the *intentions* contained in the pragmatic position (cf. Tarski 1935, p. 153).

Perhaps the most fundamental of these "intentions" was expressed by Peirce thus: "consider what effects, that might conceivably have practical bearings, we conceive the object of our conception to have. Then, our conception of these effects is the whole of our conception of the object" (Peirce 1931–35, p. 5.402). Note, first of all, that the focus here is not on propositions, per se, but on *conceptions*. Regardless of historical considerations concerning the development of Peirce's thought, we take this shift in emphasis to be fundamental. The notion of the *conception* of an object expresses an epistemic totality that goes beyond that expressed by a proposition alone. That snow is white represents only part of our conception of snow, and an analysis of truth constructed wholly in terms of the former is not going to be able to accommodate issues regarding the truth of the latter.

It may be objected that, first of all, this Peircean view nevertheless involves truth-as-correspondence with regard to the effects of our conception and that, second, such conceptions can be regarded as nothing more than complexes or sets of propositions. With regard to the first point, this is correct and, indeed, we ourselves would emphasize this as fundamental to the account we are striving to put forward. Peirce's statement can be broadly construed as saying that the pragmatic truth of an assertion depends on its practical effects, with the latter represented in terms of some "basic propositions" and accepted as "true" in the correspondence sense (Mikenberg, da Costa, and Chuaqui 1986, p. 202). It is precisely this dependence that prevents a collapse into some kind of idealism, of course, and that maintains the connection with reality, suitably construed.

The involvement with truth-as-correspondence emerges in a further way that will enable us to respond to the second objection. Switching attention from the effects of our conceptions to the final, or total, conception that is the goal of inquiry, Peirce famously claimed that different minds might set out in an investigation with diametrically opposed beliefs but will then be led to one and the same conclusion (see, for example, Peirce 1931–35, 7.187).[4] Hence, "Truth is that concordance of an abstract statement with the ideal limit towards which endless investigation would tend to bring scientific belief" (Peirce 1931–35, 5.565).

This idea has generated a huge amount of debate, but since we are not interested in exegesis we will content ourselves with two remarks before moving on to our response to the second objection. Our first comment is that one must resist the reduction of the sentiments expressed in this statement to the thesis that truth is simply identical to whatever it is that a community of inquirers agree to accept. The "agreement" and "concordance" are with an external reality, the nature of whose objects ultimately determines the form of the final conception reached by the community. And what drives the community toward this final form is, of course, the scientific method. Thus, "when one says that Peirce's proposal is to translate truth and objectivity into intersubjective agreement, this is true only when asserted in conjunction with the independent constraint exercised by the real" (Smith 1978, p. 56).

Our second remark is that Peirce's account clearly rests on a "cumulative-convergence" view of scientific progress, which has been widely criticized in the past. That such a view is in fact false is not so obvious in these post-Kuhnian times, and indeed we shall effectively be providing the elements of a defense of it in later chapters (cf., for example, Rescher 1978, esp. p. 29).

Thus truth-as-correspondence emerges again in the agreement between these "final" conceptions and reality, and in this sense it is correct to say that "Truth is defined by [Peirce] in terms of *correspondence*, the correspondence of a representation to its object" (Smith 1978, p. 51). However—and this is the all-important point—*this* correspondence is achieved only in the limit, toward which we are driven by a complex procedure involving abduction, deduction, and induction. It is only in the limit that our conception of an object can be regarded as the sum of some set of propositions, each of which is true in the correspondence sense. At any given time, as we approach that limit, our conception, expressed by some "abstract statement," cannot be regarded so straightforwardly. If the final conception is taken to be complete or total, then our conception at any given time prior to the realization of this limit may be said to be *partial*. And because it is, at any given time, partial, it is, at that time, *open* in the sense that it may be completable in a variety of ways (although as we shall argue, the space of further development is rather more circumscribed than some suppose). It is precisely this sense of partiality and openness that our account attempts to capture and further explore.

The extent to which truth-as-correspondence also features in James's account is highly debatable and turns, in part, on what is understood by this notion (see, for example, the debate between Thayer [1983] and Moser [1983]). There is, certainly, as befits an empiricist, an emphasis on the sensible constituent of knowledge, although in granting this a particular cognitive force James goes further than Peirce, for example. However, our primary interest here is with his concern with the nature of the agreement between our "ideas" (cf. conceptions) and objects, *when those ideas are not perfect copies of the objects*. This concern lies at the heart of James's form of pragmatism. Thus he insists, "Truth ... is a property of certain of our ideas. It means their agreement, as falsity means their disagreement, with reality" (James 1907a, pp. 198–199; rep. in his 1932, p. v), and then asks, "Where our ideas [do][5] not copy definitely their object, what does agreement with that object mean?" His answer is contained in the following:

> Pragmatism ... asks its usual question. "Grant an idea or belief to be true," it says, "what concrete differences will its being true make in any one's actual life? What experiences [may] be different from those which would obtain if the belief were false? How will the truth be realized? What, in short, is the truth's cash-value in experiential terms?" The moment pragmatism asks this question, it sees the answer: True ideas are those that we can assimilate, validate, corroborate, and verify. False ideas are those that we cannot. That is the practical difference it makes to us to have true ideas; that therefore is the meaning of truth, for it is all that truth is known-as. (James 1907a, pp. 200–201 and 1932, pp. v–vi)

Hence for James also, truth is to be understood in terms of a relation of "agreement" between an idea and its object. Where this idea might be said to "copy" or duplicate the object, as in "our true ideas of sensible things" (James 1907a, p. 199), agreement reduces to correspondence, and the assimilation, validation, corroboration, and verification are direct and immediate. Where, however, the idea is not a straightforward copy of the object, which is the case with the most useful and interesting of our ideas, the notion of correspondence is inadequate to express the relationship, and the full pragmatist analysis of truth must come into play.[6] In this regard, Smith notes,

> It is important to notice that there are times when James thought of "copy" as the most obvious case of agreement and he seems to have assumed that where "copying" exists agreement exists as well, so that we have a kind of immediate verification. However, where copying fails to obtain, as in the case of the idea of energy in the spring of the watch, he asked what agreement means and proposed in those cases to appeal to the pragmatic maxim—grant an idea to be true, what difference will it make, etc.—as the proper way to determine whether there is "agreement" in a non-copy sense. (Smith 1997, p. 211, fn. 39)

(The example of the energy or, as James put it, the "elasticity" in a watch spring (1907a, p. 199) is significant; it is "iconic" in senses to be explored later.)

Thus, in his response to a critique of pragmatism, James takes what Tarski called the "classical" conception of truth as stating "this simple thing that the object of which one is thinking is as one thinks it" and writes:

> It seems to me that the word "as," which qualifies the relation here, and bears the whole "epistemological" burden, is anything but simple. What it most immediately suggests is that the idea should be *like* the object; but most of our ideas, being abstract concepts, bear almost no resemblance to their objects. The "as" must therefore, I should say, be usually interpreted functionally, as meaning that the idea shall lead us in to the same quarters of experience *as* the object would. Experience leads ever on and on, and objects and our ideas of objects may both lead to the same goals. The ideas being in that case shorter cuts, we *substitute* them more and more for their objects; and we habitually waive direct verification of each one of them, as their train passes through our mind, because if an idea leads *as* the object would lead, we can say ... that in so far forth the object is as we think it, and that the idea, verified thus in so far forth, is true enough. (James 1907b, p. 467; reprinted in 1932, pp. 166–167)

In other words, for an idea to be true the object must be "as" the idea declares it, where the "as-ness" here is explicated in terms of the idea's verifiability[7] (James 1907b, p. 170). "Verifiability," in turn, signifies "certain practical consequences of the verified and validated idea" (1907a, p. 201). These consequences are then characterized in terms of correspondence: "It is hard to find any one phrase that characterizes these consequences better than the ordinary agreement-formula—just such consequences being what we have in mind whenever we say that our ideas "agree" with reality" (James 1907a, p. 201).

Hence the pragmatist account of the truth-relation is intended to cover not only "the most complete truth that can be conceived of"namely, correspondence truth—but also "truth of the most relative and imperfect description" (James 1907, p. 183). In other words, it is explicitly intended to accommodate situations in which the relation holds between "reality" and "ideas," "conceptions" and "descriptions," which are abstract, imperfect, and incomplete—that is, *partial.*

There is, of course, much more that can, and has, been said about the characteristic "functional" interpretation of this relation and the way in which James replaces the traditional static or "saltatory" relationship between an idea and its object with a dynamic or "ambulatory" one (1907c; 1932, pp. 136–161). However, we believe that we have uncovered enough of the concerns or "intentions" of the pragmatists to provide a basis for our formal account to follow. Summarizing, these intentions represent an emphasis on:

1. The nature of agreement between "imperfect" or "abstract" descriptions and reality
2. The empirical consequences of such descriptions, understood as "agreeing" with reality in the classical correspondence sense

3. "Complete" or "absolute" truth, again understood in the classical correspondence sense, as the (ideal) terminus of all inquiry

Of course, in glossing over the pragmatists' emphasis on the dynamic nature of the truth relation, we may be accused of ignoring or, at best, downplaying that which is held to be most significant and fundamental about their theory of truth.[8] Perhaps some degree of historical distortion is inevitable if a formal approach is to be allowed any grip. It may be that this dynamic element is, in fact, what puts the pragmatists' account beyond the pale of any possible formalization. Nevertheless, we feel that these brief excavations, shallow as they are, reveal a set of concerns that can be accommodated within a formal scheme—indeed, one which can be seen as a "natural" development of the Tarskian approach.

Truth and Partial Structures

> Standard model theory has concentrated on complete models, corresponding to complete constellations of facts, the limiting cases where any additions would introduce inconsistencies. A certain tradition in the philosophy of language has been cited as one influence leading to the exclusive emphasis on totality: The meaning of a sentence is equated with its truth conditions, and the truth conditions are taken to correspond with states of the whole world in which the sentence is true. (Langholm 1988, p. 1)

This concentration on complete models is an inheritance from Tarski that derives from the "intentions" underlying the "classical" correspondence view of truth and that he sought to capture. The word *correspondence* itself signifies a certain "tightness" in the relation, a lack of any looseness or vagueness or ambiguity or, indeed, anything that is not entirely straightforward. For how else could we understand the truth of the proposition "the snow is white" other than in terms of a correspondence with the relevant state of affairs?

Notice that the characterization of our representations in terms of propositions; the characterization of the relationship between these propositions and "the world," "reality," whatever, as correspondence; and the characterization of this "world" or "reality" as "complete" or "total" form an interlocking triangle. The proposition, of course, is the unit of analysis of the classical propositional calculus; shifting to the finer-grained perspective of predicate logic, the structure of these propositions is revealed, but still the structure is determinate: Any vagueness, ambiguity, or looseness of application lies in our understanding of the variables and predicate terms, not in the structure of the proposition itself. Given this rigidly determinate structure, the relationship between the proposition and that which it is about can only be that of correspondence, a relationship so transparent that capturing it formally was thought by some to be a triviality. And the thing on the other end of this relationship must be equally transparent, equally determinate, equally complete and epistemically "tight."

Yet, as we just saw, one of the primary motivations behind pragmatism was to accommodate situations—surely the more interesting if not actually in the majority—in which the relationship between the representation and that which is represented is "loose," vague, ambiguous perhaps, but certainly partial. Both our everyday and scientific beliefs concern representations that are *not* determinate, *not* tight, *not* complete; they are idealizations and approximations, they are imperfect, and they are partial, reflecting our partial knowledge and understanding of the world. The notion of a proposition is inadequate to capture these representations, as is the notion of correspondence to capture the relevant relationship, which the pragmatists called "agreement." They tried to accommodate the appropriate aspect of partiality by transforming the relationship itself into a dynamic one, thereby removing their conception from the realm of the formal, an approach that led to enormous criticism and, for some, much misunderstanding (thus James famously talked of truth "happening" to an idea, for example). Here we shall retain a static form of relationship, which will allow us to adapt a Tarski-style formalization, while still capturing the pragmatist intentions. What follows is an informal summary of technical results presented in Mikenberg, da Costa, and Chuaqui (1986).

When we say that some sentence $\mathcal{S}$ is "true," we may interpret it as strictly saying that $\mathcal{S}$ is true in a certain structure or model that represents a portion of reality. In fact, $\mathcal{S}$ can be said to "point" to the world by means of a model, which normally is not made explicit. Let us suppose that we are investigating a given domain of knowledge Δ,[9] certain aspects of which can be represented by a "data structure" $\mathcal{D}$. The manner in which $\mathcal{D}$ is obtained from Δ is, of course, complex, and here we may defer to the exponents of the recently resurgent "history and philosophy of experiment," whose accounts have shed new light on the construction of such structures. Likewise, the manner in which the various elements of $\mathcal{D}$ are related to the "objects" (for want of a better word) of Δ is also problematic; it may be that, as Wittgenstein suggested, the nature of this relationship lies beyond linguistic expression.

If our sentence $\mathcal{S}$ refers to $\mathcal{D}$, this is accomplished with the help of a structure $\mathcal{A}$, which effectively substitutes for Δ in our thought. Sometimes aspects of $\mathcal{A}$ model $\mathcal{D}$ so well that we have no doubts in accepting that there is an isomorphic correspondence between the two, as in the case, perhaps, of the relationship between the "appearances" in van Fraassen's sense and the empirical substructures of a theory (van Fraassen 1980). (Again, we wish to emphasize that, strictly speaking, such an isomorphism cannot be said to hold between $\mathcal{A}$ and Δ, since this relation is rigorously defined as holding between formal structures only. This is a point that has been obscured in recent discussions.) The further relationship between these substructures and the theoretical superstructure of $\mathcal{A}$ is, of course, much discussed in the philosophy of science, and we shall return to it in later chapters.

One of the main points of Tarski's formalization of the concept of correspondence truth is that in order to talk rigorously of truth we require a language $\mathbb{L}$ and an interpretation $\mathcal{I}$ of $\mathbb{L}$ in a structure $\mathcal{A}$. A sentence of $\mathbb{L}$, then, is true or false only with reference to $\mathcal{I}$. In other words, truth and falsity are properties of sentences of a particular language $\mathbb{L}$, in accordance with an interpretation $\mathcal{I}$ of $\mathbb{L}$ in some structure $\mathcal{A}$. Informally, of course, we may assert

$\mathcal{S}$ of $\mathbb{L}$ as being true, without explicit mention of the interpretation, and may even forget that $\mathcal{S}$ is part of $\mathbb{L}$. If we are to be rigorous, however, we must consider the interpretation.

Analogously, when we talk of "pragmatic" or partial, or "quasi" truth, we must also introduce the twin notions of interpretation and structure. In this case, however, the models underlying the interpretations are not viewed as completely or exactly mapping the given domain; rather they are regarded as only partially modeling it. Thus, the relevant properties and relations among the members of A are captured by a family of relations that may be said to be "partial" in a certain formal sense. A "simple pragmatic structure," then, is a partial structure of the form

$$\mathcal{A} = \langle A, R_k, P \rangle_{k \in K}$$

where A is a nonempty set, R_k, $k \in K$, is a partial relation defined on A for every $k \in K$, where K is an appropriate index set and P is a set of sentences of the language $\mathbb{L}$ of the same similarity type as that of $\mathcal{A}$ and which is interpreted in $\mathcal{A}$. For some k, R_k may be empty; P may also be empty. The R_k are described as "partial" because any relation R_k, $k \in K$, of arity n_k is not necessarily defined for all n_k-tuples of elements of A; also, in the general case partial functions, as well as relations, could also be included in a partial structure. In general, A denotes the set of individuals of the domain of knowledge modeled in the particular case considered (elementary particles in the case of high-energy physics, for example), and the family of partial relations R_k models the various relationships that are taken to hold among these individuals. The P is then regarded as a set of distinguished sentences of $\mathbb{L}$, which may include, for example, observation statements regarding the domain.

If $\mathcal{B}$ is a total structure, whose relations of arity n_k are defined for all n_k-tuples of elements of its universe, and $\mathbb{L}$ is also interpreted in $\mathcal{B}$, then $\mathcal{B}$ is said to be "$\mathcal{A}$-normal" if:

1. The universe of $\mathcal{B}$ is A.
2. The relations of $\mathcal{B}$ extend the corresponding partial relations of $\mathcal{A}$.
3. If c is an individual constant of $\mathbb{L}$, then in both $\mathcal{A}$ and $\mathcal{B}$ c is interpreted by the same element.
4. If $\mathcal{S} \in P$, then $\mathcal{B} \models \mathcal{S}$.

Loosely speaking, a total structure $\mathcal{B}$ is called $\mathcal{A}$-normal if it has the same similarity type as $\mathcal{A}$, its relations extend the corresponding partial relations of $\mathcal{A}$, and the sentences of P are true, in the Tarskian sense, in $\mathcal{B}$. Then $\mathcal{S}$ is said to be pragmatically or quasi-true in $\mathcal{A}$, or in the domain Δ that $\mathcal{A}$ partially reflects, if there is an interpretation $\mathcal{I}$ of $\mathbb{L}$ (and consequently of $\mathcal{S}$) in an $\mathcal{A}$-normal structure $\mathcal{B}$ and $\mathcal{S}$ is true in the Tarskian sense in $\mathcal{B}$. Clearly, $\mathcal{A}$ is not conceived of as reflecting the (total) structure of Δ via $\mathcal{D}$ but as only partially mirroring this domain; in particular, the sentences of P should be true, in the usual sense in $\mathcal{D}$. Thus the partial model $\mathcal{A}$ has to capture some fundamental aspects of Δ, or some "elements of truth," although it does not mirror Δ perfectly. Then, we can say that $\mathcal{S}$ is pragmatically or quasi-true in $\mathcal{A}$, if all logical consequences of $\mathcal{S}$ or of $\mathcal{S}$

plus the true primary statements P are compatible with any true primary statement. In other words, we say that $\mathcal{S}$ is pragmatically true in the structure $\mathcal{A}$ if there exists an $\mathcal{A}$-normal $\mathcal{B}$ in which $\mathcal{S}$ is true, in the correspondence sense. If $\mathcal{S}$ is not pragmatically true in $\mathcal{A}$ according to $\mathcal{B}$, then $\mathcal{S}$ is said to be pragmatically false in $\mathcal{A}$ according to $\mathcal{B}$. This definition of pragmatic, or quasi-, truth captures the gist of the idea of a proposition being such that everything occurs in a given domain *as if* it were true (in the correspondence sense of truth).

Let us pause to review the nature of this formal framework. First of all, it explicitly follows the Tarskian approach to truth, the fundamental difference, or addition, being the introduction of "partial structures" into the structure $\mathcal{A}$. This notion can be further elucidated as follows: The central idea is that in a partial structure, the relations and operations are defined for only some of the elements of the domain. So, if R is a n-ary partial relation and $\langle a_0, a_1, \ldots, a_{n-1}\rangle$ are n-tuples of the elements of the universe of the structure—that is, in the terms indicated, the members of A—then there are three possibilities: either $\langle a_0, a_1, \ldots, a_{n-1}\rangle$ is in R, or it is not, or it is undetermined whether it is. Thus,

> The partial models ... all specify a definite domain of individuals. Furthermore, they ... specify, for each relation symbol, a positive and a negative extension. The extensions contain n-tuples of individuals, and correspond to elements that *do* and *do not* relate to each other in the sense of the particular relation symbols. The two extensions are disjoint, but do not have to be exhaustive. About the remaining n-tuples the model does not provide information, Standard models can be viewed as partial models where for each relation symbol the positive and negative extensions together *are* exhaustive. (Langholm 1988, p. 9)

Let us give an example to illustrate what we mean. Consider a binary relation R, which can be introduced as follows: R is an ordered triple $\langle R_1, R_2, R_3\rangle$, where R_1, R_2, and R_3 are mutually disjoint sets such that $R_1 \cup R_2 \cup R_3 = A^2$. In model-theoretic terms, R_1 is the set of ordered pairs which are satisfied by those sentences expressing the relationships between the entities concerned, R_2 is the set of ordered pairs not satisfied by these sentences, and R_3 is the set of ordered pairs for which it is left open whether they are satisfied. When R_3 is empty, R constitutes a normal binary relation and can be identified with R_1. It is precisely this which is meant when we say that R_k is "not necessarily defined for all n_k-tuples of elements of A." This is the crucial formal mechanism behind our account: the idea being that the interpretation appropriate for capturing the "intentions" of "pragmatic" truth is *partial*.

That this account does indeed capture these intentions is, we hope, clear. In particular, of course, the incomplete and imperfect nature of the majority of our representations of the world is, we claim, represented by the simple pragmatic structures just provided. It is the partial relations that express what we know about the domain of knowledge concerned or represent the relevant relations between the elements of this domain in the sense of those relations we happen to be interested in. The second and third "intentions" on our list also find a formal representation within our account. Thus, the set of privileged sentences P, which include the empirical consequences of our descriptions, are taken to be true in the Tarskian correspondence sense, and the total structure $\mathcal{B}$ can be taken to represent the ideal "terminus" of enquiry.[10]

Perhaps even more fundamentally, although it is sentences such as S that are said to be pragmatically or "quasi-" true, they are so only in a simple pragmatic structure, such as $\mathcal{A}$. It is the latter that (partially) represents the world, and these partial structures are the primary locus of epistemic activity. Thus, rather than focus on propositions per se, we shall be advocating a shift to consideration of the underlying structures, complete in the Tarskian correspondence account, partial in the case of pragmatic truth, which do all the representational work, as it were. Such a shift is, we contend, crucial to obtaining a more representative characterization, in the sense of being in closer agreement with practice, of a range of notions, from belief in general to scientific theories in particular.

Finally, then, the formalization of pragmatic or quasi truth in this manner removes any last inhibitions on the use of these notions to resolve a wide range of issues, from the treatment of inconsistent belief sets to the problem of induction and the question of theory acceptance in science. It is to such issues that we now turn.

2

THEORIES AND MODELS

A possible realization in which all valid sentences of a theory T are satisfied is called a model of T.

—Tarski, "A General Method in Proofs of Undecidability"

Model Theory and the Philosophy of Science

Tarski's innovation had two fundamentally important consequences (see also Suppes 1988). First, his definition of truth-as-correspondence had an enormous impact on philosophy and, in particular, the philosophy of science. Second, by showing that not only truth but also other semantic notions, such as that of semantic consequence, could be defined in terms of a form of set theory, such as Zermelo-Fraenkel, Tarski effectively brought the notion of model within the scope of "ordinary" mathematics. These two results are both formally and historically entwined, as Vaught indicates in his brief account of the early history of model theory: In his seminars of 1926–28 at Warsaw University, Tarski expressed his dissatisfaction with the intuitive notion of "truth in a structure," which was typically left undefined in metalogical discussions (Vaught 1974).[1] Thus Tarski's analysis was driven by formal considerations, as well as philosophical ones.

Subsequently, model theory underwent an explosive development, branching off in a variety of directions to such an extent that doubts have been expressed as to whether the field can be contained under such a simple title (a summary of these developments is given in Chang 1974; for an indication of the doubts as to what, precisely, falls under "model theory," see Addison 1965). Nevertheless, we shall adopt the position that,

> to lie properly in the theory of models a study should really make "essential" use of structures as well as of languages. Vaguely, it is necessary that

> the structures (or more general kinds of entities) which underlie the interpretation of the grammars under consideration should be considered as individuals—that we construct them, perform operations on them, classify them, or determine relations that hold among them. (Addison 1965, p. 441)

It is this "essential" use of structures and their treatment as unitary individuals that we regard as fundamental; indeed, this view may be taken as the formal counterpart of the epistemic shift from the language in which our beliefs are expressed to the structural objects of those beliefs. We shall return to this issue shortly.

Returning to our historical outline, the model-theoretic aspect of Tarski's technical achievement had further fundamental consequences outside model theory itself by initiating the development of the "model-theoretic" or "semantic" approach to scientific theories.[2] The twin driving forces behind this approach were the application of model-theoretic techniques, understood as a kind of "logical analysis" in a broad sense and a concern with scientific practice. Thus Beth, one of the originators of the approach who was profoundly influenced by Tarski's work, ably expresses these motivations: "a philosophy of science, instead of attempting to deal with speculations on the subject matter of the sciences, should rather attempt a logical analysis—in the broadest sense of the phrase—of the theories which form the actual content of the various sciences. The Semantic method, which was introduced by A. Tarski towards 1930 ... is a very great help in the logical analysis of physical theories" (Beth 1949, p. 180). These twin concerns also underlie the work of Suppes, another of the originators of this approach, who learned about set-theoretic analysis from Tarski and McKinsey and applied it in the axiomatization of theories in physics (see Suppe 1989, p. 8).

The work of Beth and Suppes influenced in turn van Fraassen and Suppe, whose books and papers, together with those of Suppes in particular, can be seen as delineating the core of the model-theoretic program within the philosophy of science (Suppe 1989; van Fraassen 1980, 1989, 1991). This core is marked out by a commitment to the view that scientific theories can be characterized by what their linguistic formulations refer to when the latter are interpreted semantically, in the model-theoretic sense. Thus on this view, theories may be viewed as "extralinguistic" in a sense to be touched on shortly. Precisely how this extralinguistic nature is cashed out varies within the model-theoretic approach: thus for Beth and van Fraassen, theory structures are captured in terms of state spaces, for Suppe they are understood as relational systems, and for Suppes and Sneed, they are regarded in terms of set-theoretical predicates. The function of these different mathematical characterizations is, however, the same: It is to specify the admissible behavior of physical systems. Thus, Suppe gives the example of classical mechanics, where the state of an n-body system is determined by 6n variables of position and momentum, and the laws of the theory specify which changes of this state are allowed:

> If one represents the states as points in 6n-dimensional spaces, the theory structure is construed as a configurated state space (van Fraassen). One also

can construe the state transition structure as a relational system consisting of the set of possible states as a domain on which various sequencing relations are imposed (Suppe). By defining set-theoretic predicates (Suppes, Sneed), one can specify either a state space, a relational system, or some other representing mathematical structure or class of structures. (Suppe 1989, p. 4)

The set-theoretic approach of Suppes is in this latter sense more general, and for this reason, as well as that of personal influence, we shall follow it here.

The Axiomatization of Theories

Let us return to the issue—which we regard as absolutely fundamental—expressed by the question "what is a scientific theory?" According to the so-called Received View (the name, sounding today rather dated, was coined by Putnam in 1960), the answer is relatively straightforward: A theory is an axiomatic calculus given a partial observational interpretation via a set of correspondence rules; that is, a theory on this view is a logico-linguistic entity. The deficiencies of such a view are, by now, very well known and, of course, were a contributing factor in the development of the "Weltanschauungen" analyses of Toulmin, Hanson, Kuhn, and others (for an excellent critical account of such analyses, see Suppe 1977). The aftermath of these developments was a denigration of axiomatization in general as merely an elaborate display of technical ingenuity applied to no apparent purpose.

However, that a particular characterization of theories and their consequent axiomatization is faulty does not of course give any grounds for dismissing all such characterizations or attempts at axiomatization in general. Following Suppes, we believe that axiomatization is a fundamentally important component in the philosophy of science, for its role in introducing clarity with respect to the basic concepts of a theory, its help in theory comparison (anathema, of course, to those who believe in incommensurability), for the way it can open up a theory to potentially fruitful mathematical techniques,[3] and, yes, for its usefulness in resolving certain philosophical disputes (see, for example, Suppes 1968).[4] As Sklar has remarked, in a slightly different context,

> The choice of formalization of a theory is replete with scientific and philosophical consequences.... Our characterization of terms as primitive or defined, and our characterization of consequences as definitional or postulational, are decisions that implicitly reveal our beliefs about the limits of observational testability of theories and about the ability of these theories to outrun these limits in their content; our ideas about the meanings of theoretical terms, how they are fixed, and how they change; and our view about the place of the theory formalized both in the historical context of the theories which preceded it and from which it evolved and in the assumed future science which, we anticipate, will perhaps evolve from it under pressure of new observations and new theorizing.
>
> Given that so much hinges upon the formalization we do choose, we should not be surprised to find that this choice is no trivial matter, but one that requires the full utilization of our best available scientific and philosophical

methodology. If formalizing is to be more than mere "logicifying," it *is* theorizing, and demands, if it is to be done adequately, the full resources needed for theorizing in general. (Sklar 1977, pp. 271–272)[5]

It is clear, however, that such axiomatization cannot proceed merely linguistically or syntactically. Leaving aside such well-known issues as the lack of a sound basis for the linguistic distinction between theoretical and observational terms, for example, there is the fundamental point concerning the characterization and, hence, individuation, of theories. More specifically, as Suppe, for example, emphasizes (and a concern with this issue of theory individuation runs throughout his book), on the Received View, the correspondence rules that relate theoretical and observational terms are "individuating proper parts of theories" (1989, p. 4), as they must be on a linguistic characterization. Any change in such rules must therefore result in what is, strictly speaking, a new theory. Since such changes will typically be driven by changes in experimental procedure—by the introduction of new measurement techniques, for example—on this view a change in experimental design must formally lead to a new theory. Yet this is clearly absurd from the perspective of scientific practice.

This is part and parcel of a more general point concerning the independence of the theory, as such, from linguistic formulations. It is, perhaps, banal to say that the theory of General Relativity is the same whether it is expressed in English or Portuguese. What is not so banal is the claim that it is the same theory whether it is logico-linguistically axiomatized in one particular way or another. We stress the logico-linguistic aspect since, as Sklar points out, there is a sense in which formalizing of a certain kind is theorizing, and we shall touch on this point later. There is a further practical issue here: Although it is possible to logically axiomatize certain relatively simple theories in mathematics, for example, in terms of first-order predicate logic with identity, the task becomes "awkward and unduly laborious" (Suppes 1957, p. 248) when we face more complex theories, such as geometry, where we might want to define lines as certain sets of points, for example. Even more significantly, "it is repetitious, for in axiomatizing a wide variety of theories, it is necessary or at least highly expedient to make use of set theory; if formalization in first-order logic is the method used, then each such axiomatization must include appropriate axioms for set theory" (Suppes 1957, p. 248). With even more complex theories, such as probability theory, where it is not just notions from set theory but also results of number theory and the theory of functions that are appealed to, the task moves from the laborious to the "utterly impractical" (Suppes 1957, p. 249).

Finally and relatedly, within scientific practice itself we simply do not find axiomatizations in this logico-linguistic sense. As van Fraassen has emphasized, what are called "axioms" in a textbook on quantum mechanics, say, do not look anything like the kind of thing we expect from our logic courses, nor do they play the same role (1980, p. 65). As he further remarks, while the disparity between such practice and what is typically captured syntactically "will not affect philosophical points which hinge only on what is possible 'in principle,' it may certainly affect the real possibility of understanding and clarification" (1989, p. 211). Thus, to denigrate these points as merely concerned with issues of practicality is to precisely miss the point.

Suppe (and van Fraassen) conclude that "theories are not collections of propositions or statements, but rather are extra-linguistic entities which may be described or characterized by a number of different linguistic formulations" (Suppe 1977, p. 221). Suppes, on the other hand, understands theories in terms of set-theoretical predicates, which, of course, *are* linguistic entities. What is a theory then? This is perhaps a more delicate question than has been acknowledged until now. It seems to us that it bears on the further issue of the nature of the game that we, as philosophers of science, are involved in. And this game, at least in part, has to do with *representation*. If we want to know what a theory is, the best we can do is give an ostensive response by looking at practice and taking the examples of what science presents to us as theories. The question we have to answer, as philosophers of science, is what is the most appropriate *representation* of theories? In answering this question, we must acknowledge that a theory can be represented from various perspectives: A Suppes predicate, understood as a linguistic notion, determines a family of structures, which are nonlinguistic and (normally) conversely.

Representationally, then, theories present us with two faces: the syntactical and the semantical.[6] However, no one would think today that we should axiomatize all theories within first-order logic because this is simply not possible. If we were to insist on a "linguistic" approach, we could employ either higher order logics or set theory (which is itself a first-order theory; see appendix 2). Indeed, this might be the more appropriate approach if our aim were to prove certain metatheorems about the theory. If, however, our aim is to accommodate various aspects of scientific practice, such as the interrelationships between theories and data or between theories themselves, then axiomatization must be sought, if it is to be sought at all, in some method other than the specification and elaboration of a first-order logical language. It is precisely such an alternative method that is provided by the model-theoretic approach, where a theory is presented (and here we choose our words carefully, for reasons to be elaborated later) in terms of a description of a set of models in the sense of relational structures for which all the sentences in a particular linguistic formulation of the theory express true properties about the structure when the latter acts as an interpretation or "possible realization" (Suppes 1957) of the theory.

We declare, then, that to axiomatize a theory is to apply these model-theoretic methods. Perhaps the most famous such application is the axiomatization of classical particle mechanics by Suppes and others (Suppes 1957, 1970; McKinsey, Sugar, and Suppes 1953; da Costa and Chuaqui 1988; da Costa and Doria 1992), although suggestions of this approach can be found in von Neumann's formalization of quantum mechanics, which also influenced Beth (von Neumann 1955; for further discussion of the model-theoretic approach to quantum mechanics, see van Fraassen 1970, pp. 329–330, and 1991). More recently, one of us has suggested how Maxwell's electromagnetic theory can also be axiomatized in this way (da Costa 1987a), and van Fraassen

(1985a) and Friedman (1983) have applied these techniques in discussions of the foundations of space-time theories (see also da Costa, Doria, and de Barros 1990). That such axiomatization might have important consequences, both philosophical and scientific, is exemplified by the result that including or deleting certain set-theoretical axioms, such as the continuum hypothesis or Martin's axiom, produces definite physical results: A positive entropy shift, for example, becomes a zero-entropy shift (see da Costa and Doria 1996).

If to axiomatize a theory is to define a family of models or, more generally, a species of structures, the question naturally arises, "In what terms should such structures be defined?" Now, the three best known ways to define the concept of a mathematical structure are as follows: (1) using set theory à la Bourbaki, (2) by means of higher order logic (or type theory), and (3) via category theory. Loosely speaking, 1 and 2 can be thought of as methods for studying the "psychology" of mathematical structures, whereas 3 is more concerned with the "sociology" of such structures! The last method studies a certain kind of structure *inside* a given category. We shall briefly consider each of these in turn (for further details of 1, see appendix 1):

1. The usual concept of species of structures in general and of structure in particular is essentially that of Bourbaki (1968, chapter 4). It is this notion that is employed by Beth, Suppes, and others. In particular, a model is a structure that satisfies certain sentences of a convenient language. These can be taken as the standard concepts of (mathematical) structure and model.
2. Instead of set theory, we could use higher order logic to define the notion of structure. This is basically Carnap's approach (see Carnap 1958), and it can be shown that his definition is essentially equivalent to Bourbaki's. However, one big difference is that we may interpret the basic elements of a structure as predicates and not as sets. In this manner, we are naturally led to define not only extensional structures but also intentional ones.
3. Most properties of structures depend not only on particular structures but also on collections of structures composing mathematical objects known as categories (see, for example, Bourbaki 1968, pp. 271–288). Taking into account this fact and generalizing the situation, Ehresmann had the happy idea of defining a species of structures as a functor acting on appropriate categories (see, for example, Ehresmann 1966, pp. 1–7). This categorial or functorial concept of species of structure can be thought of as extending the set-theoretical definition, and it may be that future developments will compel us to employ a concept of structure defined along these lines.

Having sketched the three most common ways to define structures, it is worth noting that we can develop yet another. Adopting a somewhat loose way of speaking for current purposes, we proceed as follows: We begin with a Boolean algebra $\mathcal{B}$ in which we suppose that there exists an extra operator $\{\ \}$, associating with each element x of $\mathcal{B}$, another element $\{x\}$. The following axiom then has to be satisfied:

$$x = y \leftrightarrow \{x\} = \{y\}$$

Then we define "x ∈ y" as $x \leq \{y\}$, where $\leq$ is the order-relation of $\mathcal{B}$, we also suppose that $\{x\}$ is a Boolean atom.

Since we have $\{x\}$, $\varnothing$, 1, v, ′, and so on, we can introduce as new postulates those of Church (1974), as well as the necessary definitions. This gives us a "formal" set theory whose elements are any abstract objects, not necessarily sets. A definition of "formal" structure then offers no particular difficulties, and an analogous generalization can be performed on the functorial concept of structure.[7]

In what follows we adopt Suppes's declaration that "to axiomatize a theory is to define a predicate in terms of notions of set theory." A set-theoretical predicate in this sense is simply a predicate that can be defined within set theory in a completely formal way. It is important to stress that the kind of axiomatization we are talking about here is not logico-linguistic but set-theoretic; thus, as Suppes again has put it, in a slogan widely taken up by the proponents of the model-theoretic approach, the appropriate tool for the philosophy of science is mathematics and not metamathematics.

Recalling our previous brief discussion of alternatives, it can then be shown that such set-theoretical predicates are essentially identical to the species of structures described by the Bourbaki mathematicians (da Costa and Chuaqui 1988). A scientific theory may thus be characterized by a set-theoretical predicate in such a way as to connect this approach with standard model theory: "When a theory is axiomatized by defining a set-theoretical predicate, by a *model* for the theory we mean simply an entity which satisfies the predicate" (Suppes 1957, p. 253) and "Roughly speaking, a possible realization of a theory is a set-theoretical entity of the appropriate logical type" (Suppes 1961, p. 166). To give a simple example, let us consider the concept of a group (see Suppes 1961, pp. 250–259). A group can be characterized by a structure of the form:

$$\mathcal{A}_G = \langle A, \bullet, *, I \rangle$$

where A is a nonempty set, $\bullet$ is a binary operator on A, $*$ a unary operator on A, and I an element of A, such that:

1. $(x \bullet y) \bullet z = x \bullet (y \bullet z)$
2. $x \bullet I = I \bullet x = x$
3. $x \bullet x^* = x^* \bullet x = I$

The corresponding predicate is then

$$P(x) \leftrightarrow \exists A \exists B \exists C \exists D\, (x = \langle A, B, C, D \rangle \ \&\ A \text{ is a nonempty set } \&\ B \text{ is a binary operation on } A \ \&\ C \text{ is a unary operator on } A \ \&\ D \text{ is an element of } A \ \&\ \forall x \forall y \forall z ((x, y, z) \in A \rightarrow (xBy)Bz = xB(yBz) \text{etc.}))\text{.}$$

Obviously, this is a fairly simple example, but we can in fact repeat this procedure with any structure we like. As set theorists like to claim, the language of set theory is a kind of universal language, and with it we can reproduce *all* extant mathematics (and practically all of our scientific thinking as well). This aspect underlies the usefulness and importance of the semantic approach because if we axiomatize our theories in this manner, then we have the whole of mathematics "at hand," as it were (Suppes 1970).

So, to axiomatize a theory is to define a set-theoretical predicate, and the structures that satisfy this predicate are the models of the theory. That is, when a theory is formalized in this way, the mathematical structures that satisfy the predicate are the models of this predicate, or the structures of this species of structure, *P*(x), or, more simply, *P*. In what follows we use informal, intuitive set theory and restrict ourselves to first-order structures only, to which can be associated a first-order language of the same similarity type. Our intention is to take the notion of a "partial structure," defined in the characterization of pragmatic or quasi truth given in the previous chapter, and employ it within the model-theoretic approach to theories. In doing so, we shall not be particularly concerned with the relevant set-theoretical predicates—that is, our aim here is not to present model-theoretic axiomatizations of particular scientific theories—rather we shall work directly with the models in an effort to capture relevant aspects of scientific practice.[8]

Theories as Partial Structures

We recall that the central idea is to capture the fact that we may not know everything about a given domain of knowledge Δ, so we model Δ by a structure of the form:

$$\mathcal{A} = \langle A_1, R_i \rangle_{i \in I}$$

where A_1 is the set of observable individuals of Δ and R_i, $i \in I$ is a family of partial relations defined on A_1, where I is an appropriate index set (this rather simplistic initial presentation ignores the hierarchy of models between the so-called theoretical models and the "phenomenological" ones that most directly model Δ; we shall consider this aspect later). The R_i, to repeat, are partial in the sense that any relation R_i, $i \in I$, of arity n_i, is not necessarily defined for all n_i-tuples of elements of A_1.

The partial structure can then be enriched by the introduction of further individuals representing so-called unobservable entities and by the construction of new partial relations among the extended set of individuals so obtained. Denoting this new set of individuals by A_2 and the family of new partial relations by R_j, $j \in J$, we suppose that $A_1 \cap A_2 = \emptyset$, $I \cap J = \emptyset$, and we put $A = A_1 \cup A_2$. Thus Δ is modeled by a partial structure that has the general form

$$\mathcal{A} = \langle A_1, A_2, R_i, R_j \rangle_{i \in I, j \in J}$$

or, more simply, with $K = I \cup J$:

$$\mathcal{A} = \langle A, R_k \rangle_{k \in K}$$

To simplify the exposition, we have not explicitly included functions and distinguished elements.

At the end of his review of the postwar history of model theory, Chang urges his colleagues in the field to study the meaning, definition, and use of models in the various sciences and suggests that "if you like the idea of putting models to use, then you should broaden yourself and be prepared to study some very unorthodox models" (Chang 1974, p. 181). We do not claim that partial structures are so very unorthodox; nevertheless, we feel that our proposal of introducing them into the model-theoretic account is very much in the spirit of Chang's suggestion. In particular, we believe that it opens up new avenues for dealing with a range of fundamental issues, such as the relationship between these mathematical models and scientific models in general, the nature of idealization in science, and the interrelationships between models in theory development.[9]

Before illustrating the usefulness of partial structures in this respect, there is an important issue that must be tackled first. The notion of a partial structure was originally introduced in the context of a definition of pragmatic or quasi truth. Given the fallibility of scientists and the lack of complete knowledge of a given domain, it seems a straightforward move to regard scientific theories as only partially or pragmatically true in this formal sense. However, a fundamental problem now arises: If such theories are simply classes of models, understood in our case as partial structures, how can they be regarded as partially or pragmatically true in precisely this sense? Such a theory, regarded in terms of partial structures, could be considered pragmatically true only *in a partial structure*! The point is, the models themselves cannot be regarded as true, or quasi-true, in any Tarski-like account because it is precisely their role to satisfy the sentences of the theory in its linguistic formulation.[10] The problem actually involves two intertwined issues: the ontological status of theories from the model-theoretic perspective, which we have alluded to a number of times already, and the dual epistemic and representative role of models within this approach. We believe that a partial (!) resolution of this problem is possible insofar as such a resolution can be outlined without having to plunge too deeply into epistemology and the philosophy of mathematics and that it can be arrived at through a careful consideration of Suppes's distinction between the "intrinsic" and "extrinsic" characterizations of theories.

Partial Structures and Partial Truth

As we indicated in the brief history with which this chapter began, Tarski's introduction of models as the "satisfiers" of sentences brought model theory within the domain of mathematics and consequently opened up the former to all the techniques of the latter. In effect, the models themselves transcended this role in "satisfaction" to become the focus of mathematical interest; they came to be treated as "individuals" in their own right, in Vaught's words. As the author of a recent textbook on model theory puts it:

> Thirty or forty years ago the founding fathers of model theory were particularly interested in classes specified by some set of axioms in first order predicate logic—this would include the abelian groups but not the Banach algebras or the primitive groups. Today we have more catholic tastes . . . model theorists are usually much less interested than they used to be in the syntactical niceties of formal languages—if you want to know about formal languages today, you should go first go to a computer scientist. (Hodges 1993, p. ix)

The model-theoretic approach in philosophy of science rides upon this move away from formal languages to structures, but the issue now arises as to whether this move should be understood merely as a shift in focus or rather, and more radically, as a casting off of the linguistic.[11]

One adherent of this approach who has adopted the more radical stance is van Fraassen, who, immediately after noting that the essence of Suppes's insight was that "to *present* a theory, we define the class of its models directly" (van Fraassen 1989, p. 222; our emphasis),[12] asserts that, "if the theory as such, is to be identified with anything at all—if theories are to be reified—then a theory should be identified with its class of models" (van Fraassen 1989, p. 222). This brings the issue into sharp focus: whether the role of models is to be merely *representational* or to be *constitutional*, in the sense that the class of models actually constitutes the theory. The "if" in van Fraassen's remark is a big if, and reification raises concerns that we do not intend to address here. Nevertheless, if models are granted a constitutional role in this sense, the question of how we are to understand the truth, or pragmatic truth, of theories becomes crucial.

In a footnote to the previous passage, van Fraassen writes:

> The impact of Suppes's innovation is lost if models are defined, as in many standard logic texts, to be partially linguistic entities, each yoked to a particular syntax. In my terminology here the models are mathematical structures, called models of a given theory only by virtue of belonging to the class defined to be the models of that theory. (van Fraassen 1989, p. 366, note 4 of chapter 8; cf. van Fraassen 1991, p. 483, note 2 of chapter 1)

It is this lifting of the yoke of a particular logico-linguistic formulation that van Fraassen adduces as a further reason for preferring the model-theoretic approach to the syntactic in philosophy of science. Of course, that this yoke can be lifted does not imply that in set theory we cannot individualize a theory, conceived as a class of models. Indeed, given any tractable class of models, this class is individualizable via a convenient set-theoretical predicate. The truth or pragmatic truth of the theory, in this case, is reduced to the truth or pragmatic truth of the set of sentences that determine the theory or to the species of structures corresponding to the family of structures of the theory.

In a criticism of van Fraassen's claims, Friedman argues as follows:

> Let us follow van Fraassen in identifying a theory with a class of models or structures. Suppose, however, that the class of models in question is a so-called "elementary class": i.e., that it contains precisely the models of some first order theory *T*. . . . Then the Completeness Theorem immediately yields the equivalence of van Fraassen's account and the traditional syntactic account. (1982, p. 276)[13]

Likewise, Worrall (1984) points out that to every "elementary" set of models there corresponds a consistent set of first-order sentences, and thus it is merely a matter of preference whether the model-theoretic or syntactic approach is employed.[14] Although this is true for the tractable classes of models, using the language of set theory, it is not correct in general.

Van Fraassen's response is to note that the requisite class of models may not be an "elementary" class in this sense (1989, p. 211), and indeed this must be so when it comes to scientific theories that involve the real number continuum.[15] Thus, he writes,

> Suppose I present a scientific theory . . . by describing its set of models, i.e., the set of structures it makes available for modeling its domain. Call this set *M*. Suppose in addition that, in doing so, the first object I mention is the real number continuum. . . . Now, you or I may formalise what I have done in some carefully chosen or constructed artificial language. . . . we can correlate the chosen set of axioms with the class of models of the language in which these axioms are satisfied. Call that set *N*. . . . But the real number continuum is something infinite. The Löwenheim-Skolem theorems then tell us at once that *N* contains many structures not isomorphic to any member of *M*. (van Fraassen 1985a, pp. 301–302)

The claim here is that we will never be able to *categorically* designate the models we are interested in with an axiomatized first-order theory, because of the existence of nonstandard models that must arise in any application of this approach to science.[16] The point, then, is that since almost all scientific theories will need the real number continuum, the class of models of such a theory will not be an elementary class, and thus these models will not be specifiable by an axiomatization even up to isomorphism.

Of course, such nonstandard models can be ruled out by stipulating the use of a "standard" or "intended" model. However, as van Fraassen notes, this method of description is simply not available to the syntactic approach (1989, p. 211). The upshot, then, is that "descriptions of structure in terms of satisfaction of sentences is much less informative than direct mathematical description" (van Fraassen 1989, p. 212). However, even granted this point, this "direct mathematical description" must be made within set theory—at least if it is the usual mathematics we all know and love that we are talking about here—and hence must be subject to the same limitations. Furthermore, it is not clear that we can dismiss satisfaction altogether. We may allow that in their descriptive or representational role, models offer significant advantages of both practice and principle over formalized languages, but the issue that now has to be addressed is that of understanding the *truth* of theories from this perspective.

Although one reason for the recent renewed interest in the model-theoretic approach is its introduction into the realist-empiricist debate, particularly by van Fraassen but also by Giere, Friedman, and others, there has been little discussion of this issue except in rather vague and unspecific terms. Van Fraassen, again, has at least addressed it, but his view is remarkably sanguine: "Truth and falsity offer no *special* perplexities in this context" (1989, p. 226; his emphasis). By "this context," he means a broadly realist view of theories from the perspective of the model-theoretic approach, according to which a theory is true if "the real world itself" is isomorphic to one of the class of designated models. (We shall later examine in more detail the relationship between theories and "the world" and in particular this relationship of isomorphism.)

Van Fraassen takes this to be equivalent to either of two formulations familiar to philosophers in general: A theory is true if and only if one of a set of possible worlds permitted by the theory is the "real" world, or if and only if the real world is as the theory says it is. These two "formulations" are standardly given as competing accounts of the nature of propositions in philosophy, where the latter act as both the objects of belief and the bearers of truth and falsity. This therefore suggests a propositional account of belief and—if theories are accepted as *objects* of beliefs—of theories, too, but it is not clear how well such an account sits with the antilinguistic emphasis noted previously.

Of course, van Fraassen himself deviates from this context in arguing that it is the theory's empirical adequacy, rather than truth as such, that should be our central concern.[17] Nevertheless, he agrees that a theory can be the object of belief and regarded as true or false: "With the realist I take it that a theory is the sort of thing that can be true or false, that can describe reality correctly or incorrectly, and that we may believe or disbelieve. All that is part of the semantic account of theories. It is needed to maintain the semantic account of implication, inference and logical structure" (van Fraassen 1989, p. 192). Hence, if we are to maintain the model-theoretic account of logical structure, we must retain the view that theories are the objects for doxastic attitudes, as expressed *linguistically* in statements and assertions. Van Fraassen explicitly acknowledges that such objects are typically taken to be propositions or, more generally, sets of propositions, and rejects the broadly instrumentalist view that claims that a theory should not be regarded as the kind of thing that can be true or false. The latter has been advocated by the structuralists Sneed, Moulines, and Stegmüller, who reject the traditional line that a theory is nothing more than a complex statement (Stegmüller 1979; Moulines 1976). As van Fraassen notes, the structuralist analysis of theories is very similar to the model-theoretic, although notably the former retains the Received View analysis of the relationship between theory and phenomena in terms of correspondence rules.[18] To believe a theory on this analysis is not to believe that the theory is true but to believe only that it bears a certain relationship to the phenomena. According to van Fraassen, this reinterpretation of our doxastic attitudes is too high a price to pay, in particular because it renders his own constructive empiricism unintelligible![19]

Thus van Fraassen rejects this "nonstatement" view and asserts both that theories can be the objects of beliefs and that truth is a category that applies to them, while acknowledging that according to constructive empiricism, truth as such is irrelevant to *success* in science.[20] Indeed, he writes, "the content of a theory is what it says the world is like; and this is either true or false. The applicability of this notion of truth-value remains here, as everywhere, the basis of all logical analysis" (van Fraassen 1989, p. 193; cf. p. 226).[21] But if the content of a theory is to be taken to be true or false and expressed, within the "statement" view, as a set of propositions, what are we to make of the dismissal of linguistic formulations recorded previously?

When van Fraassen asserts that "the semantic view of theories makes language largely irrelevant to the subject" (1989, p. 222), what he is referring to is the *description of the structure* of theories. In attempting to describe this structure, the model-theoretic approach offers significant advantages over logico-linguistic formulations.[22] Of course, to *present* a theory (note the choice of words again), we must do so logico-linguistically. That is a fact of communication, as van Fraassen notes, but to dismiss it as "trivial" (1989, p. 222) is perhaps going too far. When we turn from a discussion of the structure of theories to a consideration of our epistemic attitudes toward them, we have no choice but to resort to a linguistic formulation of some form or other. Evidence for the attribution of belief comes in the form of belief reports that are sentential. That which occupies the "..." in "believes that..." is standardly taken to be a statement, expressing a proposition, regarded as true or false. Without a linguistic characterization of some such form, we cannot employ Tarski-like formulations of truth, and without some such formulation, the notion of truth invoked previously must remain vague and undefined.

How are we to reconcile these model-theoretic and syntactic characterizations? How, that is, are we to reconcile our account of theories in terms of (classes of) models, with our account of belief that such theories are true, or partially true, where this is understood to be truth, or quasi truth, in a model? The beginnings of an answer to these questions can be constructed by acknowledging what is implicit in the previous discussion—namely, the dual role of models as that in terms of which theories are represented and that in terms of which sentences are satisfied. This dual role is an integral aspect of the history of model theory itself, and the difference relates to that between what Suppes calls, respectively, the "extrinsic" and "intrinsic" characterizations of a theory (1967, pp. 60–62).

In the first respect, our primary concern is with the structure of the theory and with the relationships between theories themselves and between theories and "the world," understood in terms of that structure. Characterizing that structure and these relationships in purely logico-linguistic terms generates a disparity with scientific practice that verges on absurdity: Different formulations come to be regarded as different theories and, with the relationship between theories and the world characterized logico-linguistically in terms of correspondence rules, a change in theory follows on a change in experimental technique. What is required here is an "extrinsic" characterization in terms of which we can regard theories from "outside" a particular logico-linguistic formulation, as it were; indeed, it is only if we have such a characterization to hand that we can even formulate the question of whether "a certain theory" can be logically axiomatized in the first place: "To ask if we can axiomatize the theory is then just to ask if we can state a set of axioms such that the models of these axioms are precisely the models in the defined class" (Suppes 1967, p. 60).

In this respect, then, models play a *representational* role, and to *present* a theory is to define its class of models directly. However, assuming that theories are the kind of thing that can be reified, to assert that a theory *is* simply a class of models is to accede to a strain of antimetaphysical thinking, similar in kind to that which claims that a physical object *is* simply a bundle of properties. The ontological status of theories is an important issue, but, for the moment at least, we wish to take on board the minimal amount of metaphysical baggage that we can get away with! Furthermore, if theories are, ontologically, models and the latter are mathematical entities, then the question as to their status becomes subsumed under that of the status of such entities in general; further discussion requires the elaboration of a philosophy of mathematics,[23] which we do not intend to enter into here. Finally, it seems to us that there is a disadvantage to adopting such an ontological view of theories—namely, that one cannot then directly avail oneself of Tarski-like definitions of truth, and with nothing put forward to replace the latter, this becomes a serious disadvantage indeed.

So, on our view, theories—whatever they *are*, ontologically—are represented, from the extrinsic perspective, in terms of models or classes of models, and from the intrinsic perspective, can be taken to be the objects of epistemic attitudes, and in particular that of belief, and be regarded as true, empirically adequate, quasi-true, or whatever. To say that the models *themselves* can act as such objects is to switch from one perspective to the other, with problematic consequences: Set-theoretic structures themselves cannot be truth bearers. This is the problem with *identifying* theories with such structures, as, strictly speaking, we could then no longer claim that a theory is true or false. And hence we insist: Models are set-theoretical constructs, and so they are, in a certain sense, extralinguistic entities. On this point the proponents of the semantic approach are correct, and representing scientific theories using such models avoids some of the more problematic consequences of regarding theories linguistically. From this (extrinsic) standpoint, models can be treated as distinct "objects" or individuals, in Addison's sense. We can operate with models, consider families of them and so on, and manipulate them for various purposes, such as defining the ultra-product of first-order structures and isometry between metric spaces or proving a representation theorem in the theory of groups, an isomorphism theorem for higher-order geometry, and so forth. This is the stance of the mathematician.

However, in order to *talk about* models—that is, to develop mathematics—we have to use some form of language (or indeed, several), and this language must contain a goodly portion of extant set theory. In this way we can develop a general theory of mathematical structures, including partial ones.[24] In this sense, then, we cannot entirely cut models free from language, although, of course, here we are not talking about "English" or "Portuguese" but the language of set theory. This language itself may be viewed as an abstract structure (namely, an algebra), and the forms and nature of the symbols are, strictly speaking, irrelevant. We need only the symbols to talk about the algebraic language of set theory and about sets.

Nevertheless, when a model or family of models is used as a representational "device," in the manner in which they are used by the semantic approach, we can say that the models are "true" as a "façon de parler" or "abuse of language." Underlying this abuse is the following picture: When describing a particular domain Δ, one assumes that it is possible to present such a description based on the properties of the elements of Δ, on the relations holding between them, on the functions that can be defined on Δ, and so on, and that somehow such a description correctly reflects certain aspects of this domain. Of course, this is a strong assumption, which may turn out to be false, but it underlies various structural accounts, including Tarski's. Furthermore, with regard to the application of pragmatic or quasi truth, there are, roughly speaking, three alternative ways to view the relationship between some pragmatic structure $\mathcal{A}$ and the particular domain Δ that it models:

1. $\mathcal{A}$ may be suitable or appropriate with respect to Δ if (intuitively speaking) it "captures" the relevant aspects of Δ; that is, if the proposition "$\mathcal{A}$ 'captures' the relevant aspects of Δ" is true in the correspondence sense.
2. $\mathcal{A}$ is pragmatically suitable or appropriate with respect to Δ if the proposition "$\mathcal{A}$ 'captures' the relevant aspects of Δ" is pragmatically true.
3. $\mathcal{A}$ is approximately suitable or appropriate with respect to Δ if the proposition "$\mathcal{A}$ 'captures' the relevant aspects of Δ," when taken as a hypothesis, can from a loose point of view save the appearances in Δ (this might be regarded as an informal version of case 2).

Again, we insist, these are just elements of the picture that underlies our "façon de parler" in talking of models as true, quasi-true, and so forth.

Let us return to our principal issue. Since our epistemic attitudes are expressed by belief reports that are sentential in nature, we must shift to the "intrinsic" characterization of theories in order to accommodate them. Here the models play an *epistemic* role as the "possible realizations" that satisfy the sentences of the belief reports, and by playing such a role they allow truth and, significantly, quasi truth to be defined. It is important to acknowledge the point that the fact that belief *reports* are expressed in terms of sentences does not imply that the *objects* of the beliefs themselves are sentential in character. Of course, *grammatically* the verb "believe" must take as its object some sentential clause or pronoun, and this might be adduced as a reason for concluding that doxastic objects are sentential, but, as Grandy points out, "This argument is unconvincing: we could argue in parallel that since every report of what a painting or picture shows contains a sentence, that paintings and pictures consist of sentences" (Grandy 1986, p. 330). The essential point then is that "The fact that in belief reports the content sentences are linguistic reflects only the fact that reports are sentences, not on the structure of the representation being reported" (Grandy 1986, p. 323). As far as the representations of science are concerned, their structure can be described in model-theoretic terms, but to accommodate the doxastic attitudes of scientists themselves, we must shift to the intrinsic characterization in the context of which formal notions of truth and quasi truth can be defined. We shall return to a discussion of the nature of these attitudes in chapter 4, but for now we simply wish to emphasize two points:

1. Of course, it is because the theories that are the objects of scientific belief are *partial*, in the (extrinsic) sense defined previously, that the appropriate doxastic attitude is that of belief in their partial truth, understood intrinsically.
2. Both characterizations are required in order to accommodate scientific practice.

We shall explore just such an accommodation in the next chapter, where we argue that our formal notion of quasi truth provides the epistemic underpinning for a unitary treatment of models in science from the partial structures perspective.

Appendix 1: Structures, Species of Structures and Models

Our aim in this appendix is to define the concept of a mathematical structure. The exposition will be informal, but there is no difficulty in setting it in a more rigorous and formal framework.

Mathematical structures can be introduced in set theory or in higher order logic (type theory). Both ways are essentially equivalent, and one can pass easily from one formulation to the other. In general, professional mathematicians, led by Bourbaki (cf. Bourbaki 1968) and model theorists prefer the first approach, but certain logicians, such as Russell and Carnap, have explored the second (see, for example, Russell 1954 and Carnap 1949 and 1958). We shall define the mathematical structures in set theory; that is, we shall conceive of them as set-theoretical constructs. We further suppose that we are working in a standard system of set theory, such as that of Zermelo-Frankel (Bourbaki 1968; Halmos 1960).

Given the sets $A_1, A_2, \ldots, A_n$, we may form other sets from them by taking their sets of subsets or the sets of subsets of the Cartesian product of some or all of them (the Cartesian product will be denoted by x and the set of subsets of A by $\mathbb{P}(A)$). Starting again with the original and new sets, we can repeat these operations. The sets so obtained are said to belong to the type hierarchy of sets on $A_1, A_2, \ldots, A_n$ as base (this hierarchy includes $A_1, A_2, \ldots, A_n$). Thus, for example, given the two sets A and B, the following sets belong to the type hierarchy of sets on A, B as base:

(I) $A, B, \mathbb{P}(A), \mathbb{P}(A \times B), \mathbb{P}(A \times \mathbb{P}(A)), \mathbb{P}(B), \mathbb{P}(\mathbb{P}(A) \times \mathbb{P}(B),), \mathbb{P}(\mathbb{P}(A))$

The elements of the sets belonging to the type hierarchy on $A_1, A_2, \ldots, A_n$ as base can be characterized by their types. We define the type of such an element as follows:

1. The type of the members of A_i is a_i, where a_i is a fixed symbol, $1 \leq i \leq n$, $a_i \neq a_j$ for $i \neq j$;
2. If the elements of the set X_i have type x_i, $1 \leq i \leq k$, then the type of the members of $\mathbb{P}(X_1 \times X_2 \times \cdots \times X_k)$ is $\langle x_1, x_2, \ldots, x_k \rangle$.

Let us suppose that the types of the elements of A and B are *a* and *b*, respectively. The types associated with the elements of each of the sets (I) are as follows:

$a, b, \langle a \rangle, \langle a,b \rangle, \langle a,\langle a \rangle \rangle, \langle b \rangle, \langle \langle a \rangle, \langle b \rangle \rangle, \langle \langle a \rangle \rangle$

Relations are always finitary; that is, they relate a finite number of elements. A monadic relation is simply a set. Functions or operations constitute particular cases of relations, and fixed elements are identified with 0-adic constant operations. Normally an n-adic relation containing only one n-tuple is identified with its n-tuple.

To motivate the definition of structure, let us consider some examples:

1. A group $\mathcal{G}$ is a structure of the form $\langle A, \oplus, s \rangle$, where $\oplus$ is a binary operation on A, s is a unary operation on the same set, and these objects are subject to the usual axioms for groups. If we denote the type of elements of A by a, then the types of $\oplus$ and s are $\langle a, a, a \rangle$ and $\langle a, a \rangle$. The unit set of $\mathcal{G} = \langle A, \oplus, s \rangle$ belongs to the type hierarchy on A as base, and the type of $\mathcal{G}$ is by definition $\langle , \langle a, a, a \rangle, \langle a, a \rangle \rangle$.
2. A vector space $\mathcal{V} = \langle R, V, +, \cdot \rangle$ is a structure which has two base sets: R (the set of scalars) and V (the set of vectors). The operations + and $\cdot$ satisfy certain axioms, and $\mathcal{V}$ belongs to the hierarchy on R and V as base sets. In this case, R has to be a field, and consequently one of the base sets possesses a structure already. We say that R is the auxiliary set and V the principal set.[25]
3. A topological space $\mathcal{T}$ constitutes a structure of the form $\langle E, \mathcal{F} \rangle$, where E is a set and $\mathcal{F}$ is a family of subsets of E, called the family of open sets of $\mathcal{T}$ or the topology of $\mathcal{T}$. The open sets are subject to the following conditions (axioms):
   a. $\varnothing$ and E belong to $\mathcal{F}$;
   b. Any union of elements of $\mathcal{F}$ belongs to $\mathcal{F}$;
   c. If $A, B \in \mathcal{F}$, then $A \cap B \in \mathcal{F}$.

   $\mathcal{T}$ is again an element of the hierarchy on E as base.
4. The set $\mathcal{O} = \langle A, \leq \rangle$, where $\leq$ is a binary relation on A satisfying the reflexive, antisymmetric, and transitive laws, is called an ordered system. It is a structure with one base (principal) set. If the type of elements of A is a, the type of $\mathcal{O}$ is, by definition, $\langle , \langle a, a \rangle \rangle$.

Taking the preceding examples in consideration, we define the concept of a mathematical structure as follows: A structure is a finite sequence composed of certain sets, called base sets, together with certain relations of fixed types over these sets. Let $\mathcal{L}$ be the language of set theory. A predicate, in $\mathcal{L}$, is a formula with only one free variable. If $\mathcal{S}$ is a structure, there is a predicate P, which is formed by two parts: the first, P_1, shows how $\mathcal{S}$ is built from its base sets (it individualizes the type of $\mathcal{S}$); the second, P_2, is the conjunction of the axioms of $\mathcal{S}$. P is the conjunction of P_1 and P_2. Of course, in P there are parameters, corresponding to the base sets. We say that $P(\mathcal{S})$ is the species of structures on the base sets of $\mathcal{S}$. Supposing that these base sets are $A_1, A_2, \ldots, A_n$, P may be written as follows:

$$P(\mathcal{S}; A_1, A_2, \ldots, A_n)$$

Then, the species of structures corresponding to $\mathcal{S}$ is, by definition, the predicate

$$\mathcal{P}(X) \leftrightarrow \exists X_1 \exists X_2 \ldots \exists X_n P(X; X_1, X_2, \ldots, X_n),$$

whose meaning is clear.

$\mathcal{S}$ is called a structure of species $\mathcal{P}$ or a $\mathcal{P}$-structure. Species of structures are also called Suppes predicates (see da Costa and Chuaqui 1988). Obviously, the same family of structures can be characterized by numerous Suppes predicates, all equivalent to one another. By an abuse of language, we usually talk of *the* species of structures satisfied by a given structure. In principle, to axiomatize a mathematical theory is to get the corresponding species of structures. And the development of a general, rigorous theory of mathematical structures offers no particular difficulty (see Bourbaki 1968; da Costa and Chuaqui 1988). Bourbaki's treatment is purely syntactical but there is no problem to change it into a semantical presentation (da Costa and Chuaqui 1986; for more on Bourbaki's work, see Mostowski 1967).

Let $\mathcal{S}$ be a structure and $\mathcal{L}$ a language that may be interpreted in $\mathcal{S}$. Then, if s is a sentence in $\mathcal{L}$, it is easy to define when s is true in $\mathcal{S}$ (according to an interpretation). In general, if Δ is a set of sentences of $\mathcal{L}$, each of which is true in $\mathcal{S}$, then $\mathcal{S}$ is called a model of Δ. By another abuse of language, when $\mathcal{S}$ is a $\mathcal{P}$-structure, we say that $\mathcal{S}$ is a model of $\mathcal{P}$.

First-order structures are structures of the kind

$$\langle A, R_i \rangle_{i \in I}$$

where A is a set ($\neq \emptyset$) and R_i, $i \in I$ are relations over A (in particular, functions over A). First-order languages are interpretable in such structures. In this work, we usually consider first-order structures only because the theory of such structures can be extended to structures of any type whatsoever, and it exemplifies nicely the general theory (Mostowski 1967). And, of course, just as one can define partial first-order structures, so one can define partial structures of any type. In this way, we can define pragmatic structures of any type, pragmatic truth for sentences of higher order languages, and so on.

Appendix 2: Axiomatization in Mathematics

An axiomatic theory has some primitive (nondefined) concepts and a set of primitive propositions, also called axioms or postulates. The other notions are obtained by definition and the theorems by proof. From the set-theoretic point of view, the primitive concepts constitute sets, which are implicitly defined by the set-theoretic relations appearing in the axioms. In fact, to axiomatize a theory is simply to present a species of structures having the primitive sets of the theory or some sets associated with them as base sets. The principal base sets and the relations explicitly occurring in the axioms (when they are not defined relations) compose the collection of primitive concepts of the theory or of the corresponding species of structures.

So, as Bourbaki in essence showed, to axiomatize a theory $\mathcal{T}$ is to formulate a set-theoretic predicate $\mathcal{P}$. The structures that are models of $\mathcal{P}$ constitute a family $\mathcal{F}$ of structures that in a certain sense determines $\mathcal{P}$. In consequence, to study $\mathcal{T}$ we can proceed syntactically by means of $\mathcal{P}$ or semantically, taking into account the family $\mathcal{F}$. The syntactical and semantical methods appear as complementary. In general, given $\mathcal{F}$, we can in principle obtain $\mathcal{P}$, and starting with $\mathcal{P}$, the family $\mathcal{F}$ is determined.

We define $\mathcal{P}$ in set theory. In addition, this theory can be formalized—that is, transformed into a calculus whose symbols have in principle no meaning at all (cf. Bourbaki 1968). As a corollary, the entire theory of $\mathcal{P}$—the theory of $\mathcal{P}$-structures—can also be formalized. Therefore, any mathematical theory is formalizable (at least potentially). However, in normal mathematics, we usually proceed informally, handling the mathematical structures in naive set theory (see Halmos 1960 and Bourbaki 1968).

Set theory is axiomatizable (and also formalizable) in first-order logic (it belongs to the class of first-order theories). However, we usually axiomatize the mathematical theories in set theory, and the resulting axiomatic systems are not said to be first-order since they are encompassed by all set-theoretical methods and ideas.

3

Models and Models

The Models of Model Theory and the Models of Science

It has long been argued that the model-theoretic approach acquires a certain plausibility from the fact that models are also extensively used in science itself (see, for example, Suppes 1970). Consider, for example, the following passage from a textbook of nuclear physics:

> In a nucleus ... there are too few particles for a statistical treatment, and there is no overriding centre of force which would enable us to treat the forces between nucleons as small perturbations. For this reason, physicists have fallen back on the "as if" methods of attack, also known here as the method of nuclear models. This method consists of looking around for a physical system, the "model," with which we are familiar and which in some of its properties resembles a nucleus. The physics of the model are then investigated and it is hoped that any properties thus discovered will also be properties of the nucleus. ... In this way the nucleus has been treated "as if" it were a gas, a liquid drop, an atom and several other things. (Elton 1961, p. 104)

However, the sense of "model" here is apparently quite different from that discussed in the previous chapter; Suppe calls such models "iconic" since they function as "icons" of what they are modeling[1] (Suppe 1977, p. 97). Another, perhaps better known, example is the infamous "billiard ball" model of a classical gas. Focusing on such iconic models leads to what is sometimes regarded as an alternative approach to models in science that emphasizes the role of *analogy*. This particular line is exemplified in the works of Hesse (1970) and Achinstein (1968), to give two notable examples, and it may be regarded as part and parcel of a more general view that claims that the use of models in scientific *practice* is far too complex to be captured in terms of mathematical structures (see also more recently Cartwright, Shomar, and Suárez [1995] and Morrison [1999]). We shall return to a further consideration of practice in chapter 6, but at this point all we wish to show is that the model-theoretic approach *can*, in fact, accommodate many, if not all, of the apparently different kinds and uses of models in science. It can, that is, if we employ partial structures, and hence we regard this as constituting one of the most important elements in support of our position.

Let us return to the origins of the approach and to Tarski's allusion to the future of model theory in 1935: "Only the future can definitely say whether further investigation in this field will prove to be fruitful for philosophy and the special sciences, and what place semantics will win for itself in the totality of knowledge. But it does seem that the results hitherto reached justify a certain optimism in this respect" (Tarski 1935, p. 407). Tarski's optimistic outlook was notably adopted by Suppes, who begins his paper "A Comparison of the Meaning and Uses of Models in Mathematics and the Empirical Sciences" with a series of quotations referring to models, commencing with Tarski and proceeding through physics, the social sciences, and mathematical statistics (Suppes 1961). His overall conclusion is that "the concept of model in the sense of Tarski may be used without distortion and as a fundamental concept in all of the disciplines from which the above quotations are drawn. In this sense I would assert that the meaning of the concept of model is the same in mathematics and the empirical sciences. The difference to be found in these disciplines is to be found in their use of the concept"[2] (Suppes 1961, p. 165).

This is certainly a bold claim, its boldness deriving at least in part from the counterintuition that the kinds and roles of models in science are simply too diverse for there to be a unitary notion embracing them all, much less one that can be confined to a formal, set-theoretic straitjacket. Black, for example, is emphatic: "I cannot agree with Suppes that 'the meaning of the concept of model is the same in mathematics and the empirical sciences'" (1962, p. 262).[3] In his well-known taxonomy, first presented in 1958, there are scale models, covering changes of scale in any relevant dimension,[4] analogue models, "mathematical" models, and "theoretical" models, which may be regarded realistically or as "heuristic fictions";[5] Tarski's definition of model and consequently Suppes's discussion belong, he writes, to a "different conception" (Black 1962, p. 262). Moreover, models are used extensively but divergently in science, for heuristic purposes (Black 1962; Hesse 1970; Toulmin 1953, pp. 35–39; Nagel 1961, pp. 106–114; Redhead 1980; Hartmann 1995), for "existential" interpretations (Black 1962), to qualitatively "probe" a complicated theory (Redhead 1980), and to test theories when a computation gap exists (Redhead 1980). Is it possible that this plethora of different kinds and uses of models in science can be captured in terms of *any* unitary, formal framework?

Achinstein's Criticisms

The most serious attempt at tackling this question so far has been presented by Achinstein, who answers in the negative (Achinstein 1968, pp. 209–225). However, his concern is with the earlier "semantical" theory of the Received

View, which regards a model as a set of *statements* (Suppe 1977, p. 95ff.) obtained by interpreting terms in an axiomatized calculus, rather than as a *structure* in which the formulas of the calculus are satisfied (Achinstein 1968, p. 228, fn. 4).[6] It is worth taking a look at this earlier theory of models and Achinstein's criticisms of it in order to see more clearly the advantages offered by the model-theoretic view, if nothing else. This will prepare the ground for our own positive response to the question.

The technical apparatus of the Received View is well known:[7] Theories are construed as axiomatic calculi in which theoretical terms are given a partial observational interpretation by means of "correspondence rules," which function as a kind of dictionary in relating terms of the theoretical vocabulary (of the language in which the theory is expressed) with terms of the observational vocabulary. Hence, the following components of a theory can be distinguished:

1. an abstract formalism $\mathcal{F}$
2. a set of theoretical postulates (axioms) $\mathcal{T}$
3. a set of "correspondence rules" $\mathcal{C}$

$\mathcal{F}$ consists of a language $\mathcal{L}$ in terms of which the theory is formulated and a deductive calculus defined. $\mathcal{L}$ will contain logical and nonlogical terms; the latter can be divided into the set of observation terms and the set of theoretical terms.

A "partial interpretation" of the theoretical terms and the sentences of $\mathcal{L}$ containing them is provided by the theoretical postulates, which contain only theoretical terms and the correspondence rules.[8] These postulates are thus mixed sentences, containing both theoretical and observational terms. The correspondence rules effectively correlate the nonlogical, theoretical terms with observable phenomena. Thus, as Carnap noted,

> All the interpretation (in the strict sense of this term, i.e., observational interpretation) that can be given for L_T [the theoretical language] is given in the C-rules [correspondence rules], and their function is essentially the interpretation of certain sentences containing descriptive terms, and thereby the descriptive terms of V_T [the theoretical vocabulary]....
>
> For L_T we do not claim to have a complete interpretation, but only the indirect and partial interpretation given by the correspondence rules....
>
> Before the C-rules are given, L_T, with the postulates T and the rules of deduction, is an uninterpreted calculus.... Then the C-rules are added. All they do is, in effect, to permit the derivation of certain sentences of L_O [the observational language] from certain sentences of L_T and vice-versa. They serve indirectly for derivations of conclusions in L_O, e.g. predictions of observable events, from given premises in L_O, e.g. reports of results found by observation, or the determination of the probability of a conclusion in L_O on the basis of given premises in L_O. (Carnap 1956, pp. 46–47)

If $\mathcal{T}$ is the conjunction of theoretical postulates and $\mathcal{C}$ the conjunction of the correspondence rules, then a scientific theory is taken to consist of the conjunction of $\mathcal{T}$ and $\mathcal{C}$, designated by "$\mathcal{TC}$."

What, then, is the Received View of *models*? The essential idea has already been mentioned: A model for some theory is obtained by substituting interpreted predicates for certain predicate symbols, interpreted individual constants for certain individual constants, and so on, in the underlying formal calculus. By supplying "flesh" for the formal skeletal structure in this sense (Nagel 1961, p. 90), it is claimed that the model can give meaning to the symbols of the calculus (Hutten 1953–54, pp. 293–295). Since a *theory* on the Received View is just such a partially interpreted, formal (that is axiomatized) calculus, this characterization blurs the distinction between theories and models.[9] The point was well put by Braithwaite, who wrote that

> a model for a theory T is another theory M which corresponds to the theory T in respect of deductive structure. By correspondence in deductive structure between M and T is meant that there is a one-one correlation between the concepts of T and those of M which gives rise to a one-one correlation between the propositions of T and those of M which is such that if a proposition in T logically follows from a set of propositions in T, the correlate in M of the first proposition in T logically follows from the set of correlates in M of the propositions of the set in T.
>
> Since the deductive structure of T is reflected in M, a calculus which expresses T can also be interpreted as expressing M: a theory and a model for it can both be expressed by the same calculus. Thus, an alternative and equivalent explication of model for a theory can be given by saying that a model is another interpretation of the theory's calculus. (Braithwaite 1962, p. 225)

However, although the formal structure of a theory and "its" model(s) will be identical, as they must be on this view, the *interpretation* will not, of course; in particular, the individual constants in the calculus will differ in their denotation (see Braithwaite 1962). Typically, and crucially, they will denote objects or processes that are taken to be more familiar than the objects and processes originally denoted by the theory as partially interpreted.

Let us take the overused example of the billiard ball model of a gas: The individuals denoted by the relevant terms of the theory are the gas atoms, whereas those of the model are billiard balls. This is an obvious point, but what is important here is that in a model the individual terms denote objects that we already have some degree of familiarity with (that, it is claimed, is one of the reasons for proposing and using models). Thus a model may be taken to provide an *analogy* (Nagel 1961, chapter 6; Hesse 1953–54 and 1970), and in this respect the difference between "iconic" models and other kinds also becomes blurred (cf. Black 1962; again, this is something we shall return to later). Nagel, for example, took a model for a theory to be both semantic and iconic (Nagel 1961; Suppe 1977, pp. 96–98).[10] The billiard ball model of a gas is iconic in that a collection of billiard balls is obviously only similar and not identical to a collection of atoms, and it is semantic in that an interpretation is given to the theorems of the theory in terms of the system of billiard balls such that the theorems come out true under the interpretation. However, as Suppe has pointed out, if the theorems are empirically true, then under the Received View (realistically interpreted) the semantic model would be the world, yet the billiard ball model could never be such. The likes of Braithwaite and Carnap would reply that this is precisely why models can be so dangerous: They encourage a confusion of their domain with that of the theory.

According to Hesse, the mathematical structures specified by the formalism of the theory could be iconic models (Suppe 1977, p. 99) and are essential components of theories, since without them theories would have no explanatory power. The idea is that without iconic models theories would not be testable, since the theoretical terms must be interpreted via a model in such a way as to delineate the kinds of phenomena relevant to it. As Suppe points out, this merely establishes that a *semantic* interpretation is essential, as required by the later versions of the Received View. Furthermore, the claim that an iconic model is *required* for explanation either founders on the counterexample of quantum mechanics or involves a notion of iconic model that is so broad as to be indistinguishable from that of a partial interpretation. Nevertheless, iconic models do play an important role in science, and we shall return to them shortly.

Now, it is at precisely this point that opinions diverge as to the role and significance of models in science. By replacing the standard or intended interpretation of the theory with one that is articulated in more familiar terms, it has been claimed (by Campbell, Hesse, and others) that a model can contribute to our *understanding*. This "modelist" view, as it has been called, was vigourously resisted by Carnap, for example, who insisted, "It is important to realise that the discovery of a model has no more than an aesthetic or didactic or at best a heuristic value, but it is not at all essential for a successful application of a physical theory" (Carnap 1939, p. 68). Likewise, Braithwaite explicitly contrasted the "modelist" account of understanding with the "contextualist" approach: Theoretical meaning and understanding are provided contextually by the relations between theoretical terms and the correspondence rules relating the latter to observable terms (Braithwaite 1962). And that, the contextualists insist, is all you need.

The modelist can respond that what the model provides is an interpretation in *familiar* terms. However, this is not always the case, as the example of quantum mechanics reveals. Indeed, an argument can be made that the development of the "new" quantum theory of the mid-1920s and the subsequent claims that the theory could not be supplied with a "model" in terms of more familiar quantities (hidden variables) played a significant role in supporting the "contextualist's" view of models. In response to these examples, where no model articulated in "familiar" terms is available, Hesse, famously, broadened her definition of a model to include "any system, whether buildable, picturable, imaginable, or none of these" (1970, p. 19). But now we appear to have lost one of the principal advantages models were claimed to have over theories to begin with!

From a modern perspective, the "contextualist" account of models—as supplying just another interpretation—has been seen as too meager even by the Received View's own standards. If models are just interpretations, Psillos asks, "Why not abandon [them] altogether and go just for theories?" (Psillos 1995, p. 109). But that is just what Braithwaite and Carnap do, in effect. Of course, the Received View's answer seems too meager by those *current* standards that call upon models "to concretise, specify or approximately realise assumptions about the physical system described by the theory, as for instance is the case with the Bohr model of the atom and its embedding atomic theory" (Psillos 1995, p. 109). This is precisely to insist on the "modelist" view of models as *models of systems*, which the proponents of the Received View rejected. The instance given of the Bohr model is telling: As we have suggested, part of the motivation for the Received View's insistence that understanding should not be tied to the provision of a model surely lies with the perception that, as standardly interpreted, no such models and hence no such understanding could be supplied for the powerful new quantum mechanics that ultimately swept aside Bohr's attempt (we shall return to this example in chapter 5).

Not all theories are like quantum mechanics, however. Does this account put the Received View radically out of step with scientific practice? Not necessarily: As an interpretation, a model need not be sound; its initial propositions need not be true or thought to be true (Braithwaite 1962, p. 225). As Braithwaite acknowledges, "Scientists frequently use quite imaginary models" (1962, p. 226), giving as an example 19th-century mechanical models for optical theory.[11] Nevertheless, if thinking in terms of models is always "as if" thinking (Braithwaite 1953, p. 93), then it might seem that the Received View cannot adequately accommodate the role of models as "a tool of discovering the furniture of the world" (Psillos 1995, p. 110). Needless to say, the antirealist will have a definite view on this.[12] Is the Received View then committed a priori to antirealism (Psillos 1995, p. 111)? Not necessarily. A proponent might argue that insofar as an "as-if" attitude is *not* adopted with regard to those aspects of scientific practice that we are nevertheless uncomfortable in regarding as full-fledged theories, they can be called *theoruncula* or (affectionately) *theorita* and still be understood in terms of interpretations of a calculus (Braithwaite 1962, p. 225). A *theorita* can be regarded as incorporating some element of truth, while acknowledged as not quite measuring up to a complete theory. Our point is that the Received View of models does have a range of resources it can draw upon to deal with these criticisms.

Nevertheless, others arise once we begin to pay closer attention to scientific *practice*. Perhaps the most detailed set of objections to the Received View of models has been offered by Achinstein: First of all, he points out, although theoretical models might be viewed as sets of statements, the "representational" kind, such as scale models, cannot since they are *objects* (1968, pp. 230–231; cf. Downes 1992); an example here might be Crick and Watson's famous wire and tinplate model of DNA. Second, the denotation of individual terms in certain kinds of models is not different from that of such terms in the theories themselves (Achinstein 1968, p. 231); the theoretical model of a gas refers, not to billiard balls, but to gas atoms. Third, given this, regarding both theoretical models and theories themselves in terms of partial interpretations of the same uninterpreted calculus fails to capture the essential differences between the two. In particular, a theoretical model, such as the Bohr model of the atom, does not provide an interpretation of, or give meaning to, terms in some calculi but rather can be regarded as a set of simplifying assumptions that employs such terms, *as already interpreted*, in order to describe the system of interest (Achinstein 1968, pp. 233–235). Similarly, in an analogy, "the laws governing one phenomenon are not treated as mere uninterpreted formulas that the other laws are supposed to interpret" (p. 236). In the analogy between heat conduction and electrostatic attraction, for example, it is absurd to regard the term "source of heat" as providing the *meaning* for "source of electricity" or vice versa.

Fourth, it is implausible to require that the formal structure of the theory and model be *identical*, as the Received View of models does. Maxwell's "imaginary" model of the electromagnetic field, for example, is not an interpretation of the calculus of any theory that would be recognized as such (Achinstein 1968, p. 239). With regard to analogies and representational models, the "identity" at work here is typically only *partial* and may hold between aspects of the system that are broadly nonformal; thus there is an analogy between the Bohr atom and the solar system, even though the relevant mathematical expressions are formally very different (Achinstein 1968, pp. 240–241). Insisting on identity of the underlying calculi imposes a requirement of completeness with regard to similarity that is entirely inappropriate in the case of analogies (where familiarity breeds dissimilarity) and furthermore fails to capture the nonformal similarities that may be absolutely essential where both representational and analogue models are concerned (Achinstein 1968, pp. 244–248).

Finally, Achinstein also attacks the view that all models are analogies, a view he dismisses as simply implausible. Bohr's model of the atom is a theoretical model and not itself an analogy, and although the laws of heat conduction describe phenomena that are *analogous* to those of electrostatics, such laws cannot be construed as a theoretical model of the latter (Achinstein 1968, p. 248). Nor is it the case that analogies are required to formulate models, theoretical or otherwise. Hence the *identification* of the two is simply unwarranted. We shall return to this issue of accommodating analogies shortly.

Having considered and effectively demolished what he takes to be the Received View of models, Achinstein then goes on to discuss the purported role attributed to it in the *practice* of science (1968, pp. 253–256). As he notes, although such a role is often alluded to in the literature, the claim that models are important in both formulating and extending theories is rarely fleshed out. When it is, this role is seen to be dependent on identity in formal structure again: A set of empirical laws is perceived to have the same formal structure as the theorems of an already established theory, which encourages the development of a new theory explaining these laws, or a model, sharing such structure with a theory, may possess further structure that will suggest an analogous extension of the theory. However, identity of *formal* structure cannot be a bridge for conferring plausibility, whether from an established theory to a newly formulated one or from a model to a theoretical extension:[13] "the plausibility of statements must be determined by an appeal to content and not logical form alone" (Achinstein 1968, p. 254). Yet stripped of any plausibility claim, this account of the role of models can offer no heuristic at all. There is simply no such role for models to play.

These are powerful arguments. However, Achinstein's conclusion that there can be *no* "theory" of models and analogies in science (1968, pp. 256–258) is unduly pessimistic, as we shall now argue.

Icons and Analogies

Let us return to the fundamental idea that a model is an *icon*, in the sense of "literally embodying the features of interest in the original" (Black 1962, p. 221). To function appropriately, some of the features of the model will be essential to the representation of the things for which it stands, whereas others will be irrelevant: "There is no such thing as a perfectly faithful model; only by being unfaithful in some respect can a model represent its original" (Black 1962, p. 220).[14] Even analogue models may be viewed as "iconic" but in a more abstract way: "The analogue model shares with its original not a set of features or an identical proportionality of magnitudes [as in the case of scale models] but, more abstractly, *the same structure or pattern of relationships*" (Black 1962, p. 223; our emphasis). The "dominating principle" here is that of isomorphism (and here Carnap [1958] is cited), but to insist that "every incidence of a relation in the original must be echoed by a corresponding incidence of a correlated relation in the analogue model" (Black 1962, p. 222) is too strong a demand, as Achinstein indicated. What is required is some *weaker* notion that picks up just some relevant subset of the relations concerned. We shall suggest such a weaker relationship later.

Setting to one side the issues concerning understanding touched on here, let us return to the "modelist" account since it offers further illumination of analogies in this context. Hesse, in particular, famously proposed a tripartite distinction between "positive," "negative," and "neutral" analogies (1970). Thinking once again of the billiard ball model, the properties of the balls that are ascribed to gas atoms, such as motion and impact, form part of the positive analogy; properties such as, say, color, which the balls have but the gas atoms do not, form the negative analogy; and, crucially, those properties of the elements of the model of which it is not known whether they are possessed by the atoms or not constitute the neutral analogy. We say "crucially" because it is through the introduction of the neutral analogy that Hesse can defend the modelist thesis that models are essential to scientific practice in that they render theories genuinely predictive: "these are the interesting properties because . . . they allow us to make new predictions"[15] (Hesse 1970, p. 8). The crucial idea is that it is the neutral component that captures the aspect of *novelty*; because we simply do not know, of course, whether the relations expressed by the neutral analogy hold or not, any predictions based on them will be genuinely new.[16]

Dropping the negative analogy gives what Hesse denotes by "model_1" or what Achinstein calls a "theoretical" model, which, without this analogy, ceases to be regarded as referring to a system different than that which is being modeled. With the negative analogy included, we have what Hesse calls "model_2" or a form of "analogue" model. What is fundamentally important, however, is the inclusion in both cases of the *neutral* analogy; it is this which distinguishes the model from a theory as standardly understood,[17] and it is this which gives the model its "growing points": "My whole argument is going to depend on these features and so I want to make it clear that I am not dealing with static and formalized theories, corresponding only to the known positive analogy, but with theories in the process of growth" (Hesse 1970, p. 10). It is curious that the fundamental importance of the neutral analogy is so little commented upon. Yet without it models cease to be "logically essential" for theories in Hesse's version of the "modelist" approach: It is through exploration of the neutral analogy that scientists come to make predictions based on the model and use these to test the theory (Hesse 1970, p. 33).

Hesse also takes on board Achinstein's point that the similarities upon which analogies are erected may not necessarily be mathematical and goes on to make a further distinction between "formal" and "material" analogies (1970, pp. 68–69). The former are based on identities in formal mathematical structure, whereas the latter involve identity at the level of *properties*, and here, again, the existence of the neutral analogy is emphasized as absolutely crucial if models are to perform a predictive function.[18]

Returning to Black's taxonomy, "theoretical" models can also be situated in the same continuum (Black 1962, pp. 230–231) since "the relations between the 'described model' and the original domain are like those between an analogue model and its original" (p. 230). Again, "the key to understanding the entire transaction is the *identity of structure* that in favourable cases permits assertions made about the secondary domain to yield insight into the original field of interest" (Black 1970, pp. 230–231; our emphasis). Here issues of realism intrude again as Black distinguishes between "as-if" and "existential" uses of such models, the difference being exemplified by Maxwell's shift on the ether from "the purely geometric idea of the motion of an imaginary fluid" (Maxwell 1965, p. 1:159) to the "wonderful medium" that fills space (p. 2:322). In the former case, the model is regarded as a heuristic fiction, but in the latter, scientists work not *by* but *through* an underlying analogy (Black 1962, p. 229).

Partial Structures and Partial Isomorphisms

Let us now return to our original question: Can we construct an overarching account of models in science? The clue to a positive answer is given by the common element in Black's discussion—namely, the emphasis on the iconic nature of models as expressed in terms of relevant structural relationships, suitably weakened to include the more plausible similarity, rather than strict identity, and suitably broadened to cover similarities in both formal structure and material properties. It is precisely such structural relationships that can be conveniently captured by the model-theoretic approach, and in particular, Hesse's fundamental notion of positive, negative, and neutral analogies finds a natural formal home in the context of partial structures, as we shall now see.[19]

We recall that, from the "extrinsic" standpoint indicated previously, we can characterize a theory in terms of a class of models represented by partial structures of the form:

$$\mathcal{A} = \langle A, R_i, f_j, a_k \rangle_{i \in I,\, j \in J,\, k \in K}$$

We now suggest that the iconic nature of models can be similarly represented;[20] let us denote one such model as follows:

$$\mathcal{A}' = \langle A', R'_i, f'_j, a'_k \rangle_{i \in I,\, j \in J,\, k \in K}$$

Thus $\mathcal{A}$ might characterize the kinetic theory of gases and $\mathcal{A}'$ the infamous billiard ball model. Clearly the A and A′ are not the same in this case, which corresponds to that of an "analogue" model—gas atoms are not, after all, taken to be billiard balls in a literal sense. But in asserting that the billiard ball model is a *model* in the relevant sense,[21] one is asserting that the behavior of the gas atoms can be represented by the behavior of billiard balls, in certain respects and to certain degrees (or, conversely, that the behavior of billiard balls is similar to that of gas atoms).

This representative function obviously rides on the back of a relationship between the R_i in the two cases—a relationship that must be understood in terms of a correspondence between certain elements of the family R_i and certain of those in the family R'_i. In other words, there must be elements of the respective families of relations that are the same in the two cases, so that $\mathcal{A}$ and $\mathcal{A}'$ can be said to exhibit a certain kind of structural isomorphism. This relationship can be termed a "partial isomorphism": $\mathcal{A}$ is partially isomorphic to $\mathcal{A}'$ when a partial substructure of $\mathcal{A}$ is isomorphic to a partial substructure of $\mathcal{A}'$. The notion of a partial structure (or substructure) is so conceived that a total structure (or substructure) constitutes a particular case of a partial structure (or substructure). In other words, we can say that, with regard to a partial isomorphism, certain of the R_i—some subfamily—stand in a one-to-one correspondence to certain of the R'_i.[22]

This can be used to explicate what is meant by "similarity" in this context (cf. Giere 1988). The claim that $\mathcal{A}$ is "similar" to $\mathcal{A}'$ cannot be simply reduced to the claim that the A are similar to the A′ since these are just the "bare" individuals in the model and similarity must be understood in terms of a relationship between properties, monadic and otherwise. Hence, to say that $\mathcal{A}$ is similar to $\mathcal{A}'$ is to say, at least in part, that the family of relations R_i is similar to the family R'_i. To say this is to say, in turn, that some subfamily of the R_i stand in a one-to-one correspondence to some subfamily of the R'_i. It is only some of the R_i that stands in this correspondence, of course, else we would not be talking of *similarity* at all but rather of isomorphism. Furthermore, the notion of similarity coming in certain degrees and respects can be easily accommodated within this view, through a consideration of the number and kind of relations that enter into the correspondence respectively.[23] Thus our claim is that the billiard ball model is an analogue model of a gas insofar as there exists a structural similarity, expressed in terms of a "partial isomorphism," holding between the respective families of relations.[24]

This analysis can be carried further, both in this case where the A and A′ are different and where they are the same, corresponding to the case of "theoretical" models. In the latter situation, both $\mathcal{A}$ and $\mathcal{A}'$ will refer to the same set of individuals. A further variant on this theme arises when $\mathcal{A}'$ is "impoverished" in Redhead's sense of a model that is used as an approximation to a theory in order to facilitate computation (1980). In this case, the R_i' will be a suitably impoverished set of the R_i; certain properties and relationships represented in $\mathcal{A}$ might be ignored in order to give a simpler set of equations, for example. Continuing to draw on gas theory, an example here might be that of the treatment of transport phenomena in gases where a number of simplifying assumptions are made—low but not too low, density, atoms with zero volume, negligible forces of interaction, and so on—before the resulting equations are solved. Of course, as many commentators have noted, one must always treat approximations with care, since they may give radically different solutions from the original equations (see, for example, Black 1962, pp. 225–226; Redhead 1980). We can accommodate this within our account by noting that care must be taken regarding which of the R_i are to be dropped, or ignored, to give the impoverished set R_i'. This is obviously where the skill in making a good approximation comes in, and, of course, the role of background knowledge is crucial. As in the previous case of similarity, the "degree" of approximation can be measured, in a sense, by comparing the R_i and R_i' (cf. Redhead 1980, p. 151).

The contrast with "analogue" models is that, in the latter case, certain of the R_i' will also not feature in the R_i, corresponding to the difference in properties of the A and A′. Thus, referring once more to the billiard ball model, the volume of the atoms/billiard balls and the forces between them might initially be ignored, for example, and the inclusion of these factors, as in the van der Waals equation, then gives a more accurate approximation. Another example is the liquid drop model of the nucleus, where the system of nucleons is given an "iconic" representation in terms of liquid molecules. It is worth considering this particular analogue model in some detail.

As is well known the analogy between the atomic nucleus and a liquid drop was first suggested by Bohr and was subsequently used to obtain the first mass formula for atomic nuclei (von Weizsäcker 1935; see also Bohr and Wheeler 1939). The underlying motivation for exploring such models is expressed well in the passage with which we began this chapter, and in particular the analogy rests on two experimental results. The first is that, to a good approximation, the radius of a nucleus is given by:

$$r = r_0 a^{1/3}$$

where r_0 is called the nuclear unit radius and a is the mass number. This implies that the number of nucleons per unit volume ($a/4/3\pi r^3$) is a constant, and so the nucleon density is independent of the size of the structure, just like molecule density in a liquid drop.

Second, the binding energy per nucleon is approximately constant over a wide range of nuclei. Similarly with a liquid drop, the latent heat is independent of the drop size. In the latter case, this arises because of the short-range nature of the intermolecular forces that "saturate," in the sense that once enough close neighbors have been bound, the presence or absence of more distant molecules does not alter the binding of a given one. This implies that the total energy is proportional to the total number of particles in the system, since each particle makes a fixed contribution to this energy. Pursuing the analogy then leads to the suggestion that the nuclear forces also saturate in this way, which accords with the experimental result noted previously.

Obviously, what we have here is a clear case of structural, but partial, isomorphism and corresponding analogy, with the As being nucleons, the A′s liquid molecules, and the corresponding properties and relations among the R_i and R_i' being those concerned with nucleon/molecular density, respectively, and "saturation." Equally obvious is the heuristic fertility of such a model. This fertility rests, fundamentally, on the partial isomorphism between the structures, which drives the search for further correspondences between those properties and relations for which it is not known whether such correspondences exist or not. It is here that Hesse's "growing points" are located, in what she calls the "neutral" analogy, and her useful distinction finds an immediate representation within the partial structures formalism.

To explore this point further, let us distinguish among the properties and relations in the family R_i those that are *intrinsic* from those that are *extrinsic*. The former are those properties and relations that are essential, or "semantically relevant" (Achinstein 1968, p. 24), for classifying the individuals as A's, whereas the latter can be regarded as nonrelevant or "accidental." If A′ is the same as A, as in a "theoretical" model, then $\mathcal{A}'$ must differ from $\mathcal{A}$ as regards the extrinsic members of R_i' only. In this case, we may say that $\mathcal{A}'$ displays an analogy with $\mathcal{A}$, which is expressed by the correspondence between the intrinsic—and certain of the extrinsic—members of R_i and R_i'. How "close" the analogy is may be measured, again, by the difference between these families of relations, although here the difference will be expressed in terms of the extrinsic properties and relations only.[25]

In the case of "analogue" models, where the A and A′ are different, some, but not necessarily all, of the intrinsic properties in R_i and R_i' will be different. Here much, if not all, of the analogy rests on related objects in $\mathcal{A}$ and $\mathcal{A}'$ possessing related properties; that is, related members of A and A′ possess related properties and relations in the R_i and R_i', respectively. $\mathcal{A}$ and $\mathcal{A}'$ then exhibit a structural isomorphism. This form of analogy is extremely powerful, as it allows us to represent a system by a variety of models, "visualizable, picturable or none of these," as long as this kind of structural isomorphism obtains.[26] As Redhead notes, "This is the source of the very important process of cross-fertilization which plays such a conspicuous role in the overall growth of theoretical physics" (1980, p. 149).[27]

What, then, of Hesse's distinctions? With regard to "material" and "formal" analogies, our earlier discussion suggests that the distinction does not map easily onto that between "theoretical" and "analogue" models and that, in particular, demanding *identity* of formal structure is too strong a constraint. Let us consider "theoretical" models: Here the "intrinsic" properties of the theory and model are the same, so a "material" analogy clearly exists between them. The closeness of the analogy will obviously be increased by any further correspondence between the extrinsic properties of the elements concerned. Equally obviously, however, the closer the analogy, the less useful it may be. In particular, if computational tractability requires "impoverishment" in Redhead's sense, then differences must exist between such properties, the price in this case having to do with concerns regarding the applicability of the impoverished model that results. This "impoverishment" will require corresponding changes in the mathematical structure—else how is greater computational ease to be achieved?—and thus the "formal" analogy will be correspondingly weaker. Demanding identity of such a structure would rule out cases of impoverished models where we might be inclined to insist that there is nevertheless *some degree* of formal analogy involved.[28] Recasting this notion in terms of *similarity* of structure and expressing it thus in terms of a "partial isomorphism" between $\mathcal{M}$ and $\mathcal{M}'$ allows us to capture a greater variety of types and uses of models in science.[29]

Turning to "analogue" models, we can see that much of the strength of the analogy depends on the existence of this kind of formal relationship between the mathematical structures concerned. However, elements of a material analogy may also exist in terms of identity of certain intrinsic properties, although this will typically not be as close as in the previous case. Nevertheless, this, too, suggests that we should think of these analogies in terms of respects and degrees, and viewing them from the perspective of partial structures seems to us to provide a more adequate framework.[30]

Furthermore, this framework precisely captures Hesse's more fundamental distinction between positive, negative, and neutral analogies as that between the set of ordered pairs that satisfy, do not satisfy, and for which it is left open whether they satisfy or not the relation R, respectively. Notably, it is through the introduction of such families of partial relations that the model-theoretic approach can accommodate Hesse's all-important neutral analogy. It is here that science's "growing points" are situated, and it is through partial structures that the openness of scientific practice can be represented. We believe this constitutes one of the fundamental advantages of this approach, and we shall discuss it in more detail in subsequent chapters.

Achinstein's Criticisms Revisited

With these considerations to hand, let us return to Achinstein's criticisms and see if they can be met by the model-theoretic view.

First of all, it should be clear that those criticisms that concern the *syntactic* aspects of the Received View of models do not touch the account we are defending. Neither models nor theories are regarded as partial interpretations of axiomatized calculi on this account, nor are the former conceived of as determining the meaning of terms in the latter. Second, and more important, on this account the twin attributes of partiality and openness can be straightforwardly accommodated. In this respect, both the Received View and the model-theoretic approach as normally conceived are equally deficient. It is difficult to imagine how an interpreted set of sentences can capture the respects and degrees in which a relationship between one representation and another might be partial, but similarly a model as standardly understood is a "closed" structure, insofar as the members of the relevant family of relations are all known to apply or not. On neither view is there any room for ambiguity, doubt, fallibility, partiality, or openness.

Not only does the partial structures approach deal with this but also it allows for the relationship to be dependent on both formal similarities and those that obtain between the relevant sets of properties. The families of relations represented in a structure are perfectly general in this respect. This allows us to answer Achinstein's point about the motivation behind the presentation and development of such similarities: As he correctly claims, mere identity of formal structure is not enough to confer heuristic plausibility upon the use of models, but the importance of content can also be accounted for on our view. Again, the point is that both formal and physical relationships come to be included in the relevant family of relations. Perhaps the rejoinder will be that achieving such generality renders our account entirely vacuous; to that we can only reply that any unitary theory will presumably be subject to the same criticism and, furthermore, that the strength of our account will become apparent in subsequent chapters. Of greater significance is the criticism that, as in the case of the syntactic view, the model-theoretic approach illegitimately blurs the distinction between theories and models. However, for such a criticism to have any bite, the nature of such a purported distinction must be laid bare. In the former case, it has to do with the meaning supplied by the model to the terms in the theory. Such a distinction obviously makes no sense if the central terms of reference of the Received View are abandoned. Turning to our account, we are precisely rejecting any structural differences between theories and models, but we believe that this lack of distinction can be supported on epistemic grounds. We shall return to this point at the end of the chapter.

Before we do, we need to dispose of the further aspects of Achinstein's critique. With regard to models and analogies, we may agree that their *identification* is unwarranted. Nevertheless, they are not so unrelated that they cannot be incorporated within one overall framework; that they do bear connections to one another is indicated by Hesse's discussion, for example. Let us consider Achinstein's example of the Bohr model of the atom. Of course, the underlying formal structure is different *in certain respects* from that of the classical model of the solar system—thus, the relevant energy equations have very different forms (Achinstein 1968, p. 241)—but in other respects there are clear similarities that derive from the fundamental conception of the electrons as orbiting the nucleus.[31] This underpins the balance that is set between the centripetal force on the electron and the attractive radial force toward the nucleus; indeed, it underpins the very invocation of the concept of angular momentum, which then comes to be quantized! These formal similarities are fundamental enough and strong enough to support the analogy with the solar system, the role of which is to put some (complementary) flesh onto the bare formal bones and, furthermore, that thereby acquires heuristic power.[32] Interestingly enough, it is precisely in these respects that the model can be said to be inconsistent (Achinstein 1968, p. 216), an inconsistency that prefigured the collapse of further attempts to derive analogies from the quantum domain; we shall return to the treatment of inconsistency in chapter 5.

Turning now to Achinstein's second counterexample, it is again correct that heat conduction and electrostatics are analogous but unrelated as models, *at least from the perspective of the Received View of the latter*. Nevertheless, the similarities between the two may be immensely fruitful heuristically, and of course any "cross-fertilization" in Redhead's sense gains further plausibility from more fundamental underlying similarities in the microcausal accounts of the phenomena. This then is the point: the existence of similarities that can be represented in terms of (partial) mappings between partial structures.[33] It is the same point that Black alluded to in locating the various kinds of models on a spectrum of abstractness, and it concerns the iconic nature of these representations, the idea of capturing "the features of interest" and incorporating both the faithfulness and unfaithfulness of the representation. In these terms, both models and analogies find a home in our account.

The Autonomy of Models

Achinstein's claim that the Received View's account of models cannot capture what goes on in practice has been reprised in recent discussions, with the focus now shifted to the Semantic Approach.[34] Cartwright, for example, has argued that much of what goes on in science involves modeling that is independent from theory in methods and aims (Cartwright, Shomar, and Suárez 1995). In similar vein, Morrison has suggested that models may be "functionally autonomous" from theories and hence "mediate" between them and the phenomena (Morrison 1999).[35] The claim, then, is that the independence and autonomy of models cannot be accommodated by the Semantic Approach, which should therefore be rejected. However, this claim is false.

The focus of Cartwright and company's concern is the Received View's account of the relationship between theories and models, which, they claim,

> gives us a kind of homunculus image of model creation: Theories have a belly-full of tiny already-formed models buried within them. It takes only the midwife of deduction to bring them forth. On the semantic view, theories are just collections of models; this view offers then a modern Japanese-style automated version of the covering-law account that does away even with the midwife. (Cartwright, Shomar, and Suárez 1995, p. 139)

The argument can be represented as follows: According to the Semantic Approach, theories are families of mathematical models; if this approach is an adequate representation of scientific practice, then any scientific model should feature as a member of such a family; however, there are models that do not so feature, since they are developed independently of theory; hence the Semantic Approach is not an adequate representation of scientific practice. While valid, this argument is not sound because the second premise does not represent the correct understanding of the Semantic Approach's view of models. Let us suppose it is true that models exist that are developed in a manner that is in some way independent of theory.[36] Still, they can be represented in terms of structures that satisfy certain Suppes predicates.

The point can be straightforwardly extended to Morrison's account of the supposed "autonomy" of models (da Costa and French 2000):[37] Whether a model is obtained by deduction from theory or by reflecting on experiment, it can be brought under the wings of the Semantic Approach by representing it in structural terms. And there is a general point here: Surely no one in their right minds would argue that all model development in the sciences proceeds deductively!

There is a further argument, however, which is that models are *representationally* autonomous from theory, again in a way that cannot be captured by the Semantic Approach. Morrison insists that models are explanatory because "they exhibit certain kinds of structural dependencies" (1999, p. 39). Moreover, models make these structural dependencies evident in a way that abstract theory cannot, and hence, again, they act as "autonomous agents" (Morrison 1999, p. 40). Interestingly, one of the examples given is that of models of the nucleus, where the computational difficulties involved in relating the relevant high-level theory—quantum chromodynamics—to the relevant phenomena have meant that these models "provide the only mechanism that allows us to represent and understand ... experimental phenomena." (Morrison 1999, p. 40). Thus models are representational, whereas theories are not. (We shall briefly return to such claims in chapter 9.) Now, two sorts of cases need to be considered to evaluate such a claim.

In the first, a theory exists, but the relationship with the relevant model is not straightforwardly deductive. This is the case with models of the nucleus. Another example might be the models of quantum chemistry, which, because of the many-body problem, for one reason, cannot be deduced directly from Schrödinger's equation (Hendry 1998; we shall return to this example in chapter 7). Nevertheless, high-level considerations do play a role in constraining the form of the possible models at the lower, "phenomenological" level, and it is difficult to see how these higher level aspects could be straightforwardly ruled out as completely nonrepresentational. Indeed, Morrison herself admits to a spectrum of explanatory and representational power as she acknowledges that models incorporate "more detail" about structural dependencies than high-level theory.

Second, a model might be "autonomous" in the sense that it is just not clear *yet* how it might be related to high-level theory. This was actually the case with regard to models of the nucleus, such as the liquid drop model, which were developed long before quantum chromodynamics came on the theoretical scene. Nevertheless, their development was not independent of the relevant theoretical considerations at the time, and at best the "autonomy" is relative and only temporary.[38] Hartmann has explored the idea that such models are examples of "preliminary physics" (Hartmann 1995, p. 52, and also 1999; see also Redhead 1980). Of course, this preliminary stage may last many years, and, as Hartmann himself acknowledges, this is not sufficient ground for distinguishing models from theories. In this work, Hartmann draws a distinction between what he calls the "static" and "diachronic" views of models (Hartmann 1995, p. 52). The former describes what models are; the latter is concerned with their construction and development. The crux of Cartwright's and Morrison's criticisms of the Semantic Approach is that by being wedded to a particular *static* view of models—by tying them deductively to theories—a particular *diachronic* view is forced, in which model development can only proceed "from the top down." However, we insist that the Semantic Approach is not so wedded and that the diachronic aspects of actual practice can be accommodated. As McMullin has noted, it is crucial to this diachronic aspect that models contain "surplus content," which will allow for extensions that, in turn, may be both *suggested* and yet *unexpected* (1968, p. 391; cf. Hesse 1970 and Braithwaite 1962); that is, the models are heuristically fruitful. It is precisely this suggested unexpectedness that can be accommodated through partial structures.

Let us return to the example of nuclear models. Morrison has argued that there can be no single high-level theory to carry representation and understanding in this case, because the relevant models are inconsistent with each other. In the case of the liquid drop model, as we have seen, an analogy is made between the nucleus and a liquid drop, on the basis of partial isomorphism between the relevant sets of properties. In the case of the shell model, the nucleons are conceived of as distributed over energy "shells," much like electrons in "old" quantum theoretical models of the atom. Each such model captures a certain important aspect of the behavior of the nucleus—fission, say, in the case of the liquid drop model, or angular momentum in that of the shell model. As we have already noted, the role of analogy in constructing these models was fundamental. The analogy between the nucleus and a charged liquid drop cannot be pushed too far, of course: If it is to be conceptualized in this way, the nuclear interior is not a classical liquid, but, rather, a Fermi-Dirac one (for details see Heyde 1994, pp. 191–192). Here high-level theoretical aspects intrude into the model that Morrison describes as "semi-empirical," taking us up to quantum statistics and, further, to group theory.[39] And group theory features prominently in the description of the shell model as well. Here the analogy is with the atom as a whole, specifically the picture of electrons moving in the atom, and, more specifically still, it holds between the ionization energy of electrons and the neutron-proton separation energy. Again, the analogy rests on certain crucial idealizations, since nucleons do not move independently in an average field, of course, there being no "nucleus" in the nucleus (see Heyde 1990). And again very high-level theoretical considerations play an important role in articulating the model. Thus although, as Morrison suggests, the models may be functionally independent from the fundamental theory of nuclear forces, this being the result of a lack of appropriate meshing of the various energy regimes, they are not so independent from *all* theory. Again, the functional independence may be transitory as physicists explore various ways of reconciling the apparent inconsistencies between these models, by exploring the open parts of the structures and relating them to further theoretical ones.[40]

Models as Objects

We can also dispose of the claim that certain "models"—Achinstein's "representational" kind—are simply *objects* and cannot as a matter of categorical fact be represented in terms of sets of statements or otherwise.[41] Of course, at one level, these models *are* simply what they are: bits of wire and tinplate, or brightly colored balls held together by plastic rods. The obvious, but important, move is to ask what their *function* is, and this is equally clear: It is to represent. The famous Crick and Watson model represents, not just a particular example of DNA but *all* DNA (of that kind); likewise, the function of the colored-balls-and-plastic-rods model of benzene, gathering dust in the school laboratory, is to represent, not a particular molecule, but *all* molecules of benzene.[42] And this function, as should now be clear, can easily be captured by the semantic view.

What these models represent is the *structure* of the entity concerned and they can be accommodated in the same manner as iconic models described previously, through an emphasis on the relevant set of relations, holding between, for example, the appropriate set of atoms. Again, not all of the R_i' describing the structural characteristics of the model will correspond with the R_i of the entity, and vice versa, depending on the accuracy of the model. But what is important, of course, is the existence of what we have called the "partial isomorphism" between the model and the modeled.[43]

It is worth considering in this context what Griesemer calls "remnant" models, whose function is to act as a taxonomic exemplar (Griesemer 1990). Thus, when it is asked what are the characteristics of a particular species—plant or animal—the exemplar can be presented,[44] in much the same way as when it is asked what are the characteristics of a simple harmonic oscillator in physics, presentation can be made of a simple pendulum or weight on a spring. The latter can be presented in the form of a diagram or in the form of a constructed artifact—the ubiquitous simple pendulum of high school physics labs. Of course, there are certain advantages to presenting a taxonomic exemplar in three dimensions, but such advantages also accrue to holograms, for example.[45] The point, of course, is that the specific material of the model is irrelevant; rather, it is the structural representation, in two or three dimensions, that is all important.

Following Kuhn, the ubiquity and importance of exemplars, both pedagogically and heuristically, have been emphasized by Giere, precisely in the context of (his version of) the semantic view (Giere 1988). It is worth noting the functional importance of approximation in this context: Exemplars such as the simple pendulum can be regarded as essentially forms of Redhead's "impoverished" models, rendered tractable for pedagogic purposes (cf. Hutten 1953–54, pp. 287ff.). Although remnant models, like museum specimens, may not satisfy quantitative laws and so the notion of strict calculational tractability might be inapplicable, they may be regarded as satisfying qualitative biological or, more specifically, taxonomic laws, which may also express a degree of approximation. Just as certain properties of the pendulum exemplar—the color of the bob, for example—are typically left out of the representation, so certain aspects of the remnant model might be regarded as irrelevant—the particular height of a given specimen, for example. Furthermore, and perhaps more interestingly, just as frictional forces are omitted in the description of a simple pendulum, so certain properties of the species represented by a museum specimen might be ignored—at least at the simplest level of description.

What we are trying to emphasize is, again, the twin importance of structure and approximation—that is, the necessity of limited unfaithfulness—in the effective employment of such models to represent species. Both aspects are captured, we claim, by the sort of approach we are outlining here: the structure by a family of relations, the approximation by introducing appropriate partiality into such families, and relating the model to the modeled via partial isomorphisms. Perhaps we should leave the final word to Suppes:

> The orbital theory of an atom is formulated as a theory. The question then arises, does a possible realization of this theory in terms of entities defined in close connection with experiments actually constitute a model of the theory, or, put another way which is perhaps simpler, do models of an orbital theory correspond well to data obtained from physical experiments with atomic phenomena? It is true that many physicists want to think of a model of the orbital theory of the atom as being more than a certain kind of set-theoretical entity. They envisage it as a very concrete physical thing built on the analogy of the solar system. I think it is important to point out that there is no real incompatibility in these two viewpoints. To define formally a model as a set-theoretical entity which is a certain kind of ordered tuple consisting of a set of objects and relations and operations on these objects is not to rule out the physical model of the kind which is appealing to physicists, for the physical model may be simply taken to define the set of objects in the set-theoretical model. (1961, pp. 166–167)[46]

Theories and Models

Let us now turn to the further criticism of the earlier Received, or "semantical," view—namely, that it blurs the distinction between models and theories "per se." As we have already remarked, here issues of realism versus empiricism impinge, and not surprisingly it proves difficult to steer a neutral course in this debate.[47] From the realist perspective, as standardly understood, the difference between theories and models is simply that the former are, if not true per se, then candidates for truth, at least, whereas the latter are not even that (Redhead 1980; Suppe 1989, pp. 425–426).[48] The acuteness of the point can be seen by dragging out, yet again, the billiard ball model of a gas: On the one hand, it is clearly an "analogue" model, offering an iconic representation of the system of interest; on the other, it may be regarded as giving a semantic interpretation to the kinetic theory in terms of a system of billiard balls so that the theorems of the theory are *true* under this interpretation. As far as the realist is concerned, the model taken in the latter sense is supposed to say "how the world is," whereas the analogue model does not, in the usual discussions of these matters (cf. Suppe 1977, pp. 97–98).

This is no problem at all for the *empiricist*, of course, who at best adopts an agnostic attitude toward the set of individuals introduced in both theories and models and thus can maintain the identification of the latter. From this perspective, all theories can be regarded as essentially iconic models of the phenomena (see, for example, van Fraassen 1985b; Pérez Ransanz 1985; Olivé 1985). Of course, said empiricist would not take this to imply that theories are *false*, only that they remain empirically adequate. Nevertheless, the "appearances," understood as models of the phenomena, which are embedded in the theoretical structures, are to be regarded as true in the correspondence sense.[49] Similarly, Cartwright has argued that whereas low-level phenomenological models represent, and are therefore true, but do not explain, theories explain but do not represent and are therefore false (1983).[50] What is common, then, to Redhead, Suppe, van Fraassen, and Cartwright, is that differences between theories and certain kinds of models, at least, are drawn in terms of some form of epistemic distinction: In the case of Redhead and Suppe, it is a distinction between truth and falsity, whereas in the case of van Fraassen and Cartwright, it is a distinction between truth, where models of phenomena or phenomenological models, respectively, are concerned, and empirical adequacy for van Fraassen and outright falsity for Cartwright, where "theoretical" models are concerned. However, such sharp distinctions are unwarranted, as we shall now suggest.

Our discussion of the nature and diversity of models was presented within the framework of what we earlier called the "extrinsic characterization" of theories. Turning to the "intrinsic" perspective, our claim is that theories should be regarded as partially true only and that this captures the epistemic attitudes of scientists themselves. Insofar as this is one of the central claims of our work and will be further supported in subsequent chapters, what we have to say here must be viewed as a kind of promissory note only. Our central claim is that insofar as theories have empirical support, they can be regarded as quasi-true, in the sense articulated in chapter 1. The nature of this support consists of a complex series of connections between an equally, or perhaps consequently, complex hierarchy of models, running from "data" models and "phenomenological" models to their "high-level" theoretical counterparts. These connections can then be represented in terms of partial isomorphisms holding between families of partial relations in the structures constituting these models. It is precisely here that our extrinsic characterization supports our intrinsic one, and given this nexus of partial relationships, any epistemic cutoff point between phenomenological and theoretical models, whether the latter are regarded as merely empirically adequate or outright false,[51] is simply unwarranted. The openness to further development of theories can then be captured, extrinsically, in terms of these partial structures.

The same can be said of models and analogies.[52] In particular, the location of the "growing points" of a theory within the "neutral analogy" also finds a natural home in our account. Switching to the intrinsic characterization, it is implausible and unwarranted from a pragmatist perspective to dismiss analogue models and the like as literally false, for how can they be so regarded and still function as models? If the model is "parasitic" on the theory in the sense of being a model of the theoretical core, then if the theory is "true," in whatever sense, or at least partially so, the model must "pick up" some of this truth. That is, there must be *something* true about the model for it to work in the first place; it is *pragmatically true*.

The root of the difficulty here is the same as that which afflicts the realist-empiricist debate as standardly conceived: the indiscriminate use of the "literal" or "correspondence" view of truth, which ends up painting the picture in stark, black-and-white terms. Obviously, the model must be true in some respects and false in others—Gas atoms are not billiard balls, yet some of the relationships between the former can be "mapped" by relationships between the latter, at least in the context of classical mechanics—and it is precisely this which is captured by the notion of quasi truth. This new way of conceiving of the model-theoretic approach allows us to respond to a final burst of criticism.

Bunge has argued that the "metascientific" conception of model cannot be identified with the semantic one for two reasons (Bunge 1973, pp. 110–113).[53] The first is that "not all theoretical models have been subjected to tests for truth: consequently they cannot all be assigned a truth value" (p. 111). However, if a model has not yet been "subjected to a test for truth," then it must be languishing in what is sometimes referred to as the "domain of discovery," and as we shall subsequently argue, it is appropriate to regard models and theories as partially true in this case as well. This brings us nicely on to Bunge's second point, which is that "every tested theoretical model proves to be at best partially true in the sense that, with toil and luck, some of its testable statements turn out to be approximately true. Therefore, no theoretical model is, strictly speaking, a model in the model-theoretic sense, for this requires the exact satisfaction of every formula in the theory" (p. 111).[54] Of course, we agree with the premise but deny the conclusion! Broadening the "model-theoretic sense" of model through the introduction of partial structures allows us to accommodate the "partial" satisfaction of only some of the formula "in" the theory. Interestingly enough, Bunge goes on to suggest that "a theoretical model that has been given a pass mark constitutes a *quasi*model of its underlying formalism. But this semantic concept of quasimodel has yet to be elucidated" (p. 111; his emphasis; see also p. 113). To which we can only add—until now!

The upshot of our response to these criticisms is that from both the extrinsic and intrinsic perspectives, the distinction between models and theories "per se" dissolves. What we have in scientific practice are a variety of structures, supporting a complex web of relationships; some of these structures get described as theories, others as models;[55] some of the latter are described as "theoretical," others as "phenomenological"; and the differences are, at best, differences in degree of partiality only.[56] In all cases, they are representational structures, and it is this aspect which the model-theoretic approach is designed to capture.[57]

4

Acceptance, Belief, and Commitment

Belief and Acceptance

Writing in the aftermath of the collapse of falsificationism, Post insisted that "acceptance, rather than rejection is the more remarkable exercise of scientific judgement" (Post 1971, p. 253). It is, perhaps, just as remarkable that, within the philosophy of science, the nature of acceptance and, in particular, its relationship to belief have not received the analysis and consideration they deserve. Our suspicion is that it is for this reason that, in the case of many issues within the field, the participants often seem to be talking past one another. This is nowhere more evident than in the realism-empiricism debate, for example, where, as we shall see in chapter 8, the realist typically argues that to accept a theory is to believe that it is true, whereas the constructive empiricist insists that to accept a theory is to believe it to be empirically adequate only. However, the central importance of "acceptance" goes beyond this particular debate to the characterization of scientific practice itself. Typically this practice is conceptually divided into distinct domains, as the development of a scientific theory passes from the initial discovery, into pursuit, and culminating in justification. And it is in terms of "acceptance" that the boundaries between these domains are delineated. Thus, many commentators subscribe to a holdover from positivism, according to which, within the domain of justification, theories are accepted as true, or empirically adequate, or whatever, whereas within the domain (or domains) of discovery and pursuit they are, at best, merely "entertained," to use a recent suggestion. Abandoning unconditional assent in the latter domain(s) is then taken to license the importation of sociological or psychological criteria of theory pursuit and discovery.

Like many before us (including Post), we shall be questioning this distinction but in a somewhat roundabout manner that allows for the possibility that discovery and pursuit may possess sufficient structure to rule out as unnecessary nonepistemic or "cultural" factors. If, as we claim, we are justified in accepting theories, not as true in the simple correspondence sense, but as provisionally or partially true only, then the need for an alternate to denote the attitude adopted in discovery or pursuit simply evaporates. Thus, our aim in this chapter is to lay down the framework for a view of acceptance that is broad enough to accommodate the various senses in which quantum field theory is currently accepted, Newtonian mechanics is accepted within a limited domain, and Bohr's theory of the atom was accepted only provisionally.

The latter two examples are particularly interesting, given that Newtonian mechanics is strictly empirically false and Bohr's theory was, famously, inconsistent. We shall begin by attempting to articulate the sense in which they can be said to be "accepted" in terms of an attitude of commitment that distinguishes acceptance from mere "factual" belief. This will provide the philosophical "flesh" on our underlying formal framework of quasi truth. Thus, our focus will switch from the extrinsic relations between models to what we have called the "intrinsic perspective," although extrinsic considerations will not be entirely shunted aside.

The fundamental question that we are concerned with, then, is what is the nature of this attitude of acceptance? The various answers that have been given can be placed on a spectrum relating acceptance with belief. At one extreme, we have the position that to accept a theory is to believe it to be true, in the correspondence sense of truth. However, simple reflection upon the historical record suggests that even the best scientific theories currently accepted will someday come to be rejected as false, which encourages a degree of fallibilism even among the "naive" realists (we shall return to this issue in chapter 8). Indeed, the latter prefer to talk of the "approximate truth" of "mature" theories and thereby face the scorn of those who demand a formal explication of this notion (see Laudan 1996, p. 119, for example).

Alternatively, constructive empiricism retains the connection between acceptance and belief but, at the theoretical level, restricts the latter to belief in the *empirical adequacy* of the theory only (van Fraassen, 1980). Given the existence of an isomorphism between the "phenomena" and the empirical substructures, the latter can be regarded as true in the correspondence sense, but the theoretical superstructure cannot because, it is claimed, these substructures can be embedded in a variety of alternative superstructures (again, we shall subject this "underdetermination" to a more detailed analysis in chapter 8). Thus the belief in the latter is weaker and tied to their empirical adequacy only. This, too, may be understood in terms of quasi truth, but again our intention is not to go into details here.

One way of responding to constructive empiricism is precisely to question this conception of acceptance as belief in the empirical adequacy of a theory. Horwich, for example, has argued that there is, in fact, no real difference between believing a theory and accepting it in the sense of being disposed to use it to make predictions, design experiments, and so on (Horwich, 1991; see also Mitchell 1988). Since the latter is precisely the sense in which the constructive empiricists understand acceptance, the distinction on which their philosophy is based is one, he claims, that is without a difference. According to Horwich, to believe a theory is the same as being disposed to act *as if* it were true. This is to adopt a behavioristic construction of "x believes that p." But, it might be asked, why not adopt the disposition-to-explicit-assent or the commitment-to-assent constructions?[1] The former approach, in particular, can accommodate situations where a scientist might use a theory to make predictions, design experiments, and the like, and therefore accept it, but would withhold her complete assent, for various reasons, including the very good one given before in response to the first answer. Indeed, we would claim that such situations are the rule rather than the exception in science.

A more radical answer to our original question than any of these cuts the connection between belief and acceptance entirely. Cohen, for example, has drawn a sharp distinction between the two: With regard to belief, he agrees with Hume that it is essentially passive and involuntary in nature. Concerning acceptance, however, he adopts a Kantian approach in arguing that it involves a conscious and voluntary choice (Cohen 1989). Beliefs come over us "willy-nilly," whereas acceptance involves an element of commitment.[2] More specifically, when we *accept* a proposition, hypothesis, or whatever, we commit ourselves to "going along with that proposition (either for the long term or for immediate purposes only) as a premiss in some or all contexts for one's own and others' proofs, argumentations, inferences, deliberations, etc." (Cohen 1989, p. 368). With regard to the acceptance of theories in particular, Cohen counsels the scientist to dispense with belief in the truth of hypotheses completely and rest content with accepting them:

> Scientific enquiry, whether in pursuit of empirical uniformities or of theoretical explanations, is not to be regarded as a procedure that is consummated when justifiable beliefs, with appropriate content, arise in or come over those engaged, who have meanwhile been waiting patiently for this to happen to them. Guesses and hunches, welling up from the subconscious, may make a very considerable contribution to the progress of an enquiry. In some cases an early conviction that *p* may even usefully fortify a scientist's resolution to seek those research facilities that are necessary in order to test whether *p*. But the culmination envisaged—the culmination that adds to our resources for explanation, prediction, technology, or further research—is a conscious and voluntary act of appropriately reasoned acceptance that is echoed throughout the relevant scientific community. (Cohen 1989, p. 385)

Leaving aside Cohen's picture of the heuristic situation in science, if acceptance were completely divorced from belief in this manner, theory choice would seem to become wholly a matter of convention. Cohen's response is that this is not so, since the "observational element" of scientific knowledge does, in fact, require belief (1989, p. 386). But if this observational element is cited in support of accepting a theory, as Cohen admits it is, then the connection between belief and acceptance is restored, at least at the level of the empirical substructures again. Given this analysis, what sense are we to make of the claim that scientists "accept that *p* in the light of the evidence that *p*" (Cohen 1989, p. 385)?

This suggests that the link between belief and acceptance cannot be completely severed in the way that Cohen wants. But now we face something of a dilemma: At one extreme, we have the view that identifies acceptance and belief; at the other, there is the view that regards them as entirely distinct. Both views have their faults, yet both seem to capture a part of what is involved in accepting a theory: belief, of some kind, on the one hand and commitment on the other. Is there an alternative, some sort of "middle way," which embodies what is right about each of these two extremes?

We can shed some light on this dilemma by noting that both of these views suffer from a defect common to many discussions of these issues—namely, that of identifying "belief that *p*" with "belief that *p* is true," where truth is understood in the standard correspondence sense. It is not only that, as Cohen suggests, the conditions for rational belief are commonly regarded as indistinguishable from those of rational acceptance but also that the nature of belief itself is treated in too coarse-grained a manner. As we have noted, "belief," as such, is typically analyzed in terms of holding a proposition to be true in the simple correspondence sense. Indeed, it is often *identified* in this manner. Now this might be appropriate if the objects of knowledge are taken to be propositions of the form "The grass is green," but, we suggest, it is hopelessly inadequate for accommodating those forms of knowledge that have as their objects theories, "worldviews," or complex representations in general. Adopting an anthropological perspective, we might say that whereas philosophers define beliefs to be propositions accepted as true, scientists themselves use "belief" with a certain vagueness, a vagueness that reflects their epistemic fallibility (cf. Sperber 1982, pp. 174–175). In chapter 1, we suggested that this element of "vagueness" can be captured through the formalism of quasi or pragmatic truth. The central claim of the current chapter is that this approach can be expanded to accommodate the "remarkable exercise" of scientific acceptance.

Partial Truth and Representational Belief

The further question we must address, then, is what is the connection between this notion of partial truth and belief? More specifically, what is the attitude evinced in the statement that a theory is believed to be partially true only? These questions can be answered by taking on board a distinction drawn in a related context—namely, the debates surrounding cognitive relativism.

According to the cognitive relativist, different communities, whether broadly cultural or narrowly scientific, "live in different worlds" (Sapir 1929; cf. Kuhn 1970). How to understand such a claim is an interesting issue, but one possible take on it is to say that different communities hold different beliefs to be true, where such beliefs may be both "theoretical" and "observational" and "truth" here is understood in the correspondence sense, since it is the correspondence that allows for the equating of different beliefs with different "worlds." Sperber has identified the crucial weakness of such claims with a reliance—by both sides of the rationalist-relativist debate—on an inappropriate understanding of "belief" (1975, 1982). Thus, he rejects the traditional philosophical description of "belief" as a verb of propositional attitude, where a sentence is the object of the belief and the verb expresses the mental attitude to the proposition expressed by this sentential object. This description is fundamentally misleading since,

> it obscures the fact that we can have such "attitudes" to objects other than propositions in the strict sense. Propositions are either true or false. Sets of propositions are either consistent or inconsistent. Propositions, as opposed to sentences or utterances, cannot be ambiguous and hence true in some interpretations and false in others. Yet some of our so-called beliefs have several possible interpretations and we can hold them without committing ourselves to any of their interpretations. (1982, pp. 167–168)

In particular, "anthropologists . . . use 'belief' with a vagueness suited to their data" (Sperber 1982, p. 174), whereas "philosophers discussing relativism generally take for granted as a matter of mere definition that the objects of beliefs are 'propositions accepted as true'" (pp. 174–175).[3]

What is required is an account of belief that fits anthropological practice, and this can be achieved, Sperber maintains, by dropping this definition. Hence, the objects of many beliefs and, typically, those of non-Western cultures described as "irrational" should not be viewed as propositions in the standard sense but rather as "conceptual representations" *without fully fixed propositional content*. Just as many of our utterances cannot be expressed as sentences per se, but rather as semigrammatical strings of words, so many of our beliefs are "semipropositional" in that "they approximate but do not achieve propositionality" (Sperber 1982, p. 169).

Crucially, "a semi-propositional representation can be given as many *propositional interpretations* as there are ways of specifying the conceptual content of its elements. In principle, one of these interpretations is the proper one: it identifies the proposition to which the semi-propositional representation is intended to correspond" (p. 169). These semipropositional representations are "entertained," in Sperber's words, because they provide us with the means for processing information that exceeds our conceptual capacities, and in this way they can lead us to full comprehension. In particular, they provide a means of dealing with mutually inconsistent ideas by allowing such ideas to be represented in semipropositional form (Sperber 1982, p. 170). Furthermore, in suggesting a range of possible interpretations, they may be enormously fruitful in a heuristic sense.

Not all beliefs have semipropositional representations as their object, however. A further distinction can be made between "factual" and "representational" beliefs, where the difference is that for the former "there is awareness only of (what to the subject is) a fact, while in the case of a representational belief, there is awareness of a *commitment to a representation*" (Sperber 1982, p. 171; our emphasis). Holding a factual belief is rational when it is consistent with and warranted by other beliefs of "closely related content." Given the epistemic unavailability of some of the implications of a semipropositional representation, so that their consistency cannot be easily established, it is not rational to have a semipropositional representation as the object of a factual belief. Rather, such representations are the objects of representational beliefs, which are vaguer, or "fuzzier," in the sense of being conceptually incomplete. Thus, "A representation *R* is [the object of] a representational belief of a subject if and only if the subject holds some factual belief about *R* such as he may sincerely state *R*" (Sperber 1982, p. 172).

In such cases, the subject may, for example, hold a factual belief to the effect that the proper interpretation of *R* is true. Importantly, when *R* can be described as propositional, there is no difference, rationally, between holding that the proper interpretation of *R* is true and simply holding *R*. When *R* is a semipropositional representation, however, it may be entirely rational to hold a representational belief in *R*, in the sense of holding a factual belief that the proper interpretation of *R* is true, but irrational to hold a factual belief in *R*—that is, to believe that *R* itself is true (in the correspondence sense). Whereas the rationality of holding factual beliefs is cashed out in terms of evidential support directly involving observation, that of holding representational beliefs of semipropositional content is to be understood in terms of a warrant that is mediated in certain ways, such as by the beliefs of other epistemic agents, for example. Thus, it is rational to believe, "factually," a proposition such as "the snow is white," where this is understood as believing that "the snow is white" is true, in the correspondence sense, given the evidence of our senses. Likewise, it is rational to believe, "representationally," that "quarks are colored," given our epistemic dependence on, and trust in, the theoretical physicist who asserts such a sentence; this is what Sperber calls "evidence on its source" (1982, p. 173). In many cases, the warrant for representational belief in *p*, where *p* is taken to have semipropositional content, is underwritten by those with relevant expertise, whether they be scientists or tribal princes. Nevertheless, the chain of epistemic dependence terminates in a factual belief of some kind; to describe the nature of the links in such a chain is, of course, precisely to tackle the problem of the relationship between theory and evidence. (We shall return to this point shortly.)

Abandoning the propositional account of belief in favor of something that admits of doxastic differentiation in the preceding sense then allows one to accommodate apparently irrational beliefs within a broadly rationalist perspective. With belief no longer tied to correspondence truth, the relativist slogan "different cultures live in different worlds" is gutted of sense. Transferring this idea from specifically anthropological practice to scientific practice in general, we shall argue that (1) Sperber's distinctions provide a set of doxastic clothes for the body of formal results laid out in chapter 1 and (2), running in the other direction, that his notion of a semipropositional representation can be understood in terms of partial structures. We shall develop the second argument first, in order to be able to apply Sperber's framework to the issues we are concerned with.

The problem with Sperber's position as it stands is that his notion of a "semipropositional representation" is itself rather vague and nebulous. As a colleague at a seminar put it, "I know what a proposition is, but I have no idea what a '*semi*-proposition' could be!" Actually, the metaphysical status of propositions themselves is unclear, as is well known,[4] and it is perhaps an exaggeration to say that we *know* what they are, given the lack of consensus over whether they are, in fact, sentencelike objects, suitably abstract perhaps, or sets of possible worlds, or collections of actual objects and their properties. Nevertheless, although our claims to knowledge in this regard might be based on nothing more than a sense of cozy philosophical familiarity, at least one of the characteristics of propositions is well understood—namely, that of being the bearer of truth or falsity (in the correspondence sense). Thus, if propositions are to be replaced by "semipropositional representations," some account must be given of the characteristics of these entities and the kind of formal framework into which they fit.

We have seen that these representations are "conceptually incomplete" in important respects, although they do have "proper" interpretations. In other words, semipropositional representations are partial, in a fundamental sense, although they point, as it were, to a complete representation in the limit at least. If the notion of a representation can be captured in general by a model or structure, then, we suggest, the appropriate formal expression of Sperber's idea is in terms of partial structures in which the conceptual incompleteness is itself represented by a family of partial relations. The corresponding "complete" representation can then obviously be taken to be the relevant "total" structure in which this family of relations is suitably extended.

The distinction between "factual" and "representational" belief can then be understood in terms of that between correspondence and "pragmatic" or *quasi* truth. (As we noted in chapter 1, there is some exegetical justification for this shift in terminology.) As Sperber remarks, when holding a factual belief, there is awareness only of a "fact" and the factual belief that *p* is to be understood as a belief that *p* is true in the correspondence sense. In such cases, there is no difference between holding that *p* is true and holding that *p*. When holding a representational belief of semipropositional content, however, there is awareness of a commitment to a representation, and in such cases talk of correspondence truth is inappropriate, since it is not at all clear to what, if anything, a semipropositional representation "corresponds." Here the representational belief that *p* is interpreted as a belief that *p* is *pragmatically* or *quasi*-true only.

The double duty performed by partial structures becomes apparent again: In a report of a representational belief of the form "belief that *p*," the sentence *p* is to be regarded as partially or pragmatically true while the semipropositional content of such beliefs is then formally expressed by the relevant partial structure, which in representing our (partial) knowledge of the domain concerned is the *object* of the belief. As we noted in chapter 1, this double duty is also performed by the (total) models in the Tarskian account: In a report of a factual belief, *p* is regarded as true in the correspondence sense within some model that expresses the propositional content of *p*.[5]

Our central claim in this chapter is that scientific beliefs should be regarded as representational in precisely this manner (see da Costa and French 1990b; French 1990; da Costa and French 1993a and b). In particular, although certain elements of scientific practice might be regarded as the objects of factual beliefs, a scientist's belief in the theoretical model itself, taken to be quasi-true, is more properly characterized as representational. Both of these points require further elaboration.

Factual Beliefs and Data

On the Received View, a conceptual line was drawn between "theoretical statements" and "observation statements." One suggestion, then, would be to tie representational belief to the former and factual belief to the latter. But as we all know, observation statements are "theory-laden," and so the ties cannot be so straightforward; indeed, claims such as this led to the demise of this view. Nevertheless, these claims require careful scrutiny, and there may yet be a structural distinction corresponding to Sperber's doxastic one. Before we get to that point, however, we need to explore the possibility of locating such a structural distinction within the model-theoretic approach.

An obvious candidate for those elements that are the objects of factual belief are the empirical substructures of a theory. Thus van Fraassen, for example, has asserted that these empirical substructures are "candidates for the direct representation of observable phenomena" (1980, p. 64). A theory is then regarded as empirically adequate if it has some model such that "all appearances are isomorphic to empirical substructures of that model" (p. 64), where the "appearances" are themselves structures describable in experimental and measurement reports.[6] With the empirical substructures "directly representing" the observable phenomena via their isomorphic relationship with the "appearances," "factual" belief would indeed seem to be the appropriate doxastic attitude toward them.

However, the situation is rather more complicated.[7] Van Fraassen himself refers to "the point long emphasised by Patrick Suppes that theory is not confronted with raw data but with models of the data and that the construction of these data models is a sophisticated and creative process" (van Fraassen 1985a, p. 271). Thus the "appearances" are in fact models of data. More precisely, according to Suppes, these models are "empirical algebras"; more generally, they are partial algebras or just partially ordered sets with certain operations defined upon them.[8] But now talk of partiality in this context undermines the effort to view such models, or the corresponding empirical substructures, as the object of factual beliefs in Sperber's sense since we obviously no longer have "direct representation." On the other hand, giving up the latter entirely has consequences that would be unpalatable to many: As we have noted, the primary motivation behind Sperber's distinction is to respond to the cognitive relativist, and giving up the distinction would be tantamount to giving up the game. Likewise, dropping factual belief from our representation of scientific practice would have fundamental implications for issues to do with scientific objectivity, rationality, and the like, undermining our efforts to counter those who would give sociological accounts of these latter notions. Perhaps, then, we should not be so hasty in giving up on factual belief.

As van Fraassen has also acknowledged, the relationship between a given model and some "phenomenon" is a complicated one. One of the most influential recent accounts of "phenomena" and their distinction from "data" has been given by Bogen and Woodward, and we shall indicate how this distinction might be pressed into service on the side of "factual" belief (Bogen and Woodward 1988; cf. Kaiser 1991 and 1995). Their central idea is that "well-developed scientific theories . . . predict and explain facts about phenomena. Phenomena are detected through the use of data, but in most cases are not observable in any interesting sense of that term" (p. 306). Data serve as evidence for the existence of phenomena, and the latter serve as evidence for, or in support of, theories. The distinction is elaborated in terms of what constitutes the explananda in science: Theories explain phenomena and not data. Could such data be the objects of factual belief? To answer this, we need to look carefully at the characteristics of such data, to examine the theory-ladenness issue in this context, and to consider how Bogen and Woodward's distinction meshes with the model-theoretic points made previously. "Examples of data include bubble chamber photographs, patterns of discharge in electronic particle detectors and records of reaction times and error rates in various psychological experiments. Examples of phenomena, for which the above data might provide evidence, include weak neutral currents, the decay of the proton, and chunking and recency effects in human memory" (Bogen and Woodward 1988, p. 306; see also the example given in Franklin 1995, p. 453).

Data can be straightforwardly observed (Bogen and Woodward 1988, p. 305 and p. 314), are idiosyncratic to particular experimental contexts (Bogen and Woodward 1988, p. 317), are the result of a "highly complex and unusual coincidence of circumstances" (p. 318),[9] and are "relatively easy to identify, classify, measure, aggregate, and analyze in ways that are reliable and reproducible by others" (p. 320). Phenomena, on the other hand, are not observable, are not idiosyncratic, have "stable, repeatable characteristics which will be detectable by means of a variety of different procedures" involving different kinds of data (p. 317), and in general are constant and stable across different experimental contexts (p. 326). These characteristics—particularly those to do with observability and reliability—suggest that data might be candidates for the objects of factual belief, whereas phenomena should be associated with the latter's representational counterparts. But if both data and phenomena are theory-laden,[10] the distinction between the two collapses; representational elements creep into the data, and phenomena can be regarded as "observed."

Following Bogen and Woodward, we can respond to this problem by recalling that on the Received View an observation statement was something like "the melting point of lead is 327 degrees C" (1988, p. 308). Of course, this appears theory-laden, involving, as it does, the theoretical notion of a melting point. But what is actually observed in an experiment are various thermometer readings—"the scatter of individual data points" (p. 308)—from which the melting point of lead is inferred or estimated, subject to experimental error. It is such readings, and "splodges" on a screen, or drawings of damaged brains, that are the constituents of "data" and these—or rather the statements reporting them—are certainly not theory-laden in this sense. Of course, one could stipulate that the process of inferring the melting point of lead from these readings is theory-laden, but such a claim sheds no light whatsoever on the process itself or on the structural connections that subsequently come to be established (Bogen and Woodward 1988, p. 310; Suppes 1967, pp. 62–63).

But, it might be objected, the data are theory-laden in another but equally pernicious sense. Among their essential characteristics are not just observability but also reliability, and to assess this some theory that systematically explains the data must be invoked. Bogen and Woodward respond to this objection in great detail (1988, pp. 327–334), and the upshot is that reliability is typically established not by constructing an explanatory theory but by empirically ruling out or controlling sources of error and confounding factors, or by statistical arguments and so on (see also Franklin 1986). Here, general arguments about whether observation is theory-laden are of no help in understanding how reliability comes to be established; again, what is required are detailed case-by-case analyses. Furthermore, it is the theory-independent determination of reliability that underpins the phenomena's robustness and objectivity (cf. Culp 1995). If the reliability of the data were dependent on highly specific (and typically causal) theoretical explanations, any shift in the latter—something that is both common and notable in the history of science—would pull the rug out from under the former. As Bogen and Woodward note, this seldom happens, and their account helps clarify a further element of structural stability and continuity in science (1988, p. 334).

As for the other side of the coin—the view that phenomena can be observed—Bogen and Woodward contend that on the one hand, if observation is given a clear interpretation in terms of the "traditional" account of perception, then the phenomena are not observable in terms of this account, whereas on the other, attempts to broaden the notion to embrace phenomena weaken it to the point of fatal vagueness (1988, pp. 343–349). Thus, according to the traditional view, a "distal" stimulus outside the human sensory system causes a "proximal" stimulus within it, which in turn produces a "sensory event." However, phenomena do not play the role of distal stimuli in this sense, and although we can tell a theoretical story regarding the causal contribution such phenomena make to the process that eventually results in a "sensory event," it is not the case that they actually explain this event: Claiming that the melting point of lead is the distal stimulus involved in our temperature measurements, or that the neutrino flux is the distal stimulus involved in the seeing of splodges on the scintillation screen (p. 345) renders this notion useless. Neither the melting point of lead nor the neutrino flux is what the observer is looking at in these experiments. As for attempts to broaden the concept of observability to cover the "observation" of the neutrino flux from the sun, for example, the best known of these is based upon an information-theoretic view of observation that remains as little more than a promissory note (Shapere 1982). If phenomena as disparate as the melting point of lead and the neutrino flux from the sun are regarded as observable, then the term is rendered vague and uninformative.

Leaving further discussion of these particular issues aside, our intention is simply to motivate the distinction between factual and representational belief. The point we wish to stress here is that the objects of the former are propositions about data in Bogen and Woodward's sense, corresponding to van Fraassen's "raw" data—that is, the individual temperature readings from which we estimate the melting point of lead, or the "splodges" on the scintillation screen on the basis of which we infer the properties of neutrinos. It is here that we have "direct representation," if anywhere. The question now is how this—and in particular this notion of "phenomena"—fits in with our model-theoretic account.

Foreshadowing the future emphasis on the epistemology of experiment by Bogen and Woodward, Franklin, Galison, Hacking, and others, Suppes earlier noted,

> The concrete experience that scientists label an experiment cannot be connected to a theory in any complete sense. That experience must be put through a conceptual grinder that in many cases is excessively coarse. Once the experience is passed through the grinder, often in the form of the quite fragmentary records of the complete experiment, the experimental data emerge in canonical form and constitute a model of the experiment. (1967, p. 62)

Thus the "raw" data are fed through this conceptual grinder to generate a model of the experiment or data model, and from such models the "phenomena" are inferred.[11] At this point the issue of what exactly constitutes "phenomena" arises. When it comes to the ontological classification of "phenomena," Bogen and Woodward admit they are inclined to be somewhat casual: "Phenomena seem to fall into many different traditional ontological categories—they include particular objects, objects with features, events, processes and states" (1988, p. 321). Hence "phenomena" seem to span many levels of abstraction, from the melting point of lead to the flux of neutrinos; we can begin to get a grasp on the breadth of this range if we consider the complexity of Suppes's "conceptual grinder" in each case. At this point we must tread carefully, given the implications for the realism-empiricism debate as a whole.

When one talks of "inferring" the phenomena from the data, what one is actually inferring, of course, are the *properties* of the relevant objects, processes or whatever; indeed, it is through such properties that the latter are identified. These properties will then feature in an appropriate model of the objects, processes, and the like concerned. It is these *models of the phenomena* that are supported by the relevant data models, and now the question as to how this distinction meshes with the model-theoretic account outlined previously becomes acute, since the aforementioned ontological breadth of the phenomena would appear to jar with the apparently rather simplistic discussion of theoretical structures and empirical substructures.

Thus, models of the phenomena might be identified with the phenomenological models emphasised by Cartwright and her coworkers.[12] Two consequences follow from this identification: (1) Models of phenomena are regarded as "independent" of, or autonomous from, theory, and (2) models of phenomena are regarded as "true" or "true to the phenomena," whereas theoretical models "lie" and are false. Let us consider these points carefully.

As we noted in the previous chapter, the claim that such models are "relatively independent" of theory is cashed out, at least in part, as independence in terms of aims, methods, and techniques. Thus the difference between theoretical models and models of phenomena is spelled out *methodologically* (Woodward 1989; Kaiser 1991, 1995; Cartwright, Shomar, and Suárez 1995; Suárez 1999a; Morrison 1999).[13] Unfortunately, such claims fail to stand up to detailed scrutiny. The purported "relative independence" from theory disappears once the heuristic situation in all its complexity is taken into account. Furthermore, the analogies that typically are a central feature of these models can be represented model-theoretically (French and Ladyman 1997). This, together with the theoretical nature of models of phenomena, suggests that they, too, can be captured in terms of partial structures (cf. Bueno 1997).

But then what of the much vaunted "robustness" of the phenomena, which helps underpin these claims of theory independence? Again, there is a certain degree of vagueness here, corresponding to the ambiguity over what the relevant "phenomenon" is taken to be (Suárez pre-print, p. 5). Certainly, atoms are "robust" in the sense that they feature in many if not all but the highest level theories of matter, but the Bohr model could hardly be so described. Likewise, Shapere has long emphasized the "theory transcendent" status of terms like "electron," in the sense that instead of merely having theories in which such terms occur, we now have theories *of* the electron (1977).

In what sense, however, do we have robustness *here*? Again, we must pay attention to what it is about the phenomena, ontologically speaking, that is "inferred" from the data—namely, a certain set of properties. The robustness of the phenomena consists in the continuity through (high-level) theory change of this set of properties, and it is this in turn that constitutes "theory transcendence." Thus in the case of electrons, such properties would include so-called intrinsic ones such as charge and mass, in terms of which an entity is identified *as* an electron. The model of the phenomena encodes these properties and the relations between them structurally and is robust insofar as aspects of this structure are retained as high-level theoretical structure is stripped away and replaced. But then robustness reduces to nothing more than continuity through theory change, understood structurally, which we shall discuss in greater detail in chapter 8.

We stated that the "relative independence/autonomy" of phenomena from theory is to be understood "at least in part" methodologically. It is also characterized epistemically as models of phenomena/phenomenological models being true or "true to the phenomena," whereas theoretical models are not. We have already expressed our disquiet about such a distinction, and the prior considerations lead us to suggest that models of phenomena should be regarded as partially true only, just as theoretical models are (Bueno 1997). If the phenomena are as Bogen and Woodward say they are—objects such as the electron, or processes such as superconductivity—then being "true to the phenomena" can only be understood as describing correctly (or at least to within experimental error) certain properties of these phenomena. Insofar as these properties become well established and the model achieves a degree of robustness, it can certainly be described as true in some sense. But it cannot be regarded as wholly, totally, or completely true in the correspondence sense since further properties of the phenomenon may be discovered—such as spin in the case of the electron—and come to be represented within the model. The model can only be partial in capturing its domain, and thus the appropriate attitude is to consider it as quasi-true.

What we have, then, is the following picture of a hierarchy of models (Suppes 1962)—models of data, of instrumentation, of experiment—ordered as to level of abstraction, taking us from the data model, through models of the phenomena to the high-level theoretical structures. On this view, the relationship between theory and evidence is a highly complex one (see, for example, Suppe 1989, chapter 5). One way of representing it is in terms of the idea of partial isomorphism introduced previously (French and Ladyman 1999): Each model in the hierarchy can be represented in terms of partial structures, and the relationships between these structures then get represented in turn via isomorphisms that hold between subfamilies of the relevant families of partial relations R_i (see Bueno 1997). Again, what we have are interrelated levels of structure, with no dichotomies, either structural or epistemic, the only sharp break coming right at the beginning as we move from the "raw" data and factual belief, through the conceptual grinder to models understood representationally.[14]

Finally, this analysis allows us to respond to the assertion that this distinction between data and phenomena undercuts constructive empiricism (Bogen and Woodward, 1988, pp. 351–352; Suárez preprint). We recall that van Fraassen requires that theories be regarded as empirically adequate only, in that they "save the phenomena," understood in turn as observable appearances. But, of course, Bogen and Woodward's central thesis is that it is the data that are observable, not the phenomena, and requiring a theory to "save" the data is unreasonable, since it is phenomena that are the explananda. One way of dealing with this is to modify constructive empiricism through the introduction of partial structures (Bueno 1997, 1999a). The motivation for expressing the data models as partial structures has already been given—namely, that these models are "empirical algebras" that already encode a degree of partiality. As van Fraassen has already stated his acceptance of Suppes's point, it seems a plausible move to further regard the relationship between the models in the hierarchy in terms of partial isomorphisms; in place of the relatively straightforward notion of "embedding," we have that of "approximate embedding" (van Fraassen 1989, p. 366 fn. 5; Bueno 1999a).

However, if we allow a level of models of phenomena to intercede between data models and the higher level theoretical structures, then the constructive empiricist may in fact not be able to maintain her antirealist stance: "If we possess evidence and procedures which can justify belief in claims about phenomena, even though many phenomena are unobservable, it is hard to see on what grounds van Fraassen could deny that we are justified in believing as true many other typical theoretical beliefs regarding entities like atoms, electrons and neutrinos" (Bogen and Woodward 1988, p. 351). Since we are justified in believing such claims just in case they are supported by reliable data, the ontological breadth of phenomena forces the constructive empiricist to give up her antirealism. However, the fallacy in this argument is clear in the previous quotation: It is to regard "belief that *p*" as "belief that *p* is true," and we have already suggested that the appropriate attitude where claims about phenomena are concerned is to regard them as partially true only.[15] Thus the data justify us in believing as *partially* or *quasi*-true claims about phenomena, and if empirical adequacy is understood in these terms,[16] the constructive empiricist's antirealist line can be maintained. From this perspective, empirical adequacy can be interpreted as a form of representational belief.

Representational Beliefs and Theories

Let us turn now to this notion of representational belief. In a sense, there is less to establish here; we have already argued for a blurring of the apparent distinction between theories and models in science on the grounds that both should properly be regarded as quasi-true only. With regard to models, this is supported by a consideration of their very nature as representing only in part their "object," and as we have stressed, following Hesse, it is because they are "unfaithful" in this sense that they gain such heuristic power. Those examples of "structures" that are typically deigned to be theories are also incomplete in ways that allow for further growth and development; theories are "open" as recent accounts of scientific practice have emphasized.

In addition, as we indicated at the beginning of this chapter, the history of science can be turned to in support of the view being defended here: Given this history, we have no warrant for regarding current theories as true in the correspondence sense; rather, we should take them to be *partially true* and, correspondingly, the beliefs in such theories to be representational in Sperber's sense. There is a further problem, however, to do with the "final" theory: If there is an "end to science," an end that may, moreover, be in sight, then there are no grounds for projecting representational belief into the future indefinitely. That there is such an end is one of the core tenets of "convergent realism," which the appeal to the historical record was designed to attack. Furthermore, the convergent realist typically appeals to the kind of cumulative view of science based on structural correspondences between theories that we are going to defend in subsequent chapters. How then can we use the historical record in this way to defend Sperber's representational beliefs as the appropriate doxastic attitude in science? The point is particularly acute, as quasi truth has been defined relative to a Tarskian conception of a "final" truth in the correspondence sense (Bueno and de Souza 1996).

Now there is clearly a thicket of issues here. One concerns the very idea of an "end" to science. In his inaugural address as Lucasian professor at Cambridge, Hawking answered yes to the question "Is the End in Sight for Theoretical

Physics?" suggesting that a form of supergravity theory was precisely the end theory in question (Hawking 1980). However, in his own presidential address to the British Society for the Philosophy of Science, Redhead took a much more cautious view, noting that Hawking's favored theory has already been replaced by the superstring program (Redhead 1993).

Although there are aspects of this debate that we do not need to go into in detail, it is worth noting the following: First of all, the "end" in question is typically an explanatory "end," and the question arises as to how we are to characterize explanation in science (we shall touch on this again in chapter 9). As Redhead notes, if one moves from the observation that the practice of physics consists in the elaboration of a range and diversity of models and approximations, to the claim that a separation can be effected between low-level "phenomenological" models and high-level "theoretical" ones, then one is likely to reject the notion of an explanatory end theory altogether (1993, p. 330). Redhead himself is unhappy with such a move, but, as he notes, there is little support to be gained by appealing to a deductive view of explanation. Rejecting such a division between the phenomenological and theoretical on either epistemic or structural grounds as we have, one might still elaborate an alternative account of explanation in terms of partial isomorphisms holding between models at various levels and thus retain the idea of an end theory of physics.

Second, however, as Redhead also points out, such a final theory may not be regarded as a "theory of everything" if one rejects the reductive stance that alleges that all sciences can ultimately be reduced to theoretical physics (see Cartwright 1995, 1999; Dupré 1995; Ereshefsky 1995). Although we have doubts about the basis for such a rejection (Morrison 1995),[17] it could conceivably still be claimed that even though representational belief may at some point no longer be the appropriate attitude in theoretical physics, it may still be so for biology, say, or psychology. Finally, however, even within physics Redhead's own view is that the end is a receding horizon and that "to achieve the end of physics is as unrealistic as grasping the pot of gold at the end of a rainbow" (1993, p. 340). Nevertheless, he insists, this claim can be dissociated from the view that science is essentially cumulative, since the latter does not logically entail the former; progress in science could simply be open-ended, of course. While acknowledging that there is more to be said on this subject, we shall continue to maintain that representational belief offers the most appropriate doxastic framework for science, and also push forward our broadly cumulative agenda.[18]

Returning to Sperber's distinction, further support for this position can be drawn from a perhaps unexpected direction, We recall that for Sperber representational beliefs are further distinguishable from factual beliefs by their warrant: The evidence for a representational belief is typically "evidence on its source" in the sense, as we noted before, that this warrant is mediated by the beliefs of other epistemic agents. Now, it might be thought that this is appropriate for the beliefs of nonscientific laypeople, or those of scientists from one field considering the theories of another about which they have little or no professional expertise, but surely not for those of scientists working directly within a domain. However, it is precisely this sense of "directness" that is problematic.

The traditional theory of knowledge focuses on the lone thinker—one only has to recall Descartes's image of himself in the *Meditations*—and, it has been claimed, is "falsely individualistic" in ignoring our epistemic dependency on others (Hardwig 1985, 1991). There are two observations of epistemic practice that can be given in support of such a claim. The first concerns our lack of doxastic omniscience: "If I would be rational, I can never avoid some epistemic dependence on experts, owing to the fact that I believe more than I can become fully informed about" (Hardwig 1985, p. 340). We shall return to this particular fact of our doxastic circumstances in chapter 7.

Second, and significant for our present discussion, there is the element of teamwork and the role of the relevant epistemic community: "Much of what people count as knowledge, in modern societies, does rest on structures of epistemic dependence. Scientists, researchers, and scholars are, sometimes at least, knowers, and all of these knowers stand on each other's shoulders. You know only because others have direct evidence" (Hardwig 1985, p. 340). "Exhibit A" presented by Hardwig is a paper on experimental high-energy physics that has no fewer than ninety-nine authors. This huge team of experimentalists, already distinct from theoreticians, can itself be broken down into subgroups, with the lines of epistemic dependence running both within and between these smaller teams. Such developments are well known, of course, but the significant point for us is that "the layperson-expert relationship is also present *within* the structure of knowledge, and the expert is an expert partially because the expert so often takes the role of the layperson *within the expert's own field*" (1985, p. 342; Hardwig's emphasis).[19]

"Evidence on its source" is thus not limited to the beliefs of so-called primitive societies. It is ubiquitous in modern science as well. Perhaps, however, it might be argued that this epistemic dependence is limited to post–World War II high-energy experimental physics, where the complexity and cost of the experiments involved necessitates such teamwork. However, we would claim that both throughout the history of science and within the more theoretical domains, the idea of the lone researcher boldly going where no scientist has gone before is a myth. It is not just that, as Newton put it and as Hardwig echoes, scientists stand on the shoulders of others in the cumulative sense but that synchronically there is a complex network of relationships, both theoretical and experimental, in which scientists are enmeshed and on which they are dependent.

This picture slots in nicely with Suppes's view of the hierarchy between data models and theoretical ones, which replaces the more simplistic accounts of the relationship between theory and evidence. The latter tend to encourage the individualistic view by ignoring the complex network of interrelationships between the more theoretical and more phenomenological levels at each of which epistemic agents intervene. Similarly, a consideration of the structural relationships between theories and the way such relationships come to be established heuristically both allows for such epistemic dependence and suggests how it comes to be established.

Thus we have open-ended structures in science constructed and developed on the basis of "evidence on the source." Rationality is preserved, ultimately, because "others," at least, have direct evidence, although as we noted before in our discussion of factual belief, here, too, "directness" is problematic. The point we wish to emphasize is that it is this combination, of representational belief in a semipropositional representation taken as partially true only, that fully captures the vagueness, uncertainty, and fallibility of a scientist's doxastic attitudes.

Semipropositional Knowledge

This discussion suggests an interesting theory of belief and knowledge in general. In particular, it can embrace recent accounts emphasizing the "nonlinguistic" elements of scientific knowledge, as we shall now see.

We recall from chapter 2 that there are alternative theories of belief that analyze it in terms of *representations*; the central idea is that for at least some beliefs, the object of the belief is not a proposition but a representation of some kind. Here we have suggested that this is so for "representational beliefs," in contrast with "factual" ones, and that, furthermore, the representations themselves can be appropriately viewed as partial structures. Of course, the characterization of the objects of belief as structured in this manner is not particularly original.[20] Barwise and Perry's "situations," for example, are essentially fine-grained complexes of individuals and properties, which are represented set-theoretically (Barwise and Perry 1983). Notably, they avoid the terms "proposition" and "propositional attitude" altogether, declaring, "We think that propositions are an artifact of the semantic endeavor and not the sorts of things that people usually talk about, even when reporting someone's attitudes" (p. 178). However, the sorts of things that people usually talk about are not just situations, about which one might have "factual" beliefs, but also theories, models, or representations in general, for which such beliefs would be inappropriate. To reduce representational beliefs to situations would be to either beg the question against the antirealist or become involved in talk of possible situations or "ways the world could be." It is significant, we think, that Barwise and Perry stress the "external significance" of language and meaning in general, where this "external significance" is cashed out throughout their work in terms of examples that we might classify as "factual." The point is that much of our talk and many of our beliefs are not straightforwardly about situations "out there" in the world but about representations that are connected to "the world" in a much more complicated and indirect manner.

Nevertheless, it is interesting to note that "situation-types" are partial, in that "they don't say everything there is to say about everyone, or even everything about the individuals appearing in the situation-type" (Barwise and Perry 1983, p. 9; cf. the partial nature of data models)—a feature of the theory which, it is claimed, confers significant advantages over standard accounts of meaning. Perry, for example, touts this aspect as encouraging the reduction of possible world semantics to a special case of situation theory: Instead of regarding propositions as functions from possible worlds—understood as complete or "total"—to truth values, they should be regarded as *partial* functions in that in some cases they will remain undefined (Perry 1986).[21] (Here we are back to talking about propositions again, but as Schiffer has observed, the possible worlds theorist can construe a structure as *representing* the function it determines (1986, p. 87).) Langholm, whom we cited in chapter 1, has been guided by these ideas toward a notion of "partial model" that is very similar to that of a partial structure (Langholm 1988). In particular, partiality arises either because a particular situation does not contain the "right" facts to make a sentence true or false, or because there is not a sufficient body of data to decide whether the sentence is true or false. In the one case, partial models represent parts of the world, whereas in the other they represent partial information about the world (Langholm 1988, p. 3). Given the indirect nature of representation in science, we would be inclined toward the second interpretation, and in Langholm's work we may see the beginnings of an approximation between situation theory and the partial structures approach.

However, our concern here is simply to point out that regarding the objects of belief in terms of some sort of structured representation is by no means an outlandish proposal. Certainly, philosophers of science should not feel they have to be tied to the "standard" philosophical account of propositions, particularly since, as we have noted, there doesn't seem to be any such "standard" account!

The deficiencies of broadly "propositional" accounts of scientific knowledge has been increasingly emphasized in both the sociology and philosophy of science as well (see, for example, Giere 1995). Thus Cartwright and her coworkers have claimed that our scientific understanding is "encoded" as much in the models and instruments used by scientists as in the theories they develop, and they suggest that "these bits of understanding so encoded should not be viewed as claims about the nature and structure of reality which ought to have a proper propositional expression that is a candidate for truth and falsehood" (Cartwright, Shomar, and Suárez 1995, p. 138).

A similar claim is developed in rather more detail by Baird, whose waterwheel example we noted in chapter 3 (Baird 1995). Baird contrasts his "instrumental epistemology" with the standard expression of scientific knowledge in what he calls the "literary mode" and argues that instruments, as the material products of scientific practice, are capable of both containing and conveying knowledge. They do this in a number of ways. The first, and arguably most important, he says (Baird 1995, p. 446), involves analogies, as in those that hold between orreries and the motions of the planets and between the telegraph and perception (p. 445). Second, "instruments can carry meaning representatively" (p. 446), and here Baird gives the example of Smeaton's material model of the waterwheel. Third, and finally, an instrument can carry meaning "simply in demonstrating a material reality" (p. 447). What is meant by this is that an instrument, such as Faraday's electric motor, can both reduce the physical possibilities and, crucially, point to a set of other possibilities as possible extensions or refinements; there is a nonpropositional kind of implication at work here (p. 448).

Scientific instruments are, of course, material objects, just like Achinstein's "representational" models or Griesemer's "remnant" exemplars. For a scientist, there is no question of presenting them as anything but what they are; for the philosopher of science, however, there is the issue of representing them for the purposes of discussion and analysis. Baird does this linguistically, or sententially, in terms of the written word. However, given the ways in which such instruments can carry meaning, their function as carriers of meaning can be represented structurally. And to capture the manner in which this meaning can be extended or refined—that is, to capture the functional "openness" of these instruments—it is appropriate to use partial structures. Given this openness, this partiality, when we hold beliefs about these instruments, the objects of such belief should be taken to be semipropositional representations. In this manner, we can understand how this kind of knowledge goes beyond the propositional and can be regarded as representational, without abandoning the very possibility of some kind of formal framework.[22]

Representational Belief and "Partial" Acceptance

Let us now return to the distinction between belief and acceptance. We recall that factual and representational beliefs differ in that with regard to the former there is awareness only of a "fact," whereas in the case of the latter there is an awareness of a commitment to a representation (Sperber 1982, p. 171). Factual beliefs simply "come over" one, whereas their representational cousins involve a conscious and voluntary act of "throwing in one's lot," as it were, with the representation concerned. It is precisely this attitude of *commitment* that, following Cohen, we wish to focus on here as the hallmark of acceptance. To accept a theory is to be committed, not to believing it to be true, but to holding it *as if* it were true, for the purposes of further elaboration, development, and investigation (as we have already noted, this is a common position to hold). Thus, Stalnaker writes, "To accept a proposition is to act, in certain respects, as if one believed it" (1987, p. 80), where "belief" is understood as belief in the truth of the proposition. However, regarding a theory *as if* it were true, within the limited domain of knowledge it models, is precisely what is captured by the notion of quasi truth.[23] Thus, the statement that "scientist *x* accepts theory T" can be unpacked as "scientist *x* holds a representational belief in T as partially true only and, therefore, is committed to 'going along' with T for her epistemic purposes."

This view of acceptance clearly distinguishes it from *factual* belief, which can be described in terms of a disposition to hold a certain proposition as true, irrespective of whether it is used as a premise for other purposes. (In *this* respect, Cohen's analysis is correct.) Factual beliefs, being firmly and tightly linked to the data, are "sensitive" beliefs in the Humean sense, forced upon one by the evidential circumstances at hand. Attempts to block or dull the impact of the evidence in such cases very quickly lead one down the road to self-delusion (da Costa and French, 1990b). Representational beliefs, however, are very different in that, as we have said, there is an awareness, not just of a set of "facts" but also of the semipropositional representation used to model those "facts." For any such set, or data model, there will generally be many such representations, and the decision to accept one, based on its quasi or pragmatic truth, involves a commitment to that representation or model. Acceptance is not, therefore, to be divorced from belief in *this* sense. (It is in *this* respect that Cohen's account is faulty.)

This emphasis on commitment as an epistemic component of acceptance has been challenged by Maher, who claims that "our ordinary notion of acceptance allows that a scientist can accept a theory, without being committed to pursuing any particular research program" (Maher 1990, p. 386). In support of such a claim, we are asked to imagine a scientist, Poisson′, who, while believing in the corpuscular theory of light, works on the wave theory for various reasons, including the possibility that this theory may be true or that it might be a very nice theory if it were true. Now, Poisson′ does no work on the corpuscular theory, indicating a lack of commitment to it, yet, says Maher, "We would still say that Poisson′ accepted the corpuscular theory, provided his assertion of this theory appeared sincere" (1990, p. 386).

But what is meant by "sincerity" in this argument? In a footnote to his paper, Maher bluntly states, "I will not attempt to define sincerity here. I think everyone knows what counts as evidence for and against the sincerity of an assertion, and this shows that everyone understands the notion well enough for present purposes" (1986, p. 391). Given the reasons stated for working on the wave theory, one might wonder if Poisson′ is truly sincere in his espousal of the corpuscular view. If someone is working on an alternative theory because she thinks that there is some degree of probability that it is true, then this surely suggests that she has doubts about the theory she professes to believe in and that this belief is, at best, only partial. This is particularly so if the theories are ontologically at odds with one another, as in this case.

However, the conclusion of the argument follows only if Poisson′ has *never* worked on the corpuscular theory, so that there is demonstrably no commitment to it on this account. But in that case he is in the epistemic circumstances discussed by Hardwig; that is, Poisson′ is epistemically dependent on those who *have* gone along with the corpuscular theory for their epistemic purposes and who are thus committed to the theory in a first-order sense. Given this dependence, the "acceptance" of the corpuscular theory is best described as second- (or third- or nth-)hand or "parasitical." And of course there may not be the same *degree* of commitment as there is in the case of someone actively involved in a given research program. Nevertheless, there must some such degree since if we accept, in second-hand fashion, something on the word of an expert, we will, if pressed, reveal this acceptance through our actions and assertions.[24]

Thus "parasitic" acceptance also involves a voluntary act, in that one chooses whether or not to base one's future actions on the opinions of the experts. The epistemic situation is really no different for the scientist working on her analysis of the results from a particle accelerator, who must decide whether to accept the claim of the head of the team that designed the computer software that the anomaly she has discovered is not due to some software error. And Poisson′ is in precisely the same boat.

Acceptance and the General Correspondence Principle

Shifting our attention away from the element of commitment and back to the representational aspect, this conception can then accommodate two interesting scenarios: acceptance of theories that have been (strictly) refuted and acceptance of theories that are (formally) inconsistent.

The empirical inadequacy of a given model may be revealed through either an extension of the model's domain to cover new phenomena or further exploration of the domain the model was intended to cover. Nevertheless, within the limited domain in which the model remains empirically adequate, it may continue to be regarded as pragmatically true and, thus, accepted, *within that limited domain*. An obvious example of this is classical mechanics, known to be "strictly false" yet still employed, and for good reason, in the construction of bridges, buildings, etc. and as the theory of scientific instrumentation in many situations. This theory can therefore be regarded as pragmatically true within a certain delimited domain, and within this domain the theory may be accepted as if it were true for the purposes indicated previously.

Indeed, it was precisely a concern to capture the attitudes adopted toward such theories that provided one of the principal motivations for the development of the formalization of pragmatic truth in the first place (Mikenberg, da Costa, and Chuaqui, 1986). Clearly, what we have here is a situation in which acceptance is appropriate but belief (in the truth of the theory) would be quite out of place. Thus Cohen writes,

> we should be wrong to believe the theory, in the sense of feeling it to be true. But we can certainly accept it, in the sense of going along with it as a premiss, for all the purposes to which it is applicable. Similarly, even when we regard a physical or chemical law as a simplification or idealization we can use it as a premiss for calculations about the actual world, if we make relevant allowances and corrections. So in this sense we can accept the law even when we do not believe it to be true of the actual world. (1989, p. 386)

In other words, Newtonian mechanics may be regarded as accepted in the sense that it is believed to be pragmatically or partially true only, and this, in turn, is to be understood as meaning that it is not refuted within the domain it models and from which it draws its empirical support. To be more explicit, it is accepted for certain purposes and subject to certain provisions regarding the limits of its applicability. Although a more comprehensive theory is available, it would clearly be unreasonable to expect engineers and scientists to use this within the domain of applicability of Newtonian theory. There is, in these cases, some form of economy-accuracy trade-off (cf. Cherniak 1984), and the reasoning of scientists in these matters is not different in kind from that of laypersons in making decisions in general (see, for example, Giere 1988; da Costa and French 1993b). The point is a perfectly general one and applies, not just to strictly refuted theories such as classical mechanics, but also to approximations and idealizations in general (see French and Ladyman 1998; Laymon 1988).

In similar vein, de Oliveira argues that Newtonian mechanics contains an "element of truth," where this notion is explicated in terms of the "established sub-theory of a theory" (1985, pp. 133–134). However, he claims that to accept a theory "means to believe that it has not been refuted and that all predictions are true which can be deduced from its established sub-theory" (1985, p.134) and that Newtonian mechanics is "no longer accepted" (p. 131) since it has been refuted. There are, of course, well-known dangers in tying acceptance to refutability in this manner. Furthermore, the scope of acceptance in this sense is clearly narrower than that which we envisage here. In particular, it rules out the possibility of theories being accepted in a provisional or pragmatic sense, which is precisely the possibility we find interesting.

Nevertheless, de Oliveira does seem to be on the right track in pushing his "thesis of the unrefutedness of established sub-theories of accepted theories," which asserts that "no established sub-theory of an accepted theory has ever been refuted" (1985, p. 134). In developing this claim, de Oliveira draws on Post's discussion of the General Correspondence Principle (Post 1971). Presented as the most important of the heuristic constraints on the construction of theories, the idea behind this principle is that "no theory that ever 'worked' adequately turned out to be a blind alley. Once a theory has proved itself useful in some respects . . . it will never be scrapped entirely" (Post 1971, p. 237). More precisely, it states that "any acceptable new theory *L* should account for the success of its predecessor *S* by 'degenerating' into that theory under those conditions under which *S* has been well confirmed by tests" (1971, p. 228).

Thus the principle makes the strong claim that a new theory must reduce to its predecessor in the domain successfully modeled by the latter. It implies, of course, that there are no "Kuhn losses" in science, and it is important to realize that Post's work, with its wealth of historical examples, constitutes a concerted attack on the Kuhnian viewpoint. Even during a scientific revolution, the whole of the lower level structure of the theory being overthrown is not destroyed, although the top levels may be lopped off and the lower levels may be given a different interpretation (Post 1971). Bluntly put, we never lose the best of what we have, and this can be mirrored within our account by what might be called the Principle of the Absolute Nature of Pragmatic Truth: Once a theory has been shown to be pragmatically true in a certain domain, it remains pragmatically true, within that domain, for all time. It is this, of course, which lies behind the justification for continuing to use Newtonian mechanics within certain limits.

But what of other theories, such as phlogiston theory, which are now abandoned and have no such pragmatic value? The difference in this case can perhaps be most clearly identified by noting the narrower domain in which phlogiston theory can be regarded as pragmatically true as compared with Newtonian theory: "the phlogiston theory 'worked' in that it assigned consistent levels of phlogistication (explaining many features) to chemical substances related in more than one way by reactions. On the other hand it tried to establish a connection between colour and phlogistication, and this part of the theory was not successful even at the time" (Post 1971, p. 228; Koertge 1968).

Heuristically, the Correspondence Principle may be used to eliminate certain candidates as serious possibilities for the successor theory (Post 1971, p. 235). As Post explains, this use of the principle is generally post hoc in nature, in the sense that it is employed to eliminate those candidates for the successor theory that fail to explain the well-confirmed part of the predecessor. Of course, the true extent of the latter is only conjecture at the time these developments are taking place (Post 1971, p. 231), and considerable skill may be required in identifying those aspects of the predecessor theory to be retained by the successor.

Rueger has given an account of this process in the early development of Quantum Field Theory (Rueger 1990). He claims that the central question at the time was "Which traits of a quantum theoretical formalism, valid at low and moderately high energies, would remain unchanged if the high energy behavior of the formalism were, more or less radically, to be modified? Heisenberg tried to separate physical concepts which could not be applied in the future theory from those concepts which (probably) remained unaffected by the high energy difficulties" (p. 209). Any theory that failed to incorporate these latter concepts would be regarded as an unfit candidate for a successor. An earlier example is represented by Einstein's 1916 derivation of Planck's radiation law, where "there was reason to believe that the parts of classical theory used by Einstein would be retained in any more comprehensive theory that would ultimately resolve the conflict between classical electrodynamics and the quantum postulates" (Smith 1988b, p. 437).

A similar attempt to make the results independent of the details of any future theory can be seen in Bohr's use of his correspondence principle, which allowed him to obtain empirical results from a theory that was, strictly speaking, inconsistent (Rueger 1990; Smith 1988b). In this case, the problem of delineating the well-confirmed parts of the theory was particularly acute, and Bohr used the principle both as a "defense mechanism," which allowed for the temporary accommodation of the inconsistent foundations, and as a heuristic guide, which effectively helped to generate empirical support for the theory (Lakatos 1970, p. 144).

It was this support, we claim, that justified the belief that Bohr's theory, although inconsistent (Lakatos 1970, p. 140), was at least partially true and that also justified accepting the theory as such. Thus, even inconsistent theories may be accepted, in our sense. The crucial element that does the work for us is this understanding of acceptance in terms of a representational belief in a semipropositional representation understood, epistemically, as quasi-true only.[25] As we shall see, this blurs the distinction between heuristics and justification, but it is to the issue of inconsistency in science that we now turn.

5

INCONSISTENCY IN SCIENCE

> What we need is a model in which you have theoretical structures that may have erroneous assumptions, in which some of your data are false, that you may operate on with rules that can induce falsehoods, that may have contradictions and so on, because those are the kinds of problems we really face.
>
> —W. C. Wimsatt, quoted in Callebaut 1993

The Control of Anarchy

In their classic text *The Logic of Inconsistency*, Rescher and Brandom note that for Peirce the world of "signs," which is the world we inhabit, is an inconsistent and incomplete world (1980, pp. 124–125). Even beyond the confines of pragmatism, the prevalence of inconsistency or of "incongruent" suppositions[1] in both our scientific and "everyday" belief structures is something that is being increasingly recognized.[2] In the world of scientific representations, Bohr's theory, of course, is one of the better known examples, described by Lakatos as "sat like a baroque tower upon the Gothic base of classical electrodynamics" (Lakatos 1970, p. 142; see also Brown 1993); others that have been put forward include the old quantum theory of blackbody radiation (Norton, 1987), the Everett-Wheeler interpretation of quantum mechanics (Rescher and Brandom 1980),[3] Newtonian cosmology (Norton 1993), the (early) theory of infinitesimals in calculus and the Dirac δ-function (Priest and Routley 1984, p. 14), Stokes's analysis of pendulum motion (Laymon 1985), and Michelson's "single-ray" analysis of the Michelson-Morley interferometer arrangement (Laymon 1988). The problem, of course, is how to accommodate this aspect of scientific practice, given that within the framework of classical logic an inconsistent set of premises yields *any* well-formed statement as a consequence. The result is disastrous: The set of consequences of an inconsistent theory will explode into triviality, and the theory is rendered useless. Another way of expressing this descent into logical anarchy that will be useful for our discussion is to say that under classical logic the closure of any inconsistent set of sentences includes *every* sentence. It is this which lies behind Popper's famous declaration that the acceptance of inconsistency "would mean the complete breakdown of science" since an inconsistent system is ultimately uninformative (Popper 1940, p. 408; 1972, pp. 91–92).[4]

Various approaches for accommodating inconsistent theories in science can be broadly delineated. Thus one can distinguish "logic-driven control" of this apparent logical anarchy from "content-driven control," where the former involves the adoption of some underlying nonclassical logic and the latter focuses on the specific content of the theory in question (Norton 1993). Cutting across this is the syntactic/model-theoretic distinction with which we have been concerned here. Both the "logic-driven" and "content-driven" approaches are typically developed within the framework of the syntactic view, and a model-theoretic approach to inconsistency has been comparatively little explored, although Laymon has taken some interesting steps in this direction (1988).[5] Finally, there is a further distinction at the epistemic level, between those who claim that inconsistent theories can be "accepted" in the same sense as can be said of any theory, and those—typically followers of the content-driven approach—who urge a shift to some weaker attitude of "entertainment."

The view we shall advocate here can be situated within this nexus of cross-cutting distinctions: Not surprisingly, we offer a model-theoretic account in which regarding theories in terms of partial structures offers a straightforward and natural way of accommodating inconsistency. In particular, it allows us to bring to the fore the heuristic importance of inconsistency in science. Shifting to the epistemic perspective, inconsistent theories can then be regarded as quasi-true and accepted as such, just like any other theory in science. As we shall indicate, this common epistemic framework forces a blurring of the standard discovery-justification distinction. (There's another distinction to add to our list!) The final component of our view is the introduction of a form of paraconsistent logic, not at the structural level of the theory itself but at the epistemic level as a logic of quasi truth. Given the overall thrust of this book, we shall restrict our attention to the nontechnical components of our view, but if the reader is particularly keen, the more formal details can be found elsewhere (da Costa, French, and Bueno 1998a).

Inconsistency, Paraconsistency, and Truth

The existence of inconsistency poses obvious problems for any view that construes theories as sets of sentences expressed in terms of classical logic, precisely because of the explosion into triviality mentioned previously (cf. Rescher and Brandom 1980, pp. 21–23). The "logic-driven" approach typically retains this syntactic view of theories but abandons the underlying classical logic for some nonclassical counterpart. The obvious candidate for such an underlying framework would be one of the (infinitely) many systems of paraconsistent logic (Arruda 1980; da Costa and Marconi 1987). Briefly put, a logic is said to be *paraconsistent* if it can be employed as the underlying logic of inconsistent but nontrivial theories; a theory is said to be paraconsistent if its underlying logic is a paraconsistent logic (Arruda 1980, p. 2). Thus, for a paraconsistent logic, the two notions of triviality and inconsistency are no longer coextensive. In such a logic, the scope of the principle of contradiction, or certain related principles, are restricted in some way or other, so that from "p" and "$\sim p$" it is not in general possible to deduce every formula.[6] (The meaning of the term *paraconsistent logic* can then be extended as follows: A paraconsistent system of logic constitutes a system that can be used to systematize theories that may themselves contain theorems that, when conjoined to classical logic, trivialize the latter.)

The history of such systems is extremely interesting, going back, perhaps, as far as Aristotle himself, who, despite rejecting contradictions in both the ontological and practical spheres, appears to have suggested that the principle of noncontradiction is not the "highest law" (Lukasiewicz 1971).[7] However, the real forerunners of paraconsistent logic in modern times are Lukasiewicz and, independently, Vasil'év, who suggested that dropping the principle of contradiction would lead to "non-Aristotelian" logics, by analogy with non-Euclidean geometries (Arruda 1989). Prompted by Lukasiewicz, Jaśkowski constructed the first system of paraconsistent propositional logic with the explicit aim, among others, of accommodating inconsistency in science:

> It is known that the evolution of empirical disciplines is marked by periods in which the theorists are unable to explain the results of experiments by a homogeneous and consistent theory, but use different hypotheses, which are not always consistent with one another, to explain the various groups of phenomena. This applies, for instance, to physics in its present-day stage . . . we have to take into account the fact that in some cases we have to do with a system of hypotheses which, if subjected to a too precise analysis, would show a contradiction among them or with a certain accepted law, but which we use in a way that is restricted so as not to yield a self-evident falsehood. (Jaśkowski 1948; cited in Arruda 1980, p. 10)

Jaśkowski called his system "discussive" (or discursive) logic, and it has subsequently been proposed as a logic of acceptance in science by da Costa (da Costa and Doria 1995); we shall return to this proposal later.

Independently of Jaśkowski, one of us (da Costa) began his own work on paraconsistent systems in the late 1950s, developing them not only at the propositional level but also at the predicate level, with and without equality, as well as presenting corresponding calculi of descriptions and paraconsistent set theories (da Costa 1958, 1974a and b). Da Costa's best known paraconsistent logics are based on the propositional calculi "C_n," which were constructed to satisfy the following conditions (da Costa and Marconi 1987, p. 6):

1. The principle of contradiction, in the form $\sim(p \mathbin{\&} \sim p)$, should not be valid in general.

2. From two contradictory premises, p and $\sim p$, we should not be able to deduce any formula whatever.
3. They should contain the most important schemes and rules of classical logic compatible with the first two conditions.

On the basis of this third condition, Priest and Routley classify da Costa's systems as belonging to the "positive-plus" approach, in which the full classical "positive" logic is retained, at the cost of abandoning part of the theory of negation (Priest and Routley 1984); relevant and nonadjunctive approaches, on the other hand, renounce part of classical positive logic (the latter systems in particular will feature prominently in our discussion later). On the other hand, paraconsistent logics can be regarded either as *complementary* to classical logic, to be employed when we need a negation weaker than the classical form, or as a kind of *heterodox* logic, incompatible with the classical system and put forward in order to ultimately replace the latter.

A representative example of the latter approach is Priest, who talks of blowing the whole classical configuration asunder and argues for the adoption of the essentially Hegelian position that there are true contradictions (1987; cf. Arruda 1980, p. 4). The classical generation of all possible propositions from an inconsistent theory is then blocked through the adoption of a form of paraconsistent logic in which certain contradictions are tolerated. (For a discussion of the differences between da Costa's and Priest's approaches to such logics, see da Costa and French 1989c.) More important for our purposes, Priest includes among the pragmatic grounds for adopting his position the occurrence of inconsistency in the history of science, explicitly citing Bohr's theory with its inconsistent mixture of classical and quantum principles to support his contention that "whatever kind of argument it takes to make something rationally acceptable, an inconsistency can have it" (Priest 1987, p. 127). Thus, the "logic of science," for Priest, is paraconsistent. However, for scientific theories to fulfill this supporting role, they must be regarded as *true* simpliciter, which is precisely what we reject here. In particular, it is curious that Priest chooses Bohr's theory as one of his examples since, leaving aside the historical fact of its subsequent supersession and the current tendency to refer to it as a "model" precisely because of its inadequacies, at the time it was proposed there was considerable controversy over how seriously to take it (for the spectrum of reactions, see Pais 1991, pp. 152–155);[8] Bohr himself is recorded as saying, "No, you can't believe that. This is a rough approach. It has too much of approximation in it and it is philosophically not right" (cited in Pais 1991, p. 155).[9] Assertions of truth in this case seem particularly inapt.

Brown has also recently espoused the use of a form of paraconsistent logic in this context, while rejecting the position that inconsistent theories *as a whole* should be regarded as true in the correspondence sense (Brown 1990; the qualification in italics will turn out to be critical). At the heart of this view is a "contextual" characterization of acceptance, in the sense that a theory is regarded as accepted when "we choose to treat [it] as if true, observing contextual limits on that commitment so as to avoid bringing incompatible claims into play at any point" (Brown 1990, p. 285). The idea here is that the context of application of an inconsistent theory may be broken down into subcontexts in which the conflicting principles behind the inconsistency may be isolated. In this way, the conjunctive application of contradictory claims is effectively blocked (Brown 1990, p. 284). Since there is no context in which all the principles are collectively true, in the sense that the evidence supporting scientific theories is essentially local in nature, the logic of acceptance here is a paraconsistent "nonadjunctive" one, according to which the closure of a set of claims under conjunction is not implied by the set (Brown 1990, pp. 289–292). We noted before that such logics fall outside the "positive-plus" approach, and the particular form of nonadjunctive logic chosen in this case is that constructed by Schotch and Jennings, which allows an inconsistent premise set to have consequences that are not consequences of any particular member of that set (1980).[10]

As in Priest's case, Brown draws on the example of Bohr's theory to illustrate this approach (Brown 1993).[11] In particular, he claims that Bohr was very careful to distinguish the contexts in which his contradictory principles were to be applied. Thus, on Bohr's model an electron in a hydrogen atom could exist in one or another of a discrete set of orbits or "stationary states." Classical mechanics could then be applied to account for the dynamical equilibrium of the electron in one of these stationary states but not to transitions between states, where the relation between the amount of energy and frequency of radiation emitted is given by Planck's formula (Bohr 1981, p. 167). Hence Bohr effectively adopted a "patchwork" approach, and Brown quotes from a well-known letter in which Bohr eschews a general foundation for his model and suggests that "the possibility of a comprehensive picture should perhaps not be sought in the generality of the points of view, but rather in the strictest possible limitation of the applicability of the points of view" (Bohr 1981, p. 563).[12]

As Brown notes, "This combination of classical and non-classical principles is a logically risky game. Whatever form of commitment to these principles Bohr was proposing, it was not a commitment closed under the classical consequence relation" (1993, p. 399). Nevertheless, commitment there was, and Brown cites cases involving the confirmation of the theory by various pieces of evidence in which what he calls "literal commitment" played an important role (1993, p. 400).[13] "Literal commitment" involves belief in the truth of the separate components, or "patches," of the theory. According to Brown, then, Bohr was clearly committed to both the quantization of radiation energy and classical physics. Applying a classical closure relation on this set of inconsistent commitments, under which there is adjunction so that truth is effectively distributed, leads to triviality. Hence Brown advocates the imposition of a nonadjunctive logic that supplies a nontrivial form of closure: "This modest proposal regarding closure allows us to take Bohr's commitment to these inconsistent principles seriously and at face value, without regarding him as committed, implicitly, to anything and everything" (1993, p. 405).

Its modesty notwithstanding, this proposal is not unproblematic. As we shall see, Brown has chosen his case study very wisely, but it is not at all clear whether this nonadjunctive approach can be applied to other examples of inconsistent theories for which there is not such a clear separation of contexts. First of all, Brown is keen to stress how his approach accounts for the development of new applications of an inconsistent theory: "When previously accepted techniques were not sufficient, the theory was extended by drawing on related classical results, care always being taken to avoid bringing in anything that would raise the theory's degree of incoherence" (1993, p. 409). The obvious response is to interpret such extensions as being based on certain fundamental analogies, but to this Brown replies that "without some sort of commitment to classical physics, the practice of continually drawing on it for such analogies seems odd at best" (1993, p. 409). The point then is the commitment to classical physics, but *our* point precisely concerns the nature of this commitment and, in particular, its involvement with truth.

This goes to the heart of the nonadjunctive approach, as Schotch makes it clear that in the partitioning of a set of inconsistent statements, the "cells" of this partition must be regarded as *true* (Schotch 1993, p. 423; cf. Brown 1993, p. 404). Thus for this approach to be applicable to Bohr's theory, the contextually separated "principles" of this theory—that is, classical dynamics and Planck's law—must likewise be taken as true. Now one might respond that classical dynamics should not, and could not, by Bohr, be regarded as true precisely because of the development of quantum theory. But of course, Brown would respond that the point is that classical dynamics should be taken as true *in the appropriate context from which quantum theory is excluded.* Bohr's commitment to classical dynamics therefore involves belief that this aspect of classical physics is true, but the commitment is contextual, while his commitment to his atomic model nonadjunctively spans the contexts but involves only belief "as if" the model were true.

It is in this respect that Brown's proposal bears a close resemblance to ours. However, the crucial difference is that on our account Newtonian mechanics is regarded "as if" it were true, rather than true per se, *within* its limited domain, and this helps express what we mean when we say that it should be taken as pragmatically true only. And this difference reflects a more fundamental distinction with regard to the role of the domains or contexts in each case. In ours, the domain is invoked to block claims that Newtonian mechanics should not be regarded as true in any sense, pragmatic or otherwise, but should simply be dismissed as false outright. In Brown's case, the domain, or context, is structurally incorporated within Bohr's model and reflected in the division of adjuncts. This aspect then gives rise to a more profound criticism of the nonadjunctive approach as applied to scientific theories and models.

The central point of this criticism is that if inconsistent theories are to be regarded as structurally fragmented in this way, it is difficult to see how they can still be regarded as coherent *theories* (Priest and Routley 1984, p. 8). In what sense is such a collection of "contexts" or "cells" an integrated unit? One sense of understanding "integration" and "coherence" here is in terms of the "overall truth" of the theory. Thus, another striking example of inconsistency in the "old" quantum theory is Planck's original derivation of the blackbody radiation law, which involved contradictory classical and quantum hypotheses concerning the energy exchanges between the resonators of the blackbody and the radiation field itself. Norton has considered the suggestion that this derivation may also be accommodated within a nonadjunctive framework but concludes that Wien's Law, the classical expression for the energy density and the quantized formula for the average energy of a resonator were in fact *conjoined* to obtain Planck's law (Norton 1987). From the distributed truth of the former principles, he argues, the collective truth of the conjunction expressed in the latter was inferred. Hence adjunction was *not* abandoned in this case.

However, an adherent of the nonadjunctive approach might contend that coherence and integration are more profoundly understood in this context in terms of *deductive closure*. The critical standpoint from which to judge whether a set of claims constitutes a coherent theory is that which relates to the cognitive commitments of the scientific community. Brown's central point is that there was cognitive commitment to Bohr's theory, which can be understood as a closed set of claims, with the inherent inconsistency forcing this closure to be represented paraconsistently. In this essentially nonclassical sense, we have integration and coherence.

However, one may wonder if it is even possible to effect the clear-cut division between different "contexts" or "subsets" within a theory that this account requires. Brown does at least acknowledge the difficulty, noting,

> When we accept claims in this way, we are committed to giving systematic division of the contexts in which the incompatible claims will have their roles. This division must allow for closure of the contexts of acceptance under combination. And it must have some connection with the evidence we are able to give for the claims in question: the division of contexts should allow us to use particular claims in the contexts where we find evidence confirming those claims. (Brown 1990, p. 288)

As we have said, Bohr's theory or model of the atom offers the best possible case study for the nonadjunctive account because Bohr himself talked of pulling back from generality in favor of the "strictest possible limitation" of the principles in conflict. However, even here we question, first of all, whether there was quite the "systematic division" of contexts that this approach requires and, second, whether Bohr's contextualism should be understood at the level that Brown intends—that is, at the level of inconsistency in the mathematical formalism or, instead, at that of "paradox" in the interpretation. Let us now consider this example in more detail.

Conjointly with his nonadjunctive account of cognitive commitment, Brown offers a "schematic account" of how Bohr reasoned with his theory, which goes like this:

> OQT [Old Quantum Theory] involved extensive use of classical mechanics and electrodynamics together with quantum restrictions inconsistent with classical physics. It included explicit conditions restricting the application of the conflicting principles. While applications of the theory involved inconsistent principles at various points, the inconsistencies were isolated in separate sub contexts. Within these sub contexts the principles were treated as though true, and consequences deduced. Some of these consequences

served in turn as input for calculations carried out in other sub contexts, using other, incompatible principles.... The isolated commitments were kept separate within their cells, but some consequences inferred within those cells were allowed into the store of general commitments. These consequences were then available for use in other cells, or for direct empirical prediction and explanation. So long as the commitments imported into each cell are consistent with the isolated commitments within them, logical catastrophe can be avoided. (1993, p. 404)

In other words, consequences, typically involving physical quantities, deduced within one isolated context could be used as input for the deduction of consequences and calculation of quantities within another isolated context. Thus, for example, classical mechanics was employed together with quantized angular momentum to calculate the energy levels of a one-electron atom. These give rise to energy differences that determine the frequency of light emitted as an electron drops from a higher to a lower level, in accordance with Planck's law. Note that the restrictions on the import and export of consequences between contexts must be tighter than merely demanding that an exported calculation be consistent with the claims of the context it is being imported into. This is where the nonadjunctive account of the consequence relation comes into play, imposing the requirement that such import-export moves must not increase the "degree of incoherence" of a set of claims (Brown 1993, pp. 406–407).

Our concern is with the systematic division into contexts and with the question of whether it actually holds in this case. Bohr actually presented three different treatments of the hydrogen atom (see Pais 1991, pp. 146–152), but what they all have in common is the postulate that an electron in such an atom is restricted to one or other of a discrete set of orbits or "stationary states." Now Brown emphasizes the point that classical mechanics was taken to apply to the dynamics of the electron in its stationary state, while quantum theory was brought into play when the transition between discrete states was considered—this discreteness contradicting classical physics, of course—and this certainly does seem to mesh nicely with the nonadjunctive view. However, it is not only in the discreteness of the states that we have conflict between quantum and classical physics but also in what has been called "one of the most audacious postulates ever seen in physics" (Pais 1991, p. 147)—namely, the assertion that the ground state was stable, so that an electron in such a state would not radiate energy and spiral into the nucleus as determined by classical physics. *This* is the central inconsistency of the Bohr model, but it is there right in the heart of the context in which the governing principles are those of classical dynamics.[14] Of course, in applying and developing the old quantum theory, Bohr, Ehrenfest, Sommerfeld,[15] and others imported and employed claims from classical mechanics—this was characteristic of the "old" quantum theory, and indeed this was surely the only way to proceed, at least until the work of Schrödinger and Heisenberg—but Bohr's theory cannot be so easily broken down into distinct contexts or "cells," to each of which one can systematically assign principles held as true.[16]

What about the famous remark, so redolent of Bohr's later Complementarity Principle, to the effect that we must seek the "strictest possible limitation" of the applicability of contradictory points of view? From his nonadjunctive perspective, Brown reads this as a means of dealing with inconsistency by appropriately restricting the applicability of classical and quantum principles. However, throughout the further development and application of his model, Bohr was greatly concerned with its logical consistency.[17] Indeed, this appeared to be his ultimate goal, and like the majority of physicists at that time, he believed that such a goal would be reached only through a succession of approximations to his model by means of the correspondence principle. He was prepared to accept that these steps along the way might be inconsistent themselves, but this was justified as long as they were seen to be pointing in the right direction (Mehra and Rechenberg 1982, 4:273–274). However, by 1925 Bohr had concluded that his program had foundered and that this way of approaching the issue simply would not work: "He had become convinced that a major advance was necessary and that it would consist in formulating a mathematically consistent scheme, one in which the conflict between classical physics and quantum theory would be removed in one single step" (Mehra and Rechenberg 1982, p. 274).

The major advance came in the form of the "new" quantum mechanics of Heisenberg, Born, Jordan, and Dirac, which was indeed mathematically consistent but still contained incongruences at the "interpretational" level, where it was unclear how to understand the dual wave and particular aspects of quantum phenomena. Bohr referred to these incongruences as "higher" (Mehra and Rechenberg 1982) or "apparent" inconsistencies (Havas 1993) and was willing to accept them as an integral part of the "final" theory.[18] Indeed, his Complementarity Principle enshrined this interpretational inconsistency as a fundamental aspect of our understanding of quantum phenomena, deriving, as he took it, from the classical nature of measurement results in terms of which we have no choice but to communicate that understanding. Here, then, we do have a division into contexts, with the "strictest possible limitation" of the applicability of the space-time and causal pictures, but it is at the "higher" level of interpretation and one that has come increasingly under attack at that.

This separation of "higher" or "apparent" inconsistencies from the ordinary kind is interesting, and the monikers are significant. They mark a crucial distinction between the "old" quantum theory and the new quantum mechanics: In the former, the inconsistency was enmeshed not simply within the mathematical formalism but within this formalism as interpreted by a uniform ontological picture (corpuscular). In the latter, however, there was a consistent mathematical treatment—and in particular no division of "contexts" at this level of structure—but no uniform ontological picture and hence conflict at this level. It is precisely this separation, of course, that generated the positivist attitude toward the new theory.

A further way of marking the separation is to refer to these "higher" inconsistencies as "paradoxes," or, as we have termed them, "incongruences."

In their extensive history of the development of quantum physics, Mehra and Rechenberg note,

> Since the end of 1925 there had come about a clear recognition of the difference between a paradox implied in the description of quantum phenomena and an inconsistency of the mathematical scheme employed. Niels Bohr and the quantum physicists learned that an inconsistency was much worse than a paradox. A paradox might be very disagreeable, but still one could set up a theory from which the results would follow in a logical way, so that two results would never contradict each other. (1982, 4:274)

This dawning recognition can be attributed to the mathematicians at Göttingen, led by Hilbert, who played an interesting and not entirely clear role in these developments. Heisenberg, in particular, was concerned that the noncommutativity of conjugate quantum mechanical variables might lead to mathematical contradictions and is recorded as saying,

> I remember that in being together with young mathematicians and listening to Hilbert's lectures, I heard about the difficulties in mathematics. There it came up for the first time that one could have axioms for a logic that was different from classical logic and still was consistent. That, I think, was just *the* essential step [in physics]. You could think just in this abstract manner of mathematicians, and you could think about a scheme that was different from the other logic,[19] and still you could be convinced that you would always get consistent results. (Mehra and Rechenberg 1982, pp. 229–230)

He then goes on to make this distinction between paradox and inconsistency and remarks, "Only gradually did the idea develop in the minds of many physicists that we can scarcely describe nature without having something consistent, but we may be forced to describe nature by means of an axiomatic system which was thoroughly different from the old classical physics and even using a logical system which was different from the old one" (Mehra and Rechenberg 1982, p. 230).

Noncommutativity of variables was not such a big deal to the mathematicians, as Hilbert had already described axiomatic systems for noncommuting mathematical objects (Mehra and Rechenberg 1982). The precise nature of quantum logic has been the subject of a great deal of work in the foundations of physics, but the division of contexts when it comes to the application of the theory and on which the nonadjunctive approach is grounded simply does not feature in these developments. Nevertheless, the "higher" inconsistency or paradox of wave-particle duality was still bothersome: "The contradictions in the intuitive interpretations of phenomena occurring in the present scheme [of quantum mechanics] are totally unsatisfactory. In order to obtain an intuitive interpretation, free of contradictions, of experiments—which, by themselves, are admittedly without contradiction—some essential feature is still missing in our picture of the structure of matter" (Mehra and Rechenberg 1982, p. 278). This "essential feature" was subsequently provided by the uncertainty relations, which, of course, formed the mathematical core of Bohr's complementarity interpretation. At this level, we do have a kind of division of contexts, delineated by the uncertainty relations themselves, but it is not the kind that Brown seeks.

What are we to make of all this? It seems that from early on in the development of his program, Bohr was concerned with its logical consistency and came to realize that this could not be achieved by piecemeal steps but only by a revision of the very foundations of the field.[20] The "strictest possible limitation" on the applicability of classical and quantum principles was necessary within the context of the 1913 model in order to obtain meaningful empirical results, but we reject the view that these limitations correspond to a clear and well-marked division of contexts. Rather the situation appeared to be much more fluid, and it was only as agreement with experiment was obtained that it became clear which principles could be applied and where. Here the inconsistency was problematic precisely because it was situated within the mathematics as interpreted by a single ontological picture (essentially of the atom and its constituents as corpuscular). With the advent of quantum mechanics, we obtain a consistent mathematical scheme (and perhaps an underlying nonclassical logic) but with a "paradox" in the interpretation. Here the "higher" inconsistency is dealt with by the strictest possible limitations, not on the applicability of classical and quantum principles, but on that of different ontological pictures, particlelike and wavelike.

Content and Closure

The alternative to the "logic-driven control" of logical anarchy is the so-called content-driven view. According to this, the attitude of scientists to inconsistency is based simply on "a reflection of the specific content of the physical theory at hand" (Norton 1993, p. 417). Thus Norton argues that in the case of the quantum theory of blackbody radiation a consistent subtheory can be constructed from which Planck's law can be recovered. However, that this is possible with hindsight is irrelevant to the question of what attitude should be adopted toward inconsistent theories before such consistent reconstructions have been identified (see Brown 1990, p. 292; Smith 1988b, fn. 20). The difficulties are revealed when we reflect on the following comment:

> If we have an empirically successful theory that turns out to be logically inconsistent, then it is not an unreasonable assumption that the theory is a close approximation of a logically consistent theory which would enjoy similar empirical success. The best way to deal with the inconsistency would be to recover this corrected, consistent theory and dispense with the inconsistent theory. However, in the case in which the corrected theory cannot be identified, there is another option. If we cannot recover the entire corrected theory, then we can at least recover some of its conclusions or good approximations to them, by means of meta-level arguments applied to the inconsistent theory. (Norton 1993, pp. 417–418)[21]

Although we agree that inconsistent theories in science typically point the way to a consistent successor—an agreement on which we shall expand later—the delineation of this consistent successor can be an immensely complicated process involving a variety of heuristic moves. In many and perhaps most cases, picking out the subtheory from which this successor has sprung can be achieved only with hindsight and, as we have just said, this gives no clue as to how inconsistent theories should be regarded epistemically.

Norton's emphasis on a "content-driven" approach also carries with it problems concerning how we identify and understand the "specific content of the physical theory at hand." Here issues as to the metalevel delineation and characterization of scientific theories come to the fore. As well as the old quantum theory of blackbody radiation, Newtonian cosmology affords another example of inconsistency, since we can combine standard theorems of this theory to conclude that the force on a test mass is both $\overline{\mathbf{F}}$ and not equal to $\overline{\mathbf{F}}$, where $\overline{\mathbf{F}}$ is a force of some nominated magnitude and direction (Norton 1993). Responding to this, Malament has argued that the inconsistency is merely an artifact of traditional formulations of Newtonian cosmology that disappears if one adopts a "geometrized" version (Malament 1995). Crucially, this reformulation leaves all the observational consequences of the theory unaffected.

In counterresponse, Norton has accepted this "repair" of the theory but insists that, nevertheless, in its traditional form it remains "paradoxical" (Norton 1995, p. 511).[22] This further reinforces our point: That a consistent reformulation of an inconsistent theory can be subsequently achieved offers no insight into the attitude(s) adopted toward the theory at the time, except insofar as it undermines the neo-Hegelian line. Asserting that scientists reflect upon the physical content of the inconsistent theory merely expresses the heuristic aspects of the situation and says nothing about the more fundamental epistemic attitudes.

Indeed, Norton makes the standard (realist) assumption that theories are to be taken as true in the correspondence sense (cf. fn. 18); this is what he means when he talks of "distributed truth" in the example of the old quantum theory. But again, leaving aside our advocacy of quasi truth in these sorts of situations, such talk of truth appears profoundly inappropriate in the case of the Bohr model, where Landé reports that Bohr "always had the idea that it was makeshift and something provisional" (cited in Pais 1991, p. 155). Granted this point, for our purposes what is important about Norton's analysis is the tracing of the interconnections between the various components of these theories, which blur any clear-cut distinction between "contexts" or "cells" (and, of course, which distribute truth, on his account) and which encourage the perception of theories as fundamentally integrated.

Such integration also runs counter to the claims of another adherent to the "content-driven" view, whose work bears close comparison to that of Brown but who, crucially, rejects any suggestion that paraconsistent logic might be usefully employed in this context. Thus Smith has argued that, with regard to the drawing of inferences from inconsistent sets of beliefs, the latter can be broken up into (self-) consistent subsets from each of which implications can be derived in classical fashion (Smith 1988a; see also Kyburg 1987). Concerning the adjunction of these subsets, however, he rejects the use of paraconsistent logics, preferring, instead, to abandon the requirement of deductive closure in such cases altogether. As we shall see, this is tied in with Smith's view of the appropriate attitude that should be taken toward inconsistency in science, but it is worth noting here his emphasis that acceptance of closure should not be confused with inference. In particular, against attempts such as Priest's to produce convincing examples of true inconsistent theories from scientific practice, Smith argues that these have confused the relevant domains for the application of deductive and inductive inferences, respectively (1988a).

As in the case of nonadjoined contexts, the question immediately arises: If we are not committed to the deductive closure of a set of statements, in what sense does such a set constitute a coherent, unified theory? Smith avoids the problem on this point by referring to inconsistent proposals as "prototheories" (1988a, p. 246).[23] Thus, like Norton, Smith draws on the example of the "old" quantum theory; however, whereas Norton was primarily concerned with reconstructing a consistent formulation of this theory, Smith is quite clear that his focus is on the issue of the grounds for believing that such formulations even exist when all that a scientist has is the original inconsistent "proposal" (1988b, p. 435, fn. 20).[24] As Einstein clearly saw, Planck's inconsistent derivation of the blackbody radiation law was inadequate but also contained an element of truth (Smith 1988b, p. 435).[25] This was subsequently expressed in Einstein's famous 1917 paper on the spontaneous and stimulated emission of radiation as the conditions for equilibrium between emission and absorption processes that would have to be obeyed by both classical and quantum treatments (Smith 1988b, pp. 436–437). However, although Planck's inconsistent treatment contained this "element of truth," Smith prefers not to regard it as a fully fledged theory. Relatedly, he argues that we need something weaker than acceptance to refer to the attitude adopted toward such "proposals." We shall question these distinctions in what follows, but for the moment we would like to focus on the abandonment of deductive closure: From a formal perspective, it is this that marks out a "prototheory" and also prevents us from adopting an attitude of acceptance toward it.

Smith would certainly agree with Brown's requirement that the division into contexts of subsets should be such as to allow one to use certain claims in those subsets where they are supported by evidence. However, it might be asked whether, without the imposition of closure, this is sufficient to prevent the division into subsets from becoming an unsystematic affair, with "anything goes" being the order of the day with regard to the application of claims from these subsets. Certainly, abandoning closure is a radical move to make under any circumstances, and one might wonder whether the price of doing so is lower than that incurred by introducing nonclassical logics! On this issue, as he himself acknowledges, Smith is following the lead of Harman, who rejects the principle that our belief set should be closed under logical implication on the grounds that we are not logically omniscient: We are capable of holding only a finite number of "explicit" beliefs,[26] and many implications of our beliefs are not immediately obvious (Harman 1986, chapter 2).[27] This is used to support Harman's overall view that "there is no clearly significant way in which *logic* is specially relevant to reasoning"[28] (Harman 1986, p. 20). Presumably, Smith would not wish to go so far, abandoning as he does (deductive) logic for inconsistent "prototheories" only.

Logical omniscience is one of the fundamental "hidden" principles underlying philosophical accounts of belief and belief change that many have started to question and reject as hopelessly idealistic. Here we simply wish to note two things: The first is the connection with inconsistency in science. Norton puts it very well:

> In traditional philosophy of science, we routinely attribute powers to scientists that are near divine. It is only in desperate circumstances that we may even entertain the possibility that scientists are not logically omniscient and do not immediately see all the logical consequences of their commitments. The inhabitants of the grubby world of real science fall far short of this ideal. In truth they will routinely commit themselves consciously and even enthusiastically to the great anathema of philosophers: a logically inconsistent set of propositions. (1993, p. 412)

Likewise, Kyburg has bluntly asserted that the "demand for consistency is appropriate for angels, not men" (1987, p. 141).

Second, a lack of logical omniscience can, in fact, be accommodated within a coherent formal system by introducing a notion of truth related to that of "pragmatic truth," as we shall indicate in chapter 7 (da Costa and French 1988; see also da Costa and French 1991a). It does not follow, therefore, that because we are not omniscient in this respect, we must give up deductive closure. Of course, the sense of closure that is retained may not be classical, and here Smith and Harman may wish to offer further objections. At this point, the possibility of question begging by both sides looms. Smith wants to show that we can accommodate inconsistency in science without introducing an underlying paraconsistent logic; the price of such an accommodation is the twin distinctions between theories and "prototheories" and acceptance of the former and "entertainment" of the latter. Our contention is that we can avoid paying such a price by regarding theories, models, whatever, as only quasi-true, partially defined, and so on, and tying our notion of acceptance to that. We shall explain what we mean after we have considered the relationship between closure and commitment a little more closely.

Closure and Commitment

What the failure of logical omniscience says is that, given a set of beliefs that are held, there are many consequences of these beliefs that are not, and cannot be, consciously held, or of which we are not explicitly aware, due to the finite nature of our minds. However, this does not imply a lack of commitment to such consequences. If it did, the whole notion of revising our sets of beliefs in the light of their consequences would make little sense. (This is particularly crucial in the scientific context, where such consequences are not always immediately drawn; we shall draw on this point in chapter 7.) Since it is this aspect of commitment that distinguishes acceptance from (factual) belief, it would seem that our lack of logical omniscience could be acknowledged, while at the same time retaining the view that the set of propositions we accept is closed under (some form of) logical implication.[29] Thus, as Cohen remarks:

> While the direct act of acceptance involves a conscious and voluntary choice of a premiss or premisses, a person may be said to accept indirectly all the logical or mathematical consequences of any conjunction of the propositions that he accepts directly, whether or not he is himself conscious of those consequences or disposed to work them out. That is because, when *p* entails q and he decides to adopt the policy of taking *p* as a premiss, he stays effectively on course by taking *q* as a premiss, because anything that follows from *q* will also follow from *p*. So, if he announces to other people that he has accepted certain propositions, he implies acceptance also of their logically or mathematically necessary consequences or equivalents, and thus creates a presumption about how he will think and act. And, even if he does not announce to others his decision to accept that *p*, this decision certainly reaches indefinitely far. . .beyond any consequences or equivalents that he may perceive at the moment. (1989, pp. 370–371; cf. Brown, 1990, p. 290, fn. 14)

We have already noted that the failure of logical omniscience can be accommodated within formal systems, which, contrary to Harman, are deductively closed, *in some sense*. The systems we are referring to are, of course, nonclassical, and the closure is with respect to logical consequence, defined in a nonclassical fashion. Let us be more specific and consider again the relationship between belief and truth.

The work of Brown, Norton, and Smith has been developed within the warp and weft of two philosophical frameworks, one structural and the other epistemic. The first is the syntactic view of theories as sets of propositions, and the second is the standard conception of the relationship between belief and correspondence truth. From the perspective of the former, this set of propositions is taken to be closed in the sense that any subset of true propositions logically implies another true proposition (recall Smith's understanding of "theory"). The latter, we recall, holds that to believe that *p* is to believe that *p* is true. Putting the two together means the closure of the former spills over into the latter: Closure of true propositions gives closure of beliefs about those propositions. Thus Brown, who regards the propositions within particular contexts as true and strives to maintain a broadly realist attitude toward inconsistency, retains closure, even if the nonadjunctive fragmentation of a theory means it must take a paraconsistent form. Smith, on the other hand, has no such pretensions and advocates an "instrumentalist" view in which inconsistent "proposals" are not even theories at all and are "entertained" rather than accepted; thus he is quite happy to drop closure altogether.

The manner in which the properties of truth filter up through these channels to belief can be illustrated through a consideration of the question of whether belief in a contradiction implies inconsistent beliefs:

In formal terms, a belief in a contradiction can be written as

$$\mathbf{B}(p \,\&\, {\sim}p)$$

or, under natural presuppositions,

$$\mathbf{B}p \,\&\, \mathbf{B}{\sim}p$$

where **B** is the doxastic operator. If belief is taken to be "exclusive,"[30] so that believing that not-p implies not believing that p,[31] that is $\mathbf{B}{\sim}p$ implies ${\sim}\mathbf{B}p$, then we obtain

$$\mathbf{B}p \,\&\, {\sim}\mathbf{B}p$$

Hence the inconsistency spreads from the propositions to the belief states themselves, and holding a contradictory belief becomes equivalent to holding inconsistent beliefs.

Now consider truth. It is usually accepted that

$$\mathbf{T}{\sim}p \rightarrow {\sim}\mathbf{T}p$$

In other words, truth is also regarded as exclusive. If we further take "belief that p" as shorthand for "belief that p is true," we get:

$$\mathbf{BT}{\sim}p \rightarrow \mathbf{B}{\sim}\mathbf{T}p$$

from which we obtain

$$\mathbf{B}{\sim}\mathbf{T}p \rightarrow {\sim}\mathbf{BT}p$$

Dropping the **T**s gives us

$$\mathbf{B}{\sim}p \rightarrow {\sim}\mathbf{B}p$$

Thus the exclusive nature of belief seems to follow from that of truth, together with the standard understanding of "belief that p." We might say that there is nothing inherent in belief itself that causes it to be exclusive; it's all in the propositions and whether they are regarded as true or not.

Dropping this connection between belief and truth in favor of some weaker epistemic attitude, such as that associated with "partial" truth, then allows us to deny the previous equivalence and block the spread of inconsistency. A scientist who accepts an inconsistent theory, such as Bohr's, can therefore be said to believe in a (self-) contradictory theory but not to hold inconsistent beliefs. A logic of science suitable for accommodating such a situation should thus include as a principle:

$${\sim}(\mathbf{B}p \,\&\, {\sim}\mathbf{B}p)$$

It turns out that a form of paraconsistent doxastic logic can be developed that can accommodate

$$\mathbf{B}(p \,\&\, {\sim}p)$$

but has the previous principle as a theorem (da Costa and French 1989b). What this means is that when we reason about our own beliefs, our "external" logic is classical, whereas our "internal" logic is paraconsistent.[32]

The technical details are not so important at this point. What is important is that at the level of representational belief, such a belief in a self-contradictory theory can be held without degenerating into the holding and not holding of a belief. This is not the case at the level of "factual" beliefs, where we have the isomorphic representation of "the facts" and belief that p is understood as belief that p is true. In other words, factual beliefs are exclusive, and at this level the inconsistency spreads from the objects of belief through the belief set itself.[33]

The logic of scientific acceptance, then, is paraconsistent; indeed, it can be shown that it is a form of Jaskowski's discussive logic referred to before (da Costa, Bueno, and French 1998a). When we reason with consistent premises, Jaskowski's logic reduces to the classical form; with inconsistent premises, of course, it does not. Acceptance, in our terms, is closed under the Jaskowski system but not in relation to classical logic (unless the theory concerned is consistent). More precisely, there is no closure under classical conjunction and material implication, but we can define "discussive" forms of conjunction and implication with respect to which acceptance can be regarded as closed. It is worth noting that there may be inconsistent theories whose "degree of inconsistency," as it were (and it makes sense to talk in these terms when one is considering paraconsistent logics), is so great that they are worthless; one example might be a theory possessing an axiom of the form $p \leftrightarrow \sim p$. Hence the acceptance of inconsistent theories or of theories that are (mutually) inconsistent does not imply that we accept everything.

With these points in mind let us return to the issue of closure. Ullian has argued that truth and acceptance must be kept separate since truth is deductively closed, whereas "acceptability is unlikely to have any such closure property" (Ullian 1990, p. 345). The point of making this distinction is to suggest that perhaps talk of truth should be dropped altogether at the "higher theoretical levels": "Some theoretical parts of science are far from the soil of observation sentences. I have suggested that talk of truth is planted in that soil, so that the roots of that concept lie far from any points of clear contact with the trellised vines of highly theoretical physics" (Ullian 1990, p. 344). The conclusion, then, is that perhaps truth may not offer the best vantage point from which to view the scientific process (Ullian 1990, p. 345) and that what is needed is "a reasonable notion of acceptability to step into the breach" (p. 344). However, to disassociate acceptance entirely from truth, even at the level of the empirical basis, is to open the door to conventionalism. Although the theoretical parts of the models employed in science appear to reside in the treetops, they are still connected to their observational roots (even trellised vines must touch ground somewhere!). Of course, we agree that truth is not sacrosanct and that talk of it in the strict correspondence sense must be dropped as inappropriate for those beliefs we deem representational, but still *some* form of truth must be retained. As for a "reasonable notion of acceptability," if one persists in regarding this in classical terms only, then the slide into Harman's position would seem to be inevitable. To our minds, the consequences of such a view are sufficiently disagreeable as to make the alternative, nonclassical approach even more attractive.

Summing up then, we side with Brown against Smith (and Harman) in agreeing that if the cognitive commitments of scientists are to be taken seriously, some form of closure must be imposed in the context of acceptance. However, we reject the fragmentation of theories into contexts, cells, subsets, strands, or whatever, on the grounds, as we indicated, that it fails to capture the interconnections between different parts of the overall theoretical structure. Indeed, we suggest that such a division acquires the meager plausibility it has only because of the association of such contexts or subsets with "elements of truth," which are then passed on through theory change. While we are broadly sympathetic with the sentiments underlying this association, our claim is that no such division or fragmentation is necessary once we shift to regarding theories as quasi-true and as accepted as such.

Inconsistency and Partial Structures

Going back to the formal results displayed previously, we have indicated that there can be forms of doxastic logic—essentially paraconsistent forms—that can accommodate a belief in an inconsistency, without it diffusing through a belief set and rendering that entire set inconsistent. The inconsistency is contained, as it were, and localized to the *objects* of belief, rather than the beliefs themselves. This, we feel, is as it should be; in particular, rather than having a set of contradictory beliefs, each of which is a belief in some *p* taken as true in the correspondence sense, what we have is a belief in an inconsistency that can nevertheless be regarded as partially true. And the ultimate grounds for so regarding a theory are pragmatic, in the broad sense, to do with the empirical support the theory receives. From our internalist perspective, then, we claim that this approach captures the appropriate epistemic attitude toward inconsistent theories. Furthermore, by localizing the inconsistency in this way, our account encourages a focus on the *structure* of the inconsistent objects of belief, rather than the nature of the belief set.

And of course, it is such structural considerations that fall within the purview of the externalist perspective. From this standpoint, the interesting issues involve questions such as how inconsistent theories are used for computational purposes or how such theories might be heuristically fruitful. Unfortunately, there has been little discussion of these issues within the semantic or model-theoretic approach. Within this dearth of consideration, Laymon stands out for declaring that one of the advantages of this approach to theories is that it "allows for easy and revealing consideration of computational concerns and . . . allows for the coherent use within a calculation of inconsistent theories" (Laymon 1988, pp. 262–263). Laymon's primary concern is with the way in which theories that are inconsistent with one another are used to account for phenomena, rather than the explanatory role of internally inconsistent theories themselves (although Laymon has discussed Stokes's use of inconsistent principles in the latter's analysis of pendulum motion; Laymon 1985). Thus he gives a series of examples involving early analyses of the Michelson-Morley experiment in which, for example, the Lorentz-Fitzgerald contraction hypothesis was tested on the basis of a determination of interferometer arm length that assumed no contraction, or, in another case, a "hybrid" analysis was presented that assumed that the reflection and refraction of light were independent of any movement through the ether but that the velocity of light was dependent on such motion (1988; another example he mentions in passing is that of the well-known relativistic analysis of the bending of starlight by the sun in which the Schwarzschild metric is used in combination with a Newtonian analysis of the telescope and photographic plates).

Two fundamental points emerge from Laymon's analysis. The first is the importance of empirical results in overriding any concerns with consistency: "scientists were primarily interested in the computational consequences of their assumptions regardless of the consistency of those assumptions when viewed as sentences of some logical system" (1988, p. 262). However, the existence of inconsistencies can be outweighed in this manner only if the theory concerned is regarded not as true, in the literal or correspondence sense, but as partially or approximately so.

The second point concerns the role of idealizations and approximations in generating a certain "looseness of fit" between theories and these computational consequences, sufficient to allow mutually inconsistent theories to be used.[34] Essentially, Laymon's work can be seen as a further elaboration of the point emphasized by Thompson (1989)—namely, that the semantic approach allows for the explanatory deployment of theories from different domains. In the cases noted before, these domains are fundamentally incompatible with one another.

We have already mentioned that idealization and approximation can be accommodated through the introduction of "partial isomorphism" as the fundamental relationship—horizontally and vertically—between theoretical and data models. This also indicates the way in which one can understand how such "computational consequences" can be obtained from an "internally" inconsistent theory such as Bohr's.[35] It is by acknowledging that there is a certain "looseness of fit" between a representation and that which it represents, that inconsistency can be comfortably accommodated. By introducing "partial structures" into the semantic approach, this notion of a "looseness of fit" can be formally captured and in a way that reflects the doxastic attitudes of scientists themselves.[36] The belief involved is representational in character, being belief in a representation of semipropositional form. It is the conceptual incompleteness of such representations that underpins this "looseness of fit" and allows for their acceptance, whether they be internally or mutually inconsistent.[37] Furthermore, this partiality or openness encourages the view that they should regarded, in general, not merely as steps on the road toward some complete representation but also as potential heuristic sources for further development.[38]

If assenting to a set of statements is taken to warrant assent to their deductive closure, then dropping the latter in cases of inconsistent theories means coming up with something weaker than unconditional assent itself to refer to the epistemic attitude that is appropriate in these cases. Thus Smith introduces the notion of "entertaining" hypotheses for their "pragmatic" virtues and states, "When we have evidence for the truth of each of two incompatible claims, it is quite rational to *entertain* both. However, the fact that they are inconsistent means that we must mentally flag them to guard against indiscriminate future use. One or the other is false. At best, both can be "approximately true," "partially true" or "true under some disambiguation" etc." (1988a, p. 244). Obviously, we would agree with this last line but would question why an epistemic attitude different from that of *acceptance* needs to be marked out for these cases, where acceptance need not be understood as "unconditional." It might seem that we merely have a difference in terminology here—where we use "acceptance as partially true," Smith uses "entertainment"—but the difference is significant: As we have already said, dropping closure puts inconsistent "proposals" beyond the reach of logic altogether. Thus this position rests on a dichotomy, with on the one side, consistent theories, accepted as true and subject to classical logic, and on the other, inconsistent proposals, which are merely entertained and not closed under logical implication.

On this view, then, inconsistent proposals do not lie within the domain of acceptance at all but rather occupy the realm of "discovery" (this is made clear by Smith himself, 1988a, p. 244). This distinction is another well-known component of the Received View that has come under attack in the past thirty years or so. It marks a separation of discovery from justification in terms of rationality: The former "can be analysed only psychologically, not logically. . .and cannot be portrayed by a rational procedure"[39] (Reichenbach 1949, p. 434), whereas the opposite is true of the latter, which, indeed, is—or should be—the sole concern of epistemology (see Reichenbach 1938, p. 7). The distinction itself has come under pressure from two directions: Historians and sociologists of science have questioned whether so-called external, social factors can be contained within "discovery" and have argued that they spill over into justification, collapsing any distinction between the two.[40] Moving the other way, a number of philosophers of science have argued that discovery is susceptible to rational analysis, and Post in particular has carefully articulated what he calls the "theoretic" guidelines to new theories (1971).[41]

Claiming that there is a common epistemic attitude evinced on both sides blurs the distinction even further and suggests that we should instead refer to a unitary process of theory "development." We shall return to this point in the next chapter; for now, we wish to note that if inconsistent "proposals" have a heuristic role that can be considered by philosophers of science and not just psychologists, then the grounds for treating them as sharply different from theories is further eroded. Thus Smith explicitly acknowledges that his chief concern is with the question of how scientists move from a particular inconsistent theory to a consistent successor (1988b, p. 435 fn. 20). But, it might be argued, if there is a rationale to such moves and, in particular, if this rationale can be embedded within a general set of such rationales as applied to the heuristics of theory development in general—that is, covering the development of consistent theories as well—then the dichotomy between "proposals" and theories, with closure on the one hand but not on the other, begins to look distinctly shaky.

The heuristic role played by inconsistent theories clearly stands in need of further elucidation. As Post has noted, an inconsistency is usually regarded as a "formal flaw" that points the way to the construction of a consistent successor (Post 1971, pp. 221–225). These flaws, or "neuralgic points," are "useful hints of a higher and wider system in which these contradictions are eliminated" (p. 222).[42] The process is taken to be analogous to that of moving from the liar paradox to truth at the metalevel in logic. Of course, a strong realist when it comes to contradictions, such as Priest, would question precisely such a move, arguing that truth can be retained at the object level of the paradox itself.[43] An instrumentalist, on the other hand, while dispensing with truth altogether, would also reject the process wholesale, arguing that inconsistent theories should not be ruled out as an ultimate objective of science.[44] Both views disdain the elimination of contradictions that Post emphasizes, and both fall afoul of the actual practice of science, where attempts to move from inconsistency to consistency are clearly evident.

A crucial role in this process is played by the General Correspondence Principle, used heuristically in the same sense as discussed in the previous chapter. The general strategy identified here is that of "projection," in the sense that the scientist "uses the original [inconsistent] proposal along with the confirming evidence available for various parts of that proposal to give a schematic 'projection' of what the consistent replacement theory (or some fragment of that theory) will look like" (Smith 1988b, p. 438). The same strategy is exemplified in Bohr's use of the correspondence principle and provides a rational procedure for the heuristic use of inconsistency.[45]

In discussing these strategies, it is significant that Post highlights the incompleteness that is an essential aspect of the process, emphasizing,

> This incompleteness of theories, or their openness to growth beyond themselves, is a stronger and more interesting case than just filling the gaps in the data covered by a theory, such as missing elements in the periodic table. Just as the system of real numbers is not closed, in the sense that one can formulate equations in terms of real numbers only whose solutions point *outside* the system, so a physical theory may show its lack of closure. The points at which a theory is open, or anomalous, provide a strong indication of the way in which the future larger theory is to be built. (Post 1971, p. 222)

For Smith, it is the move by means of "projection" from an inconsistent "proposal" or prototheory to an incomplete but consistent "fragment" of the successor theory that constitutes an essential step in the heuristic exploitation of inconsistency. Such fragments are projections of the "elements of truth" of the prototheory and serve as guides to the completed successor. Our intention is not to dispute the rich historical analysis that Smith presents in support of these claims. Certainly, this process of "projection" helps us understand the way in which consistent theories come to be constructed from their inconsistent predecessors.[46]

However, we balk at the implication that such incomplete "fragments" effectively stood in for the inconsistent proposals, and that scientists like Einstein and Bohr "made deductions not from an *inconsistent* but from an *incomplete* theory, only parts of which were interpreted" (Smith 1988b, p. 443; his emphasis). It is true that they speculated as to which well-confirmed fragments of the old quantum theory would be retained by any future theory, but what they reasoned from and drew inferences from were the inconsistent models or theories themselves. Structurally, these were both inconsistent *and* incomplete in that with the inconsistency we do not know whether the relevant properties and relations hold in the domain or not; as Post says, the points where a theory is anomalous are also those where it is open. And, of course, this openness or lack of closure (this time, of the theory itself as viewed from our extrinsic perspective) is representable model-theoretically by a partial structure.[47] The way in which this openness is exploited heuristically in general and the manner in which structural relations are imported from one domain to another will be dealt with in the following chapter.

Returning to the intrinsic stance, at the heart of Smith's account lies a view of acceptance that we find simply unacceptable: "Acceptance of a set of statements warrants their unrestricted use in all reasoning processes. At least one member of an inconsistent set of statements must be false. So, without additional information, we cannot rationally assent to the *unconditional* use of any one of the statements from a set known to be inconsistent" (Smith 1988b, p. 432). Leaving aside the observation that this characterization of inconsistency begs the question against paraconsistent logics,[48] no scientific theory, past or present, receives such warrant; there are always hedging conditions, acknowledgments of limitations, approximations, and the like that are manifestations of the scientists' doxastic attitudes toward the theory. And these attitudes are inadequately and incorrectly characterized in terms of belief in the (correspondence) *truth* of the theory. What we are about here is precisely an attempt at the difficult trick of accommodating these weaker attitudes while retaining the connection between acceptance and belief of some form.[49] We prefer to stick to "acceptance" rather than "entertainment," since even in the case of inconsistent theories there is still the element of commitment that we (and Brown) take to be the hallmark of the former (witness Bohr's elaboration and defense of his theory).

The blurring of the distinction between discovery and justification thus proceeds on two fronts: First of all, heuristic guidelines are admitted, of the sort delineated by Post, lifting discovery, at least in part, back into the domain of rationality, and then acceptance is unbound from truth, allowing for that descriptively important element of fallibility.[50] There is then no need for an alternate term such as "entertainment" since acceptance understood in terms of partial truth is sufficiently broad to do the job in all cases.

The heuristic fertility of inconsistencies is something that a "conservative" view of scientific progress fails to acknowledge. According to this tradition, represented by Popper's remarks at the beginning of this chapter, the inconsistency must be removed before any further progress can be made, since it is irrational to build upon inconsistent foundations (Lakatos 1970, p. 144). The opposing "anarchist" position, on the other hand, elevates inconsistency to the status of a fundamental property of nature. It is in just this ground, of course, that Priest's approach has its roots. Lakatos's argument that such an approach is incompatible with the goal of science being truth (1970, p. 143) begs the question, since the existence of true contradictions is precisely what is asserted by the proponents of this position. Nevertheless, as we mentioned before, such a view fails to do justice to the doxastic attitudes of scientists themselves and, ultimately, to the practice of science. For this reason, the "rationalist" position is to be preferred, according to which the heuristic fruitfulness of an inconsistent theory may be exploited without a concomitant claim that it is the last word on the matter (Lakatos 1970, p. 145). Of course, not all inconsistencies are fertile in this sense. Bohr's "program" began to degenerate, in Lakatosian terms, as further inconsistencies were piled onto the original model in an increasingly desperate attempt to deal with the "undigested" anomalies that began to proliferate (Lakatos 1970, p. 154).[51] Such inconsistencies failed to extend the "empirical character" of the theory and can thus be regarded as sterile (p. 154).[52]

Summing up, the view we are advocating here can be situated both within the "rationalist" framework and between the positions of Smith and Brown: Unlike the former, we retain a unitary epistemic attitude of acceptance and a concomitant form of closure on the relevant set of beliefs, but we agree on the heuristic fertility of inconsistency; unlike the latter, we reject the separation of elements of theoretical structures into incompatible contexts, but we agree that the appropriate logic of acceptance is broadly paraconsistent, and this acceptance involves an underlying commitment. We recall from chapter 4 that on our account acceptance differs from factual belief in that the former involves this commitment, whereas the latter does not. It is, however, tied to a representational belief in the partial truth of what is accepted, and the commitment is to the use of the representation or model concerned. Certain inconsistent theories can then be regarded as quasi-true and accepted in this sense. And the element of truth such theories possess, which ultimately justifies their acceptance, is not lost but is incorporated within the structure of their successors in accordance with the General Correspondence Principle.[53]

This discussion of the heuristic moves behind theory development is further broadened in the next chapter, where we return to the externalist perspective and indicate how such moves can be understood within the model-theoretic approach.

6

Partiality, Pursuit, and Practice

So far we have focused on two of the central elements of scientific practice—namely, theories and models. In the last two chapters in particular, we were concerned with the attitudes toward these elements and what it is to say that they are "accepted." However, in chapter 5, we began to return to broadly structural considerations as we noted the heuristic force of inconsistencies in their role as formal flaws. As we indicated there, these considerations, in alliance with epistemic factors regarding the appropriate attitude toward such "flawed" representations, break down the traditional barrier between "discovery" and "justification." Since this distinction features in the traditional two-stage picture of scientific practice, in which theories or models are first "discovered" and then "justified," we need an alternative picture. Central to this will be the structural incompleteness of theories or, as Post puts it, "their openness to growth beyond themselves" (1971, p. 222). Understanding how this incompleteness is handled will enable us to construct a unitary account of theory "development" in which theories and models are proposed, developed, pursued, and justified in a complex web of structural engagements.

There is, however, the further question of the driving forces behind this development. The focus on structural incompleteness brings to the forefront the structural engines of theory development, contrary to, for example, the sociology of scientific knowledge, which emphasizes social interests as heuristic "causes," or more recent accounts of the "sociology of scientific practice," which abandons what is seen as "reductive determinism" entirely and reverts to some form of "agency." Here we shall follow Post again by analyzing the heuristics of scientific practice in terms of the objects of that practice and their structural qualities. The notion of partial isomorphism will be crucial to this picture of the development of new theories and models from the old. Before we get to that point, however, we need to consider and reject both the "social interests" approach and the more recent "mangle" of scientific practice, both of which downplay the heuristic force of the structural elements of theories and models themselves.

Practice Makes Science

At roughly the same time as the semantic approach emerged, the emphasis on models took a different form, which, although it can also be seen as a reaction to the syntactic account of theories and models advocated by the Received View, declined to adopt a formalized treatment of these notions. The views of two of the principal figures in this latter approach—namely, Black and Achinstein—have already been discussed, and one can further trace its effects on the recent work of Cartwright, for example. However, Kuhn, whose work—perhaps more than that of any other philosopher of science—represents the decisive break with the Received View,[1] also placed models at the heart of scientific practice. For Kuhn, a model was to be understood in the sense of an "exemplar" that effectively structures solutions to the problems of "normal" science.[2]

The extent to which Kuhn's ideas have had a positive impact on the philosophy of science is debatable (see Suppe 1977, pp. 633–649), but what cannot be doubted is their profound influence on sociological accounts.[3] More than the implicit relativism, more than the eschewing of any talk of truth, and more even than the rather confused but, according to the later Kuhn, central idea that occupants of different paradigms live in "different worlds," it is this notion of exemplars that was taken as fundamental by the advocates of the sociology of scientific knowledge,[4] perhaps because it indicated a way in which science could be taken as interestingly rule-free.[5] It is the exemplars that drive the conceptual extensions of the scientific "net" through the modeling process:[6] the problems and techniques that arise within a scientific tradition "may relate by resemblance and by modeling to one or another part of the scientific corpus which the community in question already recognizes as among its established achievements" (Kuhn 1970, pp. 45–46).[7] This process, in turn, is regarded as radically open-ended in the sense that "the extension of scientific culture. . .can plausibly proceed in an indefinite number of different directions; *nothing within the net fixes its future development*" (Pickering 1992, p. 4; our emphasis; see also p. 9). Indeed, innovation in science can be attributed to precisely this open-endedness.[8]

The all-important question now, of course, is what determines which direction is taken or, equivalently, what does fix the future development of the "net"? As far as these sociological accounts are concerned, the answer is unequivocal: The direction and future development are determined by social interests. Scientists extend the net, develop their theories or models, in ways that serve their interests. It is via the mediation of these interests that "closure" is achieved, in the sense of a consensus that closes off certain avenues of further development.[9] Barnes, for example, makes this explicit in a work whose title—*Interests and the Growth of Knowledge*—gives the game away: "new forms of activity arise not because men are determined by new ideas, but because they actively deploy their knowledge in a new context, as a resource to further their interests" (Barnes 1977, p. 78). In granting a fundamental role to interests as the causal base of scientific practice, the sociology of scientific knowledge effectively gave center stage to the role of human agency in science. Thus, in his sociologically oriented history of high energy physics, Pickering wrote, "In this book, the view will be that agency belongs to actors not phenomena: scientists make their own history, they are not the passive mouthpieces of nature" (1984, p. 8).

More recently, however, Pickering himself has begun to articulate an alternative view that differs in two crucial respects: (1) a shift away from science-as-knowledge to science-as-practice, with an attendant elaboration of concepts, which, it is claimed, speak directly to the latter and include the attribution of a form of agency to the material as well as the human realm and (2) an insistence on the multiplicity of elements that can account for closure, linked to a view of the "net" as "patchy" and heterogeneous (see, for example, Pickering 1992 and 1995a).[10] Interests, rather than underpinning practice in the sense of reducing and explaining it, are now situated within the plane of practice itself as just one of a variety of sorts of generative factors (see also Latour 1987). The social is no longer the sacred (cf. Pickering 1995b, p. 465, fn. 68) or, as Lynch puts it, from his ethnomethodologist's perspective: "Sociology's general concepts and methodological strategies are simply overwhelmed by the heterogeneity and technical density of the language, equipment, and skills through which mathematicians, scientists, and practitioners in many other fields of activity make their affairs accountable" (Lynch 1992b, p. 299).[11]

Nevertheless, and despite these differences, this new approach retains the dual emphasis on the role of models in extending the net and the open-ended nature of this process. Indeed, the importance of models is elevated even further since, with interests shoved aside, closure must be achieved in some other way, namely, by the bringing together of various disparate "extensions," or models—theoretical, phenomenological, experimental, or whatever—together into "association." This is, of course, a nontrivial achievement, and the difficulties involved fall under the heading of "resistances." These arise when models are brought into interaction in some way, when aspects of one model—certain fundamental relationships, say—are imported into another, for example. Resistance, then, is a manifestation of structural discord, although this is typically not made at all explicit.

Still, there is the question of the exact nature of the forces driving these associations. We must bear in mind that it is of the essence of this new sociology of scientific practice to reject any kind of global, homogeneous, unitary analysis; rather, the forces concerned will be highly localized and disunified, varying from situation to situation. Indeed, in summarizing the paper of a fellow traveler, Pickering expressed the central motif thus: "If practice carries within itself a teleological principle of making associations between disparate cultural elements, there is no need to look outside practice thus construed for explanations of particular closures in cultural extensions (though neither, of course, is it forbidden to do so). Practice has its own integrity, and once we have grasped this integrity, we no longer feel the need for an explanatory 'something else'" (Pickering 1992, p. 17). Such a "principle" cannot be understood or employed in isolation from the practice in which it is embedded and in terms of which it gains its sense and meaning (cf. Lynch 1992a, p. 252). Rather than look to a source of social determination to explain "closure" in this sense, we should look to the constituent features of the practice itself. Any question as to how a particular closure was achieved must be tied to the "local historicity" of the object of that practice (Lynch 1992a, p. 253). Explanation thus gives way to description: "Descriptions of the situated production of observations, explanations, proofs, and so forth provide a more differentiated and subtle picture of epistemic activities than can be given by the generic definitions and familiar debates in epistemology" (p. 258; also Lynch 1992b, p. 290).[12]

Now although we are sympathetic to certain of the consequences of this shift away from social interests as central causal elements,[13] it is clear that they are not entirely out of the frame. Scientific knowledge, on this view, "is just one part of the picture, something that is not analytically privileged in any way, but that evolves in an impure, posthuman dynamics together with all of the other cultural strata of science—material, human, social" (Pickering 1995b, p. 446). Thus we see that it is by no means a case of excising the social but of placing it on the plane of practice alongside a rich diversity of other factors and elements, no one set of which is any way privileged. As we shall see, it is not just this retention of the social but the importance of its role in goal formation within scientific practice that makes this view as much as a target of our concerns as the sociology of scientific knowledge. The inclusion of "agency" in this picture encourages—or better, demands—a further shift away from the representational idiom to the "performative," which sees the world as full of agency, both human and nonhuman (Pickering 1995a). This is another aspect that we shall return to shortly. The role of the representational is not denied, but, rather, we are encouraged to be interested in how the field of representations "is threaded through the machinic field of science, in trying to understand each in relation to the other, rather than treating the representational aspects of science in isolation" (Pickering 1995d, p. 414).[14]

There are aspects of this account that are obviously amenable to our more "traditional" way of looking at practice, most notably the emphasis on modeling and open-endedness. However, the introduction of agency obscures the role of structural considerations in guiding theory development and overemphasizes the degree of freedom and choice involved. Let us be clear here: We do not dispute that practice, construed sufficiently broadly, may include social and other elements. The issue is whether these have any epistemic bearing on the construction of representations within that practice; if they do not, then the "field of representations" can be considered in isolation. This is indeed what we shall argue, offering in the place of the performative idiom of agency and suchlike a structural account of practice. The points of contention will become clear if we look at Pickering's work a little more closely:

The twin emphases on models and open-endedness are clearly laid out in the following passage:

> A given model can be extended in an indefinite number of ways, nothing within the model itself foreshadows which should be chosen. Thus part of the problem of getting to grips with practice is that of understanding closure, of understanding why some individual or group extends particular models in particular ways. The solution to this problem appears to lie in the observation that models are not extended in isolation. Modelling typically aims at producing *associations* in which a plurality of projected elements hang together in some way. And the important point here is that the achievement of such associations is not guaranteed in advance—particular modelling sequences readily lead to mismatches in which intended associations are not achieved. *Resistances*, that is, arise in practice to the achievement of goals. Encounters with resistance set in train a process of *accommodation*, in which the openness of modelling is further exploited in trial-and-error revisions and substitutions of models, modelling sequences, and so on, aimed at proceeding further toward the intended association. (Pickering and Stephanides 1992, pp. 140–141)

There are claims here that are strikingly obvious, and there are claims that are strikingly problematic, or at least so it seems to us. That models are not extended in isolation will hardly come as a shock to the system of any but the most naive student of the field. Of course, we may disagree as to the nature and composition of the embedding context—the extent to which social factors intrude, for example—but, at least in this example, the primary concern is with the contextual role of other models. Likewise, there is little that is controversial in the assertion, as it stands, that "resistances" arise in practice and are then dealt with by a process of accommodation.[15] From the more conventional perspective, one obvious and rather stubborn piece of resistance would be the material world, a.k.a. reality itself, and although Pickering himself could hardly be expected to accept the latter statement without some qualification, here, too, there are points of contact between us. First, having said what we have in chapter 3 about the hierarchy of models inserted between the highest "theoretical" representation and the "phenomena," we would expect this form of resistance, at least, to mediated and perhaps also transformed by the intervening levels. Thus Pickering suggests that we "think of factual and theoretical knowledge in terms of representational chains passing through various levels of abstraction and conceptual multiplicity and terminating, in the world, on captures and framings of material agency" (1995a, p. 145; we shall briefly consider the notion of material agency shortly).[16]

Second, he advocates a kind of "pragmatic realism," which "recognises that the production and transformation of scientific knowledge in accommodation to resistance is inseparable from a larger process of the production and transformation of complex and heterogeneous forms of life" (1990, p. 688). Although we shall not explore the details here, Pickering inclines toward a Jamesian form of pragmatism, where the consequences of holding a belief are cashed out in terms of the differences that makes to what one does, regardless of how things "really" are, as opposed to a Peircean form, where these consequences are the differences that truth of the belief makes to what one does. The difference is crucial since, on the Jamesian reading, the holder of the belief is the ultimate epistemic authority, whereas the Peircean accepts that she may not know all the relevant facts that bear on the truth of her beliefs. Thus situating science within a Jamesian context will encourage one to adopt a disinterested, descriptivist stance in which no claims to epistemic authority over the practitioners of science will be asserted. Furthermore, whereas the Peircean will accept that reality goes beyond what she can anticipate, the Jamesian will admit the presence of the external world only in those situations where she encounters resistances in her practice, these resistances to be felicitously reinterpreted. But then, of course, if reality is so conditioned by one's beliefs, one might wonder how resistance can arise in the first place, a much worn line of criticism that was leveled at James himself. A similar concern might arise with respect to Pickering's account of resistance in practice—and in mathematical or conceptual practice, in particular—but as we shall see, the invocation of a form of nonhuman agency ensures that encountering and overcoming resistances is in fact a nontrivial matter.

Returning to the process of modeling in science, this can be further decomposed into the triad "bridging," "transcription," and "filling."[17] The first involves the construction of a bridgehead as a creative extension from some base model to the new system. In constructing such an extension, certain formal associations within the base model will be preserved and "transcribed" into the new one. However, given the differences between the models, such transcriptions are necessarily partial. As a consequence, gaps in the novel structure emerge that must then be "filled" in by drawing on the resources available. Resistances arise as a result of these differences, differences that emerge only as the new model or theory is constructed.

This brings us back to the issue of the nature and source of these "resistances." What the bridging process does, as we have said, is to import the basic structural elements of one model into another. Note that what we are talking about here is *importation* rather than creation; the domain in question must be structured to at least some minimal extent before bridging is attempted; otherwise, resistance would simply never arise. In scientific practice, we claim that the source of this minimal structure will be ultimately and broadly empirical, deriving perhaps from low-lying "phenomenological" models, although as we move up the hierarchy the "degree of contact" with this empirical source will get attenuated. From the perspective of the very highest theoretical levels, the depth of lower-lying structures at which the source of resistance can be located depends on the extent of the theory change in question. In relatively straightforward cases of theory extension, it may be the very next structures down in the hierarchy that are turned to in order to serve as the base of the bridgehead. On the other hand, in radical theory change, such as occurs in scientific revolutions, for example, continuity must be sought in structures much further down the ladder; obviously, this sort of account rejects any notion of incommensurability holding across revolutionary divides.[18] Given, then, the existence of this minimal structuring, it is not that surprising that further elaboration of the imported structure will encounter "resistance" due to structural incompatibilities.[19] Of course, if these incompatibilities are sufficiently extensive, they will prevent the given importation in the first place.

According to Pickering, seeking some "accommodation" to these resistances then involves "tinkering" with various extensions that are available in the given conceptual space in a series of "free" and "forced" moves. Thus transcription can be regarded as a sequence of forced moves, insofar as it follows "from what is already established concerning the base model" (Pickering and Stephanides 1992, p. 150). Bridging and filling in, on the other hand, are "free" moves in that they "mark tentative choices within the *indefinitely open* space of cultural extension" (Pickering and Stephanides, p. 142; our emphasis). "The outcome of particular modelling sequences is thus at once structured by choice but not determined by it; it is something genuinely to be found in practice" (p. 142).

There are two important points to note here. The first concerns this issue of the source of resistance. Pickering characterizes it in terms of the notion of "disciplinary agency": "It is, I shall say, the agency of a discipline . . . that leads disciplined practitioners through a series of manipulations within an established conceptual system" (1995b, p. 419). Discipline asserts itself in transcription and it is here "where disciplinary agency carries scientists along, where scientists become passive in the face of their training and established procedures" (Pickering 1995b, p. 420).[20] And it is here, as we have just said, that resistances arise as the disciplinary elements of the practice pushes back, as it were, against the practitioners.

The metaphor of "pushing back" is no idle one. For Pickering, disciplinary agency is the analogue within conceptual practice of material agency within experimental practice. The material world is a world of forces and elements beyond human control (cf. 1995a, p. 6), and likewise disciplinary agency cannot be reduced to the human form at one degree removed: "The notion of discipline as a performative agent might seem odd to those accustomed to thinking of discipline as a constraint upon human agency, but I want (like Foucault) to recognize that discipline is productive. There could be no conceptual practice without the kind of discipline at issue: there would only be marks on paper" (Pickering 1995b, p. 458, fn. 5).

As Pickering acknowledges, this attribution of agency to materiel and discipline owes a great debt to the "actor-network" school of Latour, Callon, and others (see for example, Latour 1987).[21] However, whereas for the latter the symmetry between human and nonhuman agency derives from a semiotic analysis that offers a definition of actant "devoid of its logo- and anthropocentric connotations" (Callon and Latour 1992, p. 350), Pickering rejects this "invocation of an atemporal science of signs" (1995a, p. 14) in favor of a view of material and disciplinary agency as "temporally emergent" in practice. Again, the openness of practice is brought to the fore as the "contours" of these agencies cannot be known in advance. Thus what Pickering is interested in is a temporally informed "real-time" understanding of science that rejects the idea that we can know in advance the shape of future machines, models, mathematical objects, or whatever.

This brings us to the second metric of disagreement between Pickering and the actor-network school, which concerns the nature of human agency. This runs parallel to the "machinic" and disciplinary forms, insofar as it, too, can be repetitive and machinelike and disciplined (1995a, p. 16). However, the symmetry breaks down over the rocky subject of *intentionality*, which is used "in an everyday sense to point to the fact that scientific practice is typically organized around *specific plans and goals*" (Pickering 1995a, p. 17). These goals and plans are not temporally emergent in the way that disciplinary and material agency are, since they are "predisciplined" by the culture in which they are situated. Nevertheless, "the predisciplining of intent by existing culture is only partial" (Pickering 1995a, p. 19), because of the open-endedness of modeling. Again, Pickering emphasizes that from a given model, one can construct an indefinite number of future variants and "nothing about the model itself fixes which of them will figure as the goal for a particular passage of practice" (1995a, p. 19).

This provides the link with his earlier and much discussed framework of opportunities-in-context (1984), which was developed with the emphasis on interests to the forefront and is now seen as covering the process of goal-formation in science:

> I believe that one can see an underlying general structure to goal-formation in science, and that modelling and coherence are important concepts in unraveling that structure. But I do not think that one can offer causal explanations of goal-formation on this basis. To the contrary, there seems to be an ineradicable element of chance here. There is an explanatory gap that I cannot see how to bridge between possessing a given range of resources and assembling them into a coherent goal. (1990, p. 684)[22]

As an example, Pickering gives Zweig's interpretation of the fundamental representation of the SU(3) group as referring to quarks, rather than purely mathematical entities, a move he claims cannot be understood or justified at all in terms of the context and resources available at the time. All that we can say by way of explanation is that "it just happened" (Pickering 1990, p. 721, fn. 16).

Coming back to the decomposition of the modeling process into "free" and "forced" moves, the existence of the former is crucial to the conclusion that the products of "practice" are to be viewed as the contingently successful end points of the dialectics of resistance and accommodation. This aspect of contingency is central, as throughout his work Pickering emphasizes the element of discretion involved in these free moves and states that, with regard to an example from mathematical practice, "*Nothing prior to that practice determined its course*" (1995b, p. 442; our emphasis) and that "we can never know ahead of practice what its products will be" (1995d, p. 415).[23] It is precisely these discretionary choices that reduce the openness inherent in the modeling process, and thus "conceptual practice . . . has . . . the form of a *dance of agency* in which the partners are alternately the classic human agent and disciplinary agency" (Pickering 1995d, p. 420).

We shall not delve any further into the "classic" notion of agency at work here,[24] except to recall that in chapter 4 we appealed to a broadly voluntarist account of acceptance in order to help underpin the distinction between "factual" and "representational" beliefs. The issue that separates our view of model construction and development from Pickering's then concerns the role that chance and unpredictability play in the latter and the concomitant refusal to acknowledge the heuristic force of the structural elements of the models themselves. This becomes clear when we look at the notion of "emergence," which is fundamental to the "real-time" understanding of science: The sense of emergence here is that of "brute chance" (1995a, p. 24), of resistances and accommodations that emerge "unpredictably," and of developments that "just happen."

Such talk is heavily redolent of the "flash of genius" view,[25] and Pickering acknowledges that it may be "offensive to some deeply ingrained patterns of thought" (1995a, p. 24). It is precisely on this point that we find this characterization of the modeling process so problematic, in particular when it comes to the claims that a "given model can be extended in an indefinite number of ways" and "nothing within the model itself foreshadows which should be chosen."[26] It is on the latter sort of claim, in particular, that we wish to focus attention here.

The Practice of Pursuit

Let us begin by considering the "pursuit" of models and theories. This has been famously described by Laudan as the "nether region" between discovery and justification (Laudan 1980, p. 174). In the previous chapter, we argued that the boundaries between these domains cannot be sharply drawn, not because sociological considerations emerge on both sides of the line, but rather because the appropriate epistemic attitude is the same across them. As we also remarked, with the positivist claim that all that is epistemically interesting can be found in the domain of justification only, the domain of "discovery," taken to cover both heuristics per se and pursuit, was left in a methodological vacuum. Obviously, one way to fill this vacuum is with assorted interests and opportunities-in-context, and if the wall between domains is toppled also, these pour over into justification as well. But this is to disregard the point that there is actually no vacuum to be filled, that the "domains" of discovery and pursuit have a rich and interesting and primarily *epistemic* structure that meshes smoothly with that of "justification." Thus epistemically and structurally what we have is not an abrupt scission but rather a continuous process of theory "development."

As Post notes, there is an ambiguity in the phrase "the construction of theories": "if there are any criteria of merit which we accept in judging the structure of given theories, then these criteria may presumably be carried into the field of building new theories. The study of the structure of existing houses may help us in constructing new houses" (1971, p. 217). The "structure" here derives from that of the "heuristic objects"—the theories and models themselves (Post 1971, pp. 216 and 218). Indeed, this can be taken as the central motif of the rest of this chapter: The driving forces of theory development lie, not within some social nexus nor with the notion of "agency," whether human or "material," but with the structures of the objects of this development.[27]

Of course, to say that heuristics is "structured" is not to say that it is algorithmic, in the sense that there is a set of rules and all that needs to be done is follow them in order to obtain a new theory.[28] Post, for example, talks of "a rationale, if not a 'logic' of scientific discovery" (1971, p. 216),[29] where this rationale will include the criteria for the a posteriori assessment of theories as indicated previously, as well as further "heuristic" criteria that we shall speak of later. These criteria constrain theory development and thus delimit its inherent openness, effectively suspending this development between the algorithmic and the totally free. Indeed, the sense of "openness" here needs to be handled quite carefully. Of course, it is one of the major themes of this book that scientific theories and models are only "partially" formed, in a sense that can be technically defined, but this does not mean that the space of further development is wide open.

Here the issue of creativity intrudes. There is a great deal of literature on this subject, but we shall simply note that perhaps Reichenbach's line of epistemic relevance can be drawn through psychology. On one side we have the private—and perhaps subconscious—circumstances that lead a scientist to an idea,[30] whereas on the other we have the connections between that idea and the relevant embedding context, which connections can be represented in structuralist terms. The field of creativity—of how it is that scientists actually come up with their ideas—is concerned, at least in part, with the former circumstances.[31] We nevertheless feel compelled to emphasize that these circumstances cannot be divorced from this context, and a great deal of what is put down to "genius" and "creativity" can be understood as the judicious perception and exploitation of a particular heuristic situation.[32] We think that an obvious case can be made for claiming that these theoretic guidelines of theory construction and development also function as reasons for theory choice in nonepistemic situations (or at least situations that are not explicitly epistemic; certain of these reasons may involve reference to other theories, suitably supported)—that is, as reasons for conferring appropriate warrant in what has been called theory pursuit. Such criteria, we recall, include the General Correspondence Principle, simplicity (problematic, granted), and the retention of well-confirmed universal invariance principles (Post 1971).

A nice example of theory pursuit in a situation where explicitly *empirical* factors had little or no role to play is that of the competition between the paraparticle and color models of quarks in the mid-1960s. This episode is one of the more important in the history of modern physics, as it resulted in the birth of quantum chromodynamics (QCD), generally accepted as the best theory of the strong nuclear forces currently available. It began with a statistics problem: There was good reason to believe that quarks should be assigned spin ½, and, therefore, according to the spin-statistics theorem, should be regarded as fermions, and their overall wave functions should be antisymmetric. On the other hand, there were cases in which the most successful model was one in which the wave function of the three quark-collective was *symmetric* under interchange of any two quarks. As in the case of the liquid-drop and shell models of the nucleus, discussed in chapter 3, what we have here are two models that are inconsistent with one another. The problem, then, was to overcome this inconsistency by reconciling the apparently fermionic nature of quarks with symmetric wave functions.

Greenberg proposed a particularly elegant solution to this problem that was derived from his long-standing investigation of the possibility of particles with mixed-symmetry wave functions (the relevant history of such possibilities is given in French 1985): He suggested that quarks just were examples of these theoretical possibilities, or paraparticles. Shortly afterwards, Han and Nambu put forward an alternative solution to the problem, in which quarks were given an extra degree of freedom by means of which they could have spin ½ yet also possess symmetric wave functions under interchange of any two. These extra degrees of freedom subsequently came to be denoted by "color." The color model not only reconciled the symmetry characteristics of quarks and their composites but also appeared to be observably equivalent to its paraparticle rival. The rest, as they say, is history. The three-triplet model with color went on to become the core of the theory of quantum chromodynamics, while paraparticle theory languished; although some further theoretical work continues to be done, it is now generally regarded as being outside the mainstream.

Pickering has considered this episode from the perspective of his earlier notion of opportunities-in-context, the central idea of which is that

> Each scientist has at his disposal a distinctive set of resources for constructive research. These may be material—the experimenter, say, may have access to a particular piece of apparatus—or they may be intangible—expertise in particular branches of experiment or theory acquired in the course of a professional career, for example. The key to my analysis of the dynamics of research traditions will lie in the observation that these resources may be well or ill matched to particular contexts. Research strategies, therefore, are structured in terms of the relative opportunities presented by different contexts for the constructive exploitation of the resources available to individual scientists. (1984, p. 11)

His conclusion is that the color model won the day because paraparticle theory was simply too obscure, too complex, and too "out of the ordinary" for most physicists (1984, pp. 94–95; private correspondence).[33] We recall that this idea of opportunities-in-context is now taken to cover only the goal-formation phase of scientific practice. Thus, it leaves a gap between the formation of goals and their achievement, and it is within this gap that "resistance" emerges. And goal formation is inherently contingent, involving an ineradicable element of chance.

However, this analysis simply cannot bear the weight of historical circumstance (for further details, see French 1995). In fact, the idea of a generalized form of quantum statistics had been around for quite some time and the details of the paraparticle model in particular had been in circulation for some months prior to the publication of Greenberg's work. It is certainly notable that the textbook on which generations of physicists cut their quantum mechanical teeth—Dirac's *Principles of Quantum Mechanics*—specifically mentioned this possibility. Furthermore, at the heart of parastatistics lies group theory, a fundamental part of quantum mechanics as applied to high-energy physics and certainly well known to physicists at that time.[34] Thus there was little that was fundamentally new or unfamiliar mathematically in paraparticle theory, and this was particularly true of later "cleaned up" versions. This disposes of the claim that the theory was "obscure" or "unfamiliar" and was not pursued for that reason.

Rather than turning to contextualized opportunities and the comparative unfamiliarity of the paraparticle model to explain the scientific community's preference, we should focus on the objective characteristics of the models themselves.[35] In fact, the decisive factor was that the color model was able to be "gauged," whereas the paraparticle one was not.

Classical electromagnetism is said to be "gauge invariant" in the sense that the empirical consequences of the theory are unaffected by certain transformations, which vary from one space-time point to another, applied to the potentials (whose derivatives express the electric and magnetic fields). In quantum electrodynamics (QED)—generally acknowledged as one of the most (empirically) successful theories in the history of physics—the electron and photon fields can also be transformed in this manner, without changing the physical predictions of the theory.[36] Thus gauge invariance "inherited" the immense empirical success of QED and came to be elevated to the status of a fundamental principle. So we see that there was *good reason* to pursue a model that could be shown to similarly incorporate this invariance principle, over one that could not (pursuit being differential in this way).[37] The point is made by Greenberg himself, who also acknowledges the relative familiarity of the color model:

> The SU(3) color theory became more popular than the parastatistics version because a) the former is more familiar and easy to use, and b) up to now nobody has been able to gauge the parastatistics theory, while the gauging of the SU(3) color theory gives quantum chromodynamics. Let me be explicit, the two theories are equivalent quantum mechanically, but they are apparently not equivalent from the standpoint of quantum field theory. (Greenberg, private correspondence)

We shall return shortly to a consideration of the precise nature of this similarity between QED and QCD and, relatedly, that of the "objective characteristics" of the models involved.[38]

Before we do so, let us consider this issue of the relative familiarity of models a little further. This concept is also a central feature of Giere's account of scientists' choices regarding which model to pursue (1988). One case study he presents is that of the decision to pursue Dirac (relativistic) models of the nucleus rather than Schrödinger (nonrelativistic) ones (1988, pp. 216–218). As the names suggest, members of the latter family are characterized by some form of Schrödinger's equation, with suitably approximate interaction potentials obtained by ignoring or downplaying relativistic effects, whereas the former are all based on the Dirac equation with relativistic effects included. Although admitting that it was new experimental data that lead to a resurgence of interest in the Dirac approach, Giere notes that not all the scientists involved were convinced of the usefulness of such models as applied to medium-energy interactions. At this point, Giere claims, nonepistemic considerations must come into play, and he argues for the importance of "acquired cognitive resources" in such situations, in particular those that refer to the scientists' "familiarity" with the model concerned.

This familiarity with a particular conceptual element, such as a model, is not to be regarded as abstract but rather in terms of a particular skill of applying and extending the model to new situations. Experimentalists typically share many of the same cognitive resources with the theoreticians but, in addition, possess a wide range of skills necessary for working with the material elements of scientific practice. Giere gives a number of illuminating examples, including the case of a young theorist who was originally put onto the Dirac approach by his thesis advisor. After a couple of hiatuses, he decided to return to his work on relativistic models on noticing the general excitement they were generating at the time. According to Giere, what the theorist was primarily concerned with was his short-term professional interest, and the upshot is that "The question of whether the Dirac or the Schroedinger approach is correct seems to have played little role in these decisions" (1988, p. 218).

Now the issue, of course, is what is meant by "correctness" here. If truth, in the correspondence sense, is meant, then, of course, we would be inclined to agree that whether a model is true, in this sense, plays little if any role in deciding which theory to pursue. However, even granted that a young researcher, eager to publish and act in her best short-term professional interest, might prefer to work with that model with which she was more familiar, the question arises as to how this familiarity was obtained in the first place. Giere attributes the original impetus for working on these models to a combination of the authority of the advisor and the student's theoretical interests. However, one must go back even further and ask, What was it about that particular model that made it seem worth pursuing to the advisor and attractive to the student, even before it gained any predictive success? The answer, we suggest, lies with certain objective characteristics of the model, such as, in this particular case, those that cause it to satisfy the requirement of Lorentz invariance. This is, of course, a fundamentally important requirement, embedded as it is within Special Relativity, itself supported by a variety of epistemic considerations. Indeed, we get a hint of this in the theorist's responses as recorded by Giere (1988, p. 217), and he himself notes the crucial importance of subsequent predictive success in maintaining interest in this kind of model.[39]

In response, it might be asked, Why, then, did anyone bother with the nonrelativistic approach? Again, we would point to objective features of the model: those that confer upon it a certain mathematical simplicity giving rise to greater computational tractability.[40] In this case, the model is regarded as an approximation, useful where relativistic processes are not extensively involved in the interactions concerned. Giere himself emphasizes this aspect in his example, and we have indicated how such approximations can be suitably characterized within the partial structures approach. And, of course, given the pragmatic value of such models, the greater computational complexity associated with the alternative, and the justification of the approximation concerned in terms of what is understood to be going on in the domain under consideration, their employment may be entirely rational.

Nevertheless, as Giere himself notes, "No nuclear physicist seems to doubt that, in principle, the correct model of the nucleus would be a relativistic model based on the Dirac equation" (1988, p. 184). The reason for such lack of doubt, of course, lies with the success (ultimately empirical) of Special Relativity, and here we see a well-regarded invariance principle—such regard being grounded, we stress again, in the empirical success of a particular theory—being invoked to establish overarching criteria of "correctness" for those models associated with another.

It is, perhaps, one of the more noteworthy features of modern physics that symmetry principles have come to play this heuristic role.[41] Nonetheless, invariance criteria are not infallible, as the case of parity conservation so clearly indicates (Post 1971).[42] The selection of which particular symmetry principles are to serve as heuristic guidelines is not, of course, determined a priori.[43] As science progresses, certain of these principles are so elevated, and ultimately experiment must play a crucial role in this rise to prominence, although the exact nature of this role may not be clear and the interconnections between theory or model, symmetry principle, and experiment may be tortuous and entangled. We learn what principles of invariance to apply in theory development by noting which are well supported and incorporated into our most successful theories. And this importation proceeds from one (local) domain of inquiry into another and essentially piggybacks on the structural analogies between the two.

The kind of analysis we have presented obviously drives against the sociology of scientific knowledge, with its emphasis on social interests as fundamental and explanatory. It also militates against the sociology of scientific *practice*, where, as we have noted, social interests also feature, although now they are situated within the "plane of practice" rather than causally underlying it; in particular, these interests inform the aims and goals of science. Although, as we have said, there is much in Pickering's decomposition of the modeling process that a more traditionally minded philosopher of science would agree with, his concern with agency and "temporal emergence" and the concomitant overlooking of heuristic factors lead him to see explanatory lacunae that in fact can be bridged. Let us take another example where the crucial development supposedly "just happened"—namely, Zweig's interpretation of the representation of the SU(3) group as referring to quarks.

An alternative understanding of this move can be achieved via Redhead's account of supplying a physical interpretation for surplus mathematical structure within a theory (Redhead 1975). This is related to Post's heuristic guideline of "Adding to the Interpretation" (closely connected with "Taking

Models Seriously"): "We take a hitherto incompletely interpreted part of the abstract formalism of the theory, and give it a tentative interpretation of our own, at some level" (Post 1971, p. 241). This interpretation, tentative as it is, is still some way from being put forward in the state of unfettered freedom that Pickering suggests—Zweig himself was not operating in a vacuum. Indeed, Pickering's own earlier work spells out just how detailed the context actually was:

> The idea of reducing the number of elementary particles by explaining some of them as composite was by no means new. As early as 1949, C. N. Yang and Enrico Fermi had written a paper entitled "Are Mesons Elementary Particles?," in which they speculated that the pion was a nucleon-antinucleon composite. When the strange particles began to be discovered, the Fermi-Yang approach was notably extended by the Japanese physicist S. Sakata in 1956. (1984, p. 86)

Suitably formulated in group-theoretic terms, the Sakata model became the main rival to Gell-Mann's Eightfold Way, until the 1964 discovery of the Ω^- particle resolved the competition in the latter's favor. As Pickering notes, representing the hadronic symmetries in group-theoretic terms "positively invited" the view that hadrons were composite, and the best known and developed position along these lines was the SU(3) tradition, whose fundamental representation—from which all others can be mathematically derived—contained three members: "From there it was but a small step to the identification of the fundamental triplet representation of SU(3) with a triplet of fundamental entities: the quarks" (Pickering 1984, p. 87).

Zweig studied both the Sakata model and the reviews of SU(3) and, independently of Gell-Mann, came to the view that the observed multiplet hadronic structure could be accounted for in terms of two or three hadronic constituents. The next step was not only "simple and direct" but obvious: "As one would expect for a neophyte in theoretical physics, he treated quarks as physical constituents of hadrons and thus . . . derived all of the predictions of SU(3) plus more besides" (Pickering 1984, p. 89). Thus scientific practice is essentially different from that of some unfortunate rat in a maze, randomly trying one possibility after another.[44] And what makes it different are the restrictions on this practice that are essentially "theoretic" in nature, in the sense that they involve an "internal" analysis of the theory or model in question.

The point of these historical excursions is twofold: (1) that the space of possibilities might be drastically more limited than the sociologists of scientific practice seem to appreciate, once we accept that even apparently "free" moves in pursuit are structured by objective considerations and (2) that, relatedly, it is the structure of the model itself that plays a crucial role in "foreshadowing" which is chosen. This is not to deny the "openness" of theories to further development, of course: It is via partial structures that this openness can be captured, and this also offers a suitable framework for considering the structural characteristics of models and thus for accommodating the points touched on here.[45]

There are two things to note immediately. The first is simply that from the perspective of partial structures the process of theory development can be seen as one of defining and redefining the relationships between the appropriate set of elements, be they genes, elementary particles, or whatever. In the initial stages, some subset of the R_i will be laid down according to the sorts of heuristic criteria previously suggested. Perhaps the most important concerns the Generalized Correspondence Principle, which can be taken as expressing the idea that this subset, or portions of it, will be bequeathed by the theory's predecessor(s). Adherence to well-confirmed symmetry and invariance principles will then constrain this set of relations in further ways, as well as causing the element of partiality and hence openness to be further reduced. In all cases, it is the objective qualities of the models that are important, as expressed in terms of their relational structure. Hence, whereas logically the space of opportunity may be infinitely wide, in practice it is severely delimited in terms of heuristic structure.

The second point to note is that in the practice of theory development these criteria may be applied sequentially in a different order on different occasions, with empirical considerations also added to the mixture.[46] It is here again that we see the distinction between the kind of reasons adduced under justification and those put forward as heuristic criteria begin to blur and crumble.[47]

The introduction of partial structures thus allows a treatment of evolving theories within the semantic approach, as Suppe has noted (1989, p. 427). There is a problem of theory individuation here or, more specifically, whether a change in a theory is regarded as a further development of that theory or as leading to a different one, albeit related. In the case of the latter, we should not properly talk of theories "evolving" but rather of a succession of descendants, related in terms of some kind of Lakatosian "hard core" perhaps (Suppe 1989, pp. 426–427). Alternatively, "if alterations in theories do not produce new descendant theories, but rather theories are things that can grow, evolve, and change without becoming a different theory, then this suggests that many versions of the Semantic Conception are inadequate by virtue of improperly individuating theories" (Suppe 1989, p. 428). Perhaps this problem of individuation can be circumvented by the partial structures approach as Suppe suggests.[48] Characterizing theories in this fashion allows for changes in the family of relations R_i without thereby necessitating that we regard the structure as referring to a different theory entirely.

We have already indicated how the partial structures account gives a way of treating, in fairly general terms, theories, models, and the diverse relationships between them, embracing idealizations, approximations, iconic models, and, of course, Kuhnian exemplars.[49] Our concern, then, is with the explicit representation of the structural relationships between the elements of the domain under consideration. The various relationships that hold between models (and, more generally, theories) represented in this way can be analyzed in terms of the relationships that hold between such families of relations. And, of course, it is precisely the nature and characteristics of these relationships that we are interested in from the point of view of theory pursuit.

Thus, to return to our earlier historical example, the analogy between QED and QCD, which both encouraged and permitted the expression of the latter in gauge invariant form, can be expressed in terms of an appropriate relationship of similarity holding between the respective R_i in the two cases.[50] Leaving it at that, of course, adds little to the discussion, either historically or philosophically, so let us elaborate a little further.

Taking the former perspective first,[51] particles that interact via the strong nuclear force can be grouped into multiplets of "isospin," a spinlike vector property. The isospin symmetry of the strong nuclear force can then be represented by the SU(2) group. Regarding the gauge invariance of QED from a group theoretical perspective,[52] Yang attempted to construct a similar theory for the strong interactions that would be invariant under local gauge transformations of SU(2). This attempt led to the famous Yang-Mills gauge theory of the strong interactions, which, crucially, differed from its electrodynamic analogue in containing nonlinear terms referring to the self-interaction of particles "carrying" the strong force. Formally speaking, this difference derives from the difference in group structure involved, which, in turn, can be traced to the difference in the physical properties represented.

Yang-Mills gauge theory came to be regarded as the exemplar for gauge theories in general and was extended to apply to the electroweak interaction. However, the negative analogy with QED turned out to be fundamental, as Yang-Mills theory could not be shown to be renormalizable, a feature that allows calculations to be made to arbitrarily high orders of approximation.[53] As is well known, the crucial breakthrough came with the work of 't Hooft (Pickering 1984, pp. 173–180) and his invocation of spontaneous symmetry breaking; the similarity with quantum electrodynamics was then restored. With the problem of renormalizability resolved, theorists returned to the possibility of constructing a gauge field theory of the strong interactions, and their interest focused on the SU(3) group of quark colors as the appropriate gauge group. The result was QCD, and it is precisely at this point that interest in the paraparticle model waned. What we have here is a complex web of similarities and dissimilarities existing between models, which, indeed, were constructed on the basis of the former, the ultimate ground for this incorporation of similarity, and also for the dissimilarities that arise, being empirical (albeit appropriately mediated as we have said).

Turning to the philosophical perspective, "similarity" can be decomposed into partial isomorphisms holding between certain members of the families of relations concerned (again see French and Ladyman 1999). The "degrees and respects" characteristic of this notion can then be captured in terms of the number and kinds of relations included in the relevant subsets. In the case of QED and QCD, it is in terms of the correspondence between certain members of the R_i, which refer to the relationships between the particles expressed by the relevant Lagrangians, that we can say that the theories are *similar* with respect to gauge invariance.[54] The R_i may refer not only to "physical" relationships in this manner but also to mathematical ones.[55] Thus the fundamental role of group theory, both heuristically and formally, can be accommodated within this approach: The group-theoretic similarities between

U(1) and SU(2), say, establish the bridge between domains, heuristically, and the partial isomorphism between models, formally.

Likewise, Pickering's notion of "association" can also be understood in terms of such structural analogies. In the case of scientific practice, the ultimate source of the analogy, partial as it is, lies in the properties of the particles described by the theories, these properties, such as charge and isospin, being introduced to accommodate *experimental* results. Where the analogy breaks down, "accommodation" might be sought for in terms of changes in the relevant family R_i. The move from a positive to a negative analogy in the face of resistance gets expressed in a shift of the partial isomorphism across the R_i.

The "resistance," then, is manifested in the structures or, even more explicitly, in the characteristics of the R_i.[56] Ultimately, its source is empirical. Thus, the resistance represented by the occurrence of nonlinear terms in the field equations of Yang-Mills theory can also, of course, be expressed structurally and, to repeat, derives ultimately from the properties of the interaction particles, which, in turn, were introduced to accommodate the well-supported "phenomenological model"—so called—of the strong force. The renormalizability problem was then overcome through the introduction of spontaneous symmetry breaking—that is, through an accommodation expressed in terms of changes in the R_i. New members of this family were introduced to overcome the problem, thereby restoring the partial isomorphism between the relational predicates in terms of which the renormalizability of QED and QCD is expressed.

What we have, then, is the following sort of process in this case: On the basis of some perception of structural correspondence between elements of different domains, broadly and perhaps crudely understood (such as that holding between the properties of sets of elementary particles, as in our example), a family of (partial) relations, found to hold good in one domain, typically well understood, is projected into another, about which knowledge is even less complete or, rather, even more partial. Certain of these relations may fail to hold in the new domain, as resistances arise. Others may come to be established as accommodations are made. The correspondence, then, is with only a subset of the original set, but nevertheless a (positive) analogy is established. The initial bridge may then serve as the basis for further exploration, as further relations are brought into the correspondence. The bridge itself is an "open" structure, projecting into a domain about which knowledge is only partial.[57] What we have when we compare different models as to their similarity is a time-slice of a heuristic process: The family of relations projected into the new domain forms the basis of the partial isomorphism formally holding between the relevant models. And the nature of this set—*which* relations to project—is delineated by objective heuristic criteria.[58]

Conclusion: Exemplars and Representational Belief

Let us return to the Kuhnian exemplars with which we began this discussion. In chapter 5 of *The Structure of Scientific Revolutions*, as is very well known, Kuhn grants these both pedagogic and epistemic priority over rules, in the sense that although rules are abstracted from exemplars, exemplars can guide research even in the absence of rules. How? By "direct inspection" (Kuhn 1970, p. 44). What does this mean? Kuhn's answer is famous: He refers to Wittgenstein's "family resemblance" analysis of terms like "game" or "chair." An activity is tagged with the word "game" not because it satisfies some necessary and sufficient set of criteria for what constitutes a game but because it bears a close "family resemblance" to well-established activities that have come to be called games. Likewise, the activities of normal science "may relate by resemblance and modelling to one or another part of the scientific corpus which the community in question already recognises as among its established achievements" (Kuhn 1970, pp. 45–46).

Again, an obvious way of considering family resemblance is in terms of a "partial isomorphism" holding between members of the respective R_i. The degree of closeness of the resemblance may then be crudely characterized through both the kind and number of matching relations between the models. This achieves an element of objectivity only by pushing the possibility of subjective judgment one step back. The question arises again: On what grounds are certain kinds of relations judged more important in establishing resemblance than others? The question becomes particularly acute in the context of the history previously outlined, where the resemblance between QCD and the established achievement of QED is established through the medium of group theory. One very obvious answer can be spelled out in terms of broadly sociological conventions negotiated within the relevant scientific community. (This might form part of the sociological response to the question as to why gauge invariance is taken to be so important, for example.) The equally obvious alternative brings us back to the nature and role of heuristic guidelines: Those kinds of relations that hold between the elements of empirically successful models will tend to be imposed on the elements of successor models (interpreting the General Correspondence Principle in terms of preserving structure) and will tend to be imported across domains in situations where analogies are being developed.

At first blush, it might seem misleading to introduce consideration of exemplars in a nonsociological discussion of pursuit. The standard examples (exemplars of exemplars?) include the pendulum and the weight on a spring, and although these have a fundamental pedagogical role to play in introducing new recruits to the paradigm (according to Kuhn), it is hard to see what role they might play in pursuit once professionalization has been achieved. Surely a research worker in some area of interest—quantum field theory, say—does not resolve issues of pursuit by directly inspecting the simple pendulum.

But of course, exemplars are typically arranged in hierarchies of abstraction, leading the initiate from, say, the pendulum, to the simple harmonic oscillator and beyond. Thus, a graduate textbook on quantum field theory might begin with the example of a vibrating string. Here again one might be tempted to allow nonepistemic considerations to enter the picture: Those exemplars at the higher levels of the hierarchy can be regarded as effectively delineating the space of opportunities in such a way that only those models that are appropriately or sufficiently similar to these exemplars are pursued. And the *choice* of exemplar to invoke has more to do with who one's supervisor is than with the standard epistemic virtues. There are two things to be said here: One has been said before—namely, that this account can plausibly be regarded as nonepistemic only on the surface. If we think a little about why particular exemplars are given in the relevant textbooks or pressed upon the hapless graduate student by the overbearing supervisor, it's not difficult to see that epistemic considerations might have something to do with it. Why is the example of the pendulum so ubiquitous? Because it is simple, it is computationally tractable, *and* it does seem to describe what happens in reality, given certain constraints, more or less.

The "fit" is, of course, only approximate, and it is the characteristics of this approximation that render the pendulum example simple and computationally tractable—that render it, indeed, an exemplar. As we have indicated, approximation can be accommodated within the semantic approach, and the accommodation runs along the same lines as given in the discussion of similarity: The relationship between the approximate model and that of the data is one of an isomorphism holding between portions of the relevant families of relations in each case. As the approximation is improved and the model becomes less ideal, of course, the match between the relevant R_i increases.

Talk of unanalyzed resemblances between exemplars distracts attention from the underlying dynamics of theory development. Resemblances are considered to hold between items that are already established, as it were, but in the context of pursuit and "guiding research," the focus is on the process of establishing resemblances itself. Kuhn's picture of "normal science" at this point appears to be a curiously static one: The question at the forefront is "what binds the scientist to a particular normal-scientific tradition?" (1970, p. 44)—that is, what holds the paradigm (understood in the later sense of a disciplinary matrix) together?—and it is in terms of this frozen tableau that the priority of exemplars over rules and guidelines is established. (It is only during crisis, when the binding loosens, that, according to Kuhn, a concern with rules manifests itself.)

However, this picture is not the whole picture! In Kuhn's later "Postscript," he attempts to make clear what he is about: In explicating the role of exemplars, Kuhn talks of embedded "tacit" knowledge (the term is attributed to Polyani), which would be misconstrued if taken to be represented by rules that are first abstracted from exemplars and then taken to function in the latter's stead:

> Or, to put the same point differently, when I speak of acquiring from exemplars the ability to recognize a given situation as like some and unlike others that one has seen before, I am not suggesting a process that is not potentially fully explicable in terms of neuro-cerebral mechanism. Instead I am claiming that the explication will not, by its nature, answer the question, "Similar to what?" That question is a request for a rule, in this case for the criteria by which particular situations are grouped into similarity sets,

and I am arguing that the temptation to seek criteria (or at least a full set) should be resisted in this case. (1970, p. 192)

This seems to fit the static picture delineated previously, with similarity taken as primitive and exemplars accorded priority over rules. However, it soon becomes apparent that here Kuhn is talking of the most basic level of perception of similarity in terms of *shared stimuli*. Members of a group or scientific community must see things and process stimuli in much the same ways, else how are we to understand the group's cohesion and communality? And one of the fundamental techniques by which the members of the community achieve this cohesion and learn to see things in the same way is through sharing examples of situations that other members of the group have already learned to view as similar to certain others (Kuhn 1970, pp. 193–194). Of course, Kuhn grants, this perception of similarity must be the result of appropriate neural processing and hence governed by physical and chemical laws. In this sense, he writes,

> once we have learned to do it, recognition of similarity must be as fully systematic as the beating of our hearts. But that very parallel suggests that recognition may also be involuntary, a process over which we have no control. If it is, then we may not properly conceive it as something we manage by applying rules and criteria. To speak of it in those terms implies that we have access to alternatives, that we might, for example, have disobeyed a rule, or misapplied a criterion, or experimented with some other way of seeing. Those, I take it, are just the sorts of things we cannot do. (Kuhn 1970, p. 194)

Thus the priority of exemplars over rules is a consequence of this involuntary perception of similarity, which just steals over us, as it were.[59] Yet—and this is the crucial point—Kuhn does not leave it there; this recognition of similarity over which we have no control is not ubiquitous in the sense of being applicable to all levels of exemplars. Immediately after the quoted passage, Kuhn continues, "Or, more precisely, those are things we cannot do until after we have had a sensation, perceived something. Then we do often seek criteria and put them to use. Then we may engage in interpretation, a deliberative process by which we choose among alternatives as we do not in perception itself" (Kuhn 1970, p. 194). Once we move away from the level of perception and begin to interpret, then we do have access to alternatives, we do have choice, and we must invoke criteria.

This difference is fundamentally important, and we shall now offer an interpretation of it in terms of Sperber's distinction between factual and representational beliefs.[60] We recall that the former are likewise generated involuntarily, whereas the latter, which is where acceptance is epistemically situated, are not. Our suggestion then is that it is at the level of factual belief that the perception of similarity is involuntary, and it is here that exemplars have priority over rules. We also recall that factual beliefs arise at the lowest level of the hierarchy of models, at the level of the "data" in Bogen and Woodward's sense. It is here, if anywhere, that we have "direct representation," and this provides the most basic structure of the relevant domain. And it is here, if anywhere, that the perception of similarity between domains structured at this lowest level is direct and involuntary—we just "see" it, in Kuhn's terms. This perception provides the fundamental bridge between the domains, on the back of which associations can then be made and structural relationships from one domain imported into another.

Once these associations start to be established, we have already moved up to representational belief, and at this level heuristic criteria, such as the preservation of well-supported symmetry and invariance principles, may be invoked as model construction gets under way. Here, as we said, there is an element of choice, both as to which criterion to invoke and which model to pursue and develop. At this level, exemplars lose their priority to rules, as they must, since we are no longer concerned with the static situation of perceiving similarity between minimally structured domains but rather with the dynamic situation of establishing similarities by importing structure.

The sociologists' fundamental error, then, is to assume that this priority holds at all levels, not just across but "up" the board, as it were. They fail to take note both of its basis in this involuntary perception of similarity and of Kuhn's acknowledgment of the intrusion of rules when it comes to what he calls "interpretation." Their error, ultimately, is to take *all* scientific belief as factual, and it is from this that their whole Wittgensteinian slant, with its rejection of rules and criteria, derives. With exemplars prioritized and family resemblances (mistakenly) perceived even at the highest levels, rather than established, the dynamic extension of the scientific "net" cannot be driven by structural associations but by external, nonepistemic factors. In other words, we are suggesting that the invocation of social interests is the result of the failure to see beyond the static picture of interexemplar relations.

Models do guide research but only insofar as they are heuristic objects in Post's sense. Direct inspection may indeed reveal that structure that is empirically supported and that is to be preserved according to the Correspondence Principle, but simply noting the existence of a "family resemblance" between models cannot in any way account for their development. Kuhn's apparent omission was inherited by the sociologists of scientific knowledge who filled the gap with talk of "interests." With the sociology of scientific *practice*, we see a shift to a concern with the establishment of relations between models, to a concern with "bridging" and "accommodation," but now the openness of the "net" is taken too extensively, and the element of freedom is pushed too forcefully. It is simply not the case that *nothing* within the model itself foreshadows which line of development should be pursued.

To conclude, then, exemplars might indeed be said to guide theory development, but putting it down to what the student learned from some text or advisor pressure is too glib. At some point, empirical factors intrude, even if it is through layers of abstraction or simplification. And, of course, iconic models may also function as exemplars in this way. One of Post's heuristic criteria that we have already mentioned is that of "taking models seriously," to which we would add, yes, of course, but not just in terms of the elements A, representing the entities concerned. The set of relations R_i should also be "taken seriously," and in the case of iconic models and exemplars in particular, the structural characteristics expressed by these relations may have enormous heuristic significance. Indeed, it might perhaps be said that underlying all of these considerations is the most fundamental heuristic guideline, which is also the final "message" of this chapter: Take the structure seriously.

7

Quasi Truth and the Nature of Induction

Having discussed the factors involved in the pursuit of scientific theories, we now turn our attention to the issue of justification. From the structural perspective, the higher level models come to be justified through their relationships with other models in the hierarchy, descending to the data models at the bottom. Tracking these justificatory routes through the hierarchy of partial structures is precisely what the notion of partial isomorphism is supposed to do. From the epistemic perspective, however, it is our *beliefs* that we talk of being justified, and the issue now is whether the notion of quasi truth can shed further light on this talk in the scientific context. We believe it can.

Shifting to this epistemic, or "internal," perspective, we begin by noting that inferences, or "reasonings" in general, are linguistically expressed by arguments. Any argument is composed of a set of premises and a conclusion. An argument is valid if the conclusion cannot be false when the premises are true; otherwise, it is said to be invalid. Valid arguments are the object of study in deductive logic, of course. However, in both everyday life and, critically, science, we often make inferences that are considered to be reasonable but that are invalid. Such inferences and the corresponding arguments are called inductive (or inductions in general). We are going to take this idea of induction in a very broad sense as covering all kinds of nondemonstrative inferences. The logic of such inferences is, of course, inductive logic.

The nature and role of inductive logic in scientific reasoning has, of course, been the subject of enormous discussion. Following the decline of logical empiricism and the program of inductive logic associated with it—and in particular the failure of Carnap to extend his form of inductive logic beyond the confines of certain artificial languages—many philosophers of science have concluded that the idea of such a logic is some kind of "philosophers' invention," a "make-believe" theory that has mired scientific methodology in a morass of philosophical difficulties (see van Fraassen 1985a, pp. 258–281 and 294–296, for example). However, we believe that such conclusions are based on either a conception of induction which is too restrictive, or on arguments that are simply unacceptable (see, for example, Salmon 1978, pp. 10–12). Our bold claim is that in everyday life, science, and technology, we use certain inductive techniques to make predictions and forecast future experience, among other things. Induction therefore imposes itself upon us as a significant actuality, and some form of inductive logic should be developed to investigate and systematize it. The form we have in mind, of course, incorporates the notion of quasi truth at the heart of the program.

Forms of Inductive Argument

Among the various forms of induction, we can distinguish the following (this list is by no means intended to be exhaustive):

1. Induction by simple enumeration
2. Induction based on Mill's canons (elimination)
3. Analogy
4. Direct inference (from the frequency of an attribute in the parent population to the frequency of the same attribute in a sample of that population (cf. Carnap 1963)
5. Indirect inference (from sample to population (cf. Carnap 1963)
6. Predictive inference (from sample to sample (cf. Carnap 1963)
7. The strict hypothetico-deductive method
8. The generalized hypothetico-deductive method
9. The statistical syllogism
10. The argument from authority
11. Inference from perception
12. Inference from memory
13. Testimony
14. Statistical inference in general

An argument may be conveniently and generally expressed as follows:

$$\frac{\alpha_1, \alpha_2, \ldots, \alpha_n}{\alpha}$$

where $\alpha_1, \alpha_2, \ldots, \alpha_n$ are the premises and α is the conclusion. When $\alpha_1, \alpha_2, \ldots, \alpha_n \rightarrow \alpha$ is logically true, the argument is valid; otherwise, it is invalid.

We know that a valid argument is unconditionally valid, in the sense that its validity depends only on the meanings of its premises and conclusion, other circumstances being irrelevant. On the other hand, an inductive inference is made in the presence of a set of pertinent conditions, which confer more or less plausibility upon it. Equivalently, we can say that the plausibility of an induction—that is, the plausibility of the conclusion, as supported by the premises—is also a function of certain extra conditions, which we call "side conditions." Thus the logical treatment of a form of induction reduces to the inventory and analysis of its pertinent side conditions, which, when added to the premises, contribute to the likelihood of the conclusion.

Hence, there are two basic differences between deduction and induction: Deduction is truth-preserving, while induction is not and deduction does not depend on side conditions, while the latter are essential to induction. There is a further distinction we would like to press; namely, to approximate to our inductive reasonings in even a rough and qualitative manner, it is convenient to consider as the aim of such reasonings, quasi or pragmatic truth, rather than truth *tout court*, as it were. In other words, induction arrives at "the truth," in the correspondence sense, only indirectly or in special cases.

Clearly, then, in the case of inductive inferences, we must modify our argument scheme to accommodate the set of side conditions Γ. Γ consists of the propositions expressing the pertinent system of knowledge, as well as those specific bits of evidence that contribute to the plausibility of the inference. As the standard texts of logic and statistics make clear, the set Γ is of the utmost importance: If, for example, we are dealing with indirect inference, the sample examined must be random, reasonably large, and sufficiently representative of the parent population, and all relevant facts relative to it must be critically assessed. Consequently, an induction may be appropriately represented by the following schema:

$$\frac{\alpha_1, \alpha_2, \ldots, \alpha_n}{\alpha} \qquad \Gamma$$

Of course, the distinction between premises and side conditions may sometimes appear somewhat arbitrary. This point notwithstanding, the distinction can be drawn in most cases and is absolutely essential for the development of inductive logic.

It is convenient to classify inductions into two groups: (1) those inductions that do not employ statistical techniques either directly or indirectly, explicitly or implicitly; thus their premises, side conditions, and conclusions cannot involve probabilistic concepts, such as, those of degree of belief, limits of frequencies, or probabilistic propensities, and (2) statistical inferences, strictly speaking, or those inductions that are not included in the first group. Simple induction and analogy constitute examples of the first group, whereas predictive inference, direct inference, and, in general, statistical inference belong to the second. Although probabilistic notions are not intrinsically involved in those inductions that belong to the first group, we can utilize a particular category of probability to evaluate them, as we shall see later.

We shall therefore divide inductive logic into two disciplines: (1) simple inductive logic, devoted to the study of inductions of the first category, and (2) statistical inductive logic, in which we treat inductions of the second kind. The latter will not be considered from a formal point of view here,[1] although we have discussed elsewhere the problematic nature of such inferences as they feature in "everyday" reasoning (da Costa and French 1993b). For the rest of this chapter, when we talk of "inductive logic," we shall mean simple inductive logic only.

The Generalized Hypothetico-Deductive Method

In what we shall call the "generalized hypothetico-deductive method," we have that certain propositions (the premises or bits of evidence) make plausible, in the light of the side conditions, a new proposition, the conclusion, or hypothesis. In certain cases, such as occur with certain statistical inferences, the plausibility under consideration may even be commonly estimated by a probability assignment. When the conjunction of the premises is logically implied by the hypothesis and the side conditions, we have the particular instance of the strict hypothetico-deductive method.

The generalized hypothetico-deductive method may be envisaged as, in a certain sense, the basic form of inductive inference: Any induction whatsoever can be viewed as an application of this method. It suffices, for this reduction, to regard the conclusion of the initial induction as a hypothesis made plausible by the premises plus the side conditions. Thus, the reformulation of an induction as a case of the hypothetico-deductive method does not constitute a profound logical or philosophical result but simply reflects the question of emphasis in the appraisal of the original induction. Since inductions are not truth-preserving, it is clear that we are unable to guarantee, with absolute certainty, the truth of their conclusions—even when the corresponding premises are known to be true. Hence, we may suppose that the principal aim of any induction is normally to achieve some tentative statement, something that is evident from the practice of science itself. This characteristic of induction supports the suggestion that it is natural to regard all inductions as applications of the method in question. Such a reduction, although logically possible, in fact twists the psychological intention of the first induction, but it also opens up the possibility of systematically employing the probability calculus, either in a qualitative or quantitative form, in order to evaluate inductive inferences in general.

Pragmatic Probability

Let us begin by considering subjective probability.[2] This comes in two forms: qualitative and quantitative, both of which are relevant for our purposes (cf. Koopman 1940; de Finetti 1970; da Costa 1986). However, to keep things comparatively simple, we shall limit our discussion to the quantitative form only.

The subjective probability attributed by a person S to a proposition α is usually defined as the rational degree of belief that S has in the truth of α. It is measured by a bet coefficient, as is well known. In order that bet coefficients have an immediate sense, the propositions to which a probability is assigned must be decidable; that is, we have to be able to settle, at least in principle, who, among the persons placing the bets, is the winner. Starting from decidable propositions, the subjectivist probability calculus can be developed in a straightforward manner (see de Finnetti 1970; Lindley 1965). However, when we try to apply the probability calculus to scientific theories, for example, certain difficulties are encountered. For instance, the only reasonable degree of belief in the (correspondence) truth of a universal law is zero, since their scope covers a potentially infinite set of objects. (We shall return to this issue shortly.) Furthermore, even if such a theory were to be regarded as nothing more than a set of propositions or, from the "intrinsic" perspective, were to be so regarded, it could not be considered decidable; its truth, or even its falsehood, cannot be definitively settled. In contrast, we are completely sure that a good and well-tested theory, given certain conditions, remains pragmatically true forever, as we have discussed in chapter 4. Given this, it seems to us that it is better to utilize degrees of belief not in the truth per se of theories but in their pragmatic or quasi truth. In other words, we believe one should consider, in this context, the employment of a form of "pragmatic" probability (da Costa 1986). This seems particularly natural in the case of induction.

The significance of the reduction of all inductions to the hypothetico-deductive method thus derives from the fact that it allows us to probabilize, in a uniform manner, all inductions. Of course, it is still the case that there is no a priori evaluation of inductive inference as there is of deductive reasoning. The reliability of an induction depends on the plausibility conferred by the set of premises on the conclusion, in the presence of the underlying conditions. This kind of plausibility can only be estimated, we repeat, in an appropriate way, by means of pragmatic probabilities, whether qualitative or quantitative. Logical or frequency notions do not work in this context. With regard to the former, no one has yet been able to formulate a reasonable system of logical probabilities capable of coping with complex inductions. As for frequency notions, the application of these to induction by simple enumeration and analogy is problematic. More fundamentally, perhaps, the use of frequencies presupposes the acceptance of certain theoretical constructions and hence already presupposes some form of inductive logic in our sense.[3] What we will really be concerned with, then, is not the truth of the conclusion, given the truth of the premises and underlying conditions, but instead its quasi truth, assuming the quasi truth of the premises and underlying conditions. For the effects of probabilization, the inductive procedure does not seek immediately for truth but only, in the first place, for pragmatic or quasi truth. However, when quasi truth reduces to (correspondence) truth, as in the case of observation reports (see our discussion in chapter 3), our interpretation of the probability calculus coincides with the standard one.

Our previous list of inductions evidently shows that the analysis of an individual induction has to be undertaken in a concrete fashion, taking account of all its significant peculiarities. Thus our probabilistic treatment of an induction has to be *local*: We are unable to weigh all extant pieces of evidence and side conditions, except those known to be pertinent. Our treatment is also *instrumental*, in the sense that induction may be regarded as simply a device for arriving at quasi truth, and it is from this perspective that it should be judged. Finally, the acceptance of an inductive conclusion is in principle only *provisional*, in the sense that we have delineated in chapter 4.

From this intrinsic perspective, one of the fundamental concepts of the pragmatic probability calculus is that of a pragmatic proposition. A pragmatic proposition α is a statement that asserts that another proposition β is quasi-true, in a simple pragmatic structure $\mathcal{A}$, in a domain Δ. Normally we say only that α is quasi-true or pragmatically true, leaving $\mathcal{A}$ and the domain that it models implicit. For some propositions, such as those that are decidable, α is pragmatically true if, and only if, it is true. If α is a pragmatic proposition affirming that β is pragmatically true, then the pragmatic probability of β is the standard degree of rational belief in α. By the introduction of an appropriate time parameter, we can transform a pragmatic proposition α, asserting that β is quasi-true, and that is not decidable, into a decidable proposition α'. In effect, it suffices to take α' as the proposition that says, essentially, that β is quasi-true during some time interval. By means of this device, we can attribute pragmatic probabilities even to nondecidable propositions; when the time interval is large enough, we can identify α and α'. Of course, we acknowledge that this strategy involves an idealization, but it does seem to be in accord with the pragmatist (particularly Peircean) origins of our program. It is by means of this device that we may confer nonzero prior probabilities upon scientific hypotheses. The underlying idea is that a good scientific hypothesis or theory that is pragmatically or quasi-true in $\mathcal{A}$ is in a sense "absolutely" pragmatically or quasi-true, when α is well chosen. Thus, for example, classical mechanics, Bohr's theory of the atom, and Maxwell's electrodynamics are all quasi-true within the sorts of limitations we have already discussed and will remain so for the time being. Hence, we are allowed to attribute to them the pragmatic probability 1.

When we are attributing pragmatic probabilities to certain propositions $\beta_1, \beta_2, \ldots, \beta_n$, it is natural to deal with the corresponding pragmatic propositions $\alpha_1, \alpha_2, \ldots, \alpha_n$ and then close the set $\{\alpha_1, \alpha_2, \ldots, \alpha_n\}$ by the operations defined by the usual group of connectives. The propositions thus obtained constitute a Boolean algebra, and in this manner the pragmatic probability calculus can be embedded in the ordinary subjective probability calculus (for further details, see da Costa 1986b; what we get is a finite-additive measure in a Boolean algebra, in conformity with the subjectivist approach).[4]

Inductive Logic

On the basis of these foundations, we are now able to construct our system of inductive logic (see da Costa 1987b). We have argued here that to probabilize induction is, in a certain important sense, to probabilize the general hypothetico-deductive method and that what we are essentially concerned with in such a procedure are pragmatic probabilities. Indeed, given what we have noted before, it seems to us that this is the only secure probabilistic basis for a system of this type.

We begin by noting that, given what we have just said, the inductive argument

$$\frac{\alpha_1, \alpha_2, \ldots, \alpha_n}{\alpha} \quad \Gamma$$

means that from the quasi truth of α_1, $\alpha_2, \ldots, \alpha_n$ and of the underlying conditions Γ, we infer the quasi truth of α. Thus the plausibility of inductive inferences is judged with respect to the quasi truth of the statements concerned. Furthermore, to make the application of the hypothetico-deductive method more reliable, overall, several hypotheses should initially be formulated and compared.[5] Thus, in certain cases, for example, we compare a particular hypothesis with a range of alternatives, including the negation of the statement expressing the quasi truth of the hypothesis.

The primary instrument of such comparative procedures is, of course, Bayes's Theorem. This has been the subject of extensive discussions in the philosophy of science (Rosenkrantz 1977; Earman 1992), and Bayesians claim that it provides an effective mechanism for describing changes in probability attributions. Indeed, it might be said that the essence of the hypothetico-deductive method in general lies in exactly such transformations from prior to posterior probabilities (cf. Shimony 1970, p. 85). We shall apply the theorem to pragmatic probabilities and suggest how a focus on quasi truth will allow us to deflect certain objections that have been raised against it.

Bayes's Principle

Bayes's Theorem serves as a basis for the following rule, which can be called Bayes's Principle: Let α_1, $\alpha_2, \ldots, \alpha_n$ be pragmatic propositions whose truth is involved in an investigation connected with a pragmatic structure $\mathcal{M}$, which systematizes a determined domain of knowledge Δ, and let us further suppose that every α_i, $1 \leq i \leq n$, has a prior probability different from zero (if α_i asserts that β_i is pragmatically true in $\mathcal{M}$, then β_1, $\beta_2, \ldots, \beta_n$ are the pertinent hypotheses that save the appearances). Then, given a new piece of evidence α that is also a pragmatic proposition, we should (temporarily) accept the α_i, $1 \leq i \leq n$, whose posterior probability is the highest. In many cases, we have simply $n = 2$, with α_2 being $\sim\alpha_1$.

Two important considerations should be noted at this point. The first is that, as well as the conditional probability of the evidence relative to each hypothesis, we also obviously have to be able to estimate the prior probabilities of the hypotheses being compared, *and this estimation does not take place in vacuo, as it were*. That is, the prior probabilities are not evaluated in the absence of any background knowledge. On the contrary, their computation is normally made relative to a given set of side conditions. Thus, "absolute" prior probabilities, assigned without regard to any previous body of knowledge, are not given any significant role in our system.[6]

Second, our pragmatic probability measures are not defined over universal languages or even very powerful ones. Rather, they are introduced in restricted languages that are still sufficiently rich to ensure the application of Bayes's Theorem (cf. Garber 1983). This will become apparent when we consider the problem of logical omniscience later.

It should also be emphasized that the usual confirmation theorems can easily be interpreted in pragmatic probability terms. These results then show how the use of the probability calculus, under this interpretation, renders our choices between competing hypotheses both more rational and more "organic."[7]

Simple Induction and Analogy

By way of example, let us consider how simple induction and analogy can be reduced to the general hypothetico-deductive method.

Induction by simple enumeration proceeds as follows: From the premises that some x_i, $1 \leq i \leq k$, which belongs to the class A, also belongs to the class B, we induce the conclusion that $\forall x(x \in A \rightarrow x \in B)$. Of course, such inferences are also based on a certain set Γ of side conditions, which may include such information as "no x is known that is a member of A, but not of B" or "the connection between A and B does not appear to be purely accidental."

Simple induction may then be symbolized by the schema:

$$\frac{x_1 \in A \rightarrow x_1 \in B,\ x_2 \in A \rightarrow x_2 \in B, \ldots, x_n \in A \rightarrow x_n \in B}{\forall x(x \in A \rightarrow x \in B)} \quad \Gamma$$

where, to make things easier, we may admit that the sentence $x_i \rightarrow A$ belongs to Γ, for all $i = 1, 2, \ldots, n$. In this schema, the premises are logical consequences of the conclusion, and thus an induction by simple enumeration may be logically converted into an instance of the strict hypothetico-deductive method.

We have already considered analogy from the "external" perspective in chapter 3. In its simplest form, it consists in inferences of the following kind:

The object x, which belongs to the class A, also belongs to the class B.
The object y belongs to A.
Therefore, y belongs to B.

Again, such reasoning is normally accompanied by a certain set of underlying conditions Σ, containing plausible reasons to support the inference. We can therefore convert analogy into the following inferential form:

$$\frac{y \in A}{y \in B} \quad \Sigma'$$

where Σ' is Σ plus the statement that the object x, which belongs to A, also belongs to B′. In this manner, analogy may also be reduced to the generalized hypothetico-deductive method.

To probabilize these two forms of inductive inference, so as to be able to apply Bayes's Theorem and so forth, certain pragmatic probabilities must be evaluated. With regard to analogy, for example, we have to evaluate the probability that $y \in B$, given that $y \in A$ and the statements of Σ'. It should be noted, however, that normally the probabilities involved are only roughly calculated; that is, we often proceed qualitatively. This leads us to recall the point made in the previous section, that in many cases qualitative, rather than quantitative, probabilities are all that are needed. The required theorems can then be obtained on the basis of a suitable axiomatization of this notion, such as is given in Koopman (1940).

Let us conclude this section by noting, first of all, that not all changes in our probability measures proceed according to Bayesian conditionalization, an aspect that can be regarded as a consequence of their local character. We shall return to this point shortly (see also da Costa 1986b). Second, we recognize that inductive logic as a whole is somewhat richer than the system outlined here; it includes, for example, the entire theory of elimination, in particular, Mill's methods (cf. von Wright 1951). However, we believe that our account constitutes a useful first step toward a more inclusive characterization.

Characteristics of Our System

We begin by recapitulating the essential characteristics of our system of inductive logic: It is *tentative*, *local*, and *instrumental*, three characteristics it shares with Shimony's "tempered personalist" view (Shimony 1970).

It is because the principal objective of any induction is to achieve some kind of *tentative* judgment, expressed in the form of a hypothesis, that induction as a whole can be visualized as an application of the hypothetico-deductive method and probabilized in terms of the subjective probability calculus. Furthermore, this tentative nature of our inductive inferences is expressed in the claim that such inferences do not aim for the truth, as such, but for "pragmatic" or "quasi" truth. This aspect is then reflected in our attitudes toward the acceptance of theories.

Our treatment is also *local*, in the sense that the analysis of an individual induction must be made in concrete terms, trying to take account of its relevant peculiarities. In other words, the application of pragmatic probabilities is circumscribed by the relevant conditions of a particular investigation, and, in particular, such conditions change when the set of hypotheses under consideration changes, leading to a related change in the assignment of our a priori probabilities.

Finally, our position is essentially *instrumental*, in the sense that induction is regarded merely as a device for achieving quasi truth and must be evaluated from this perspective. In particular, other alternative systems may also be constructed, with the choice among these various possibilities being dictated by pragmatic as well as logical considerations.

We have reasserted these characteristics because it is through them that our system can overcome the more serious objections usually leveled against the idea of inductive logic in general.

The Problem of Universal Laws

The first objection we shall consider relates to the problem of universal laws (see, for example, Nagel 1963 or, for a more general account, Gillies 1987). The problem is more or less the following: What probability should we assign to universal laws that have the form $(\forall x)Qx$, where the quantifier ranges over a potentially infinite set of objects? As is well known, Popper argued that the prior probability of such a law is always zero, giving a degree of confirmation also equal to zero on a probability-based confirmation theory (Popper 1972, appendix 8). Given the ubiquity of such laws in science and their positive confirmation, this presents a serious difficulty for any inductivist account.

One response to this objection is to accept that the a priori probability of a universal law is zero but assert that, pragmatically speaking, science proceeds directly from particulars to particulars without the mediation of such laws (Carnap 1963, section 10; Hesse 1974). Thus, according to this view, what is confirmed in science are not universal laws but their particular instances, with the laws possessing merely a heuristic value in terms of discovering predictions. Unfortunately this approach cannot account for the use in physics, for example, of complex, high-level theories—containing various explicitly theoretical terms—to make predictions. As Putnam has noted, there do exist examples in the history of science where there is no direct inductive evidence for a particular prediction *r* but where the evidence in conjunction with a certain group of relevant theoretical propositions gives *r* (Putnam 1975a, chapter 17).

The challenge, therefore, is to show how putative universal laws may be accommodated within a particular system of inductive logic without eliminating theories as a crucial part of the inductive process. That this challenge can be met by our system is obvious if we consider the view of truth that is employed in most accounts of the problem. The usual, if implicit, assumption as regards the (complete) truth of the laws concerned is clearly revealed by Gillies's reformulation of the problem in subjectivist terms (Popper having established his conclusion on the basis of the logical interpretation of probability): "Take, for example, I = All Ravens are Black, and suppose A is forced to bet on whether I is true. A can never win the bet, since it can never be established with certainty that all ravens are black. However, A might lose the bet if a non-black raven happens to be observed. Thus the only reasonable betting quotient is $q(I) = 0$, which indeed can be considered a kind of refusal to bet" (Gillies 1987, pp. 24–25). The crucial point is that A is forced to bet on whether the hypothesis is *true*. However, Gillies's conclusion does not follow if scientific propositions are regarded as quasi-true only, with our belief in such a proposition treated not as a belief in its truth but in its quasi truth and evaluated via the pragmatic probability calculus. It is then clear that nonzero a priori probabilities can be assigned to supposedly universal hypotheses (da Costa 1986b, p. 147). In this example, A would then be forced to bet on the quasi truth of "All ravens are black," which, of course, allows her to take account of any uncertainty involved, and the betting quotient would not then necessarily be zero.

But now the question arises: What exactly is meant by "uncertainty" in this context? That is, how do we explicate the quasi truth of universal hypotheses of the type "All ravens are black"? One possible answer is to say that such hypotheses are regarded as if they were true at, or up to, a certain time only, the uncertainty involved being expressed through the possibility of potentially refuting instances being discovered at some future time. Thus our subject A bets on "All ravens are black" being quasi-true in the sense that he or she accepts the hypothesis as if it were true within the domain of knowledge that is accessible (and here the background conditions play an important role) and up to the time of the bet, and is then betting on it continuing to be as if it were true in the future, up to some time to be specified. In other words, a temporal element is introduced into the pragmatic structures in this case. We shall consider this modification in more detail later.

This "temporal element" is obviously connected to extensions of the domain in question, in the sense that any refuting instances will arise through such extensions. On this point, the "intrinsic" and "extrinsic" perspectives again come together. Let us be clear: We are not saying that universal theories can be eliminated in the Carnap-Hesse manner, with probabilities being assigned to particular instances only. Rather, we are insisting that such theories should not be regarded as "universally"—in the sense of *absolutely*—true and that our degrees of belief should be taken to apply to partially true propositions. We recall Dorling's words: "What is important is not how philosophers construe physicists' theories but how physicists construe them" (Dorling 1972, p. 183). But it is philosophers and not physicists or, more generally, scientists themselves who assert the "literal"—that is, absolute and universal truth of laws and the theories that contain them.

Furthermore, by not eliminating such theories from the inductive process, our system allows us to retain them in establishing the plausibility of a prediction r on the basis of the evidence e. It is therefore entirely capable of meeting Putnam's challenge: What we are concerned with is the degree of belief in the quasi truth of r, given the empirical premise e and a certain set of underlying theories included in Γ, which are themselves only quasi-true as well.

The Problem of Prior Probabilities

This question of the a priori probability of universal laws is just a special case of the problem of the assignment of prior probabilities in general: Since these probabilities are essentially undetermined within the subjectivist theory itself (see, for example, Howson 1985, pp. 305–309), there arises the possibility of different (ideal) subjects making radically different assignments for the same initial hypothesis.

One possible response to this problem has been set out by Shimony, and we can adopt it here (Shimony 1970, pp. 102–103). We begin by recalling the requirement that various hypotheses concerning the subject matter under consideration should be formulated in order to increase the reliability of our method. In other words, we begin with not one but a *set* of hypotheses, and the limits on this set depend on the context of the investigation (Shimony 1970, pp. 110–114) and general heuristic principles of the sort delineated by Post, for example. If we then allow that some a priori probability can be assigned to the hypotheses—and our discussion here suggests that it always can, even in the case of universal hypotheses—and, significantly, that one and only one of the set of hypotheses is significantly more quasi-true than the alternatives, so that the likelihood p(e/h) of a piece of evidence upon this hypothesis is much greater than for the others, then any difference in a priori probabilities will be effectively "swamped," as far as a consideration of the posteriori probabilities is concerned (Shimony 1970, pp. 102–103). In other words, given that we are typically concerned with the formulation and comparison of various hypotheses and not just one, it can be shown that differences in a priori probabilities become insignificant in the application of the probabilistic approach, and (social) consensus may then be achieved.[8]

These considerations imply that a nonnegligible prior probability must be given to any new hypothesis that happens to be suggested during the investigation. This will lead to a redistribution of the probabilities already assigned and thus to a possible violation of the probability axioms (Shimony 1970, p. 104). However, the pragmatic probability interpretation, like the tempered personalist view, is *local*, in the sense given previously. Thus, when the set of alternatives is augmented in this way, the conditions of the investigation are changed, and our prior probabilities must be reevaluated. It is important to note for what follows that the use of the probability calculus to infer the prior probabilities associated with the new conditions from those associated with the old can lead to some bizarre results.

The "Hacking Problem"

This again is a special case of a general problem, referred to by Gillies as "the Hacking problem" (Gillies 1987, pp. 19–23), and the point just made provides an answer to this also.

The nature of the problem can be made clear in the following way: At some time t, before evidence e has been collected, our subject S assigns betting quotients $q_t(h)$ and $q_t(h/e)$ to h and h given e, respectively. At time t', where $t' > t$, e is known to be the case, and this is the only extra information S has acquired since t. S now assigns a quotient of $q_t(h)$ to h. Then our subject has changed her belief according to Bayesian conditionalization, provided she sets

$$q_{t'}(h) = q_t(h/e)$$

Hacking terms this the "dynamic assumption" and argues that the framework of the Bayesian view does not compel S to satisfy it (Hacking 1967, pp. 313–316).

That this assumption might indeed be implausible is claimed by Gillies on the basis of the following example:

> If [S] is asked to bet on some random process [π], it would be quite reasonable to assume at first that [π] consists of independent events, and to calculate his or her betting quotients accordingly. This amounts, within the subjective theory, to making the assumption of exchangeability, that is, the assumption that the order of the events is of no significance. The observation of a few hundred results of [π] may, however, convince [S] that order is relevant after all, that the sequence exhibits dependencies and after-effects. At this stage, if asked to bet again, [S] might want to abandon the assumption of exchangeability and use quite a different scheme for calculating his or her betting quotients. But these new betting quotients will not then be obtained from the old ones by Bayesian conditionalization (the dynamic assumption). (Gillies 1987, p. 22; Gillies uses the symbol p to denote the process, rather than π, but this becomes confusing as we shall see)

However, if the probabilistic scheme is local, as ours is, then this objection loses its force. As we noted, any change in the hypotheses under consideration implies a change in the underlying conditions of the investigation and a reassignment of our a priori probabilities (or betting quotients). That such a reevaluation can be accommodated within our approach can be seen if we recall our other requirement that, for reliability's sake, we should consider more than one hypothesis. Thus in Gillies's example, we would ask *S* to first of all consider at least two hypotheses:

> $h = \pi$ consists of independent events (exchangeability).
> $h' = \pi$ consists of dependent events (some assumption other than exchangability).

These form part of the "background conditions."

Given that she is told that π is random (more background), *S* would then reasonably assign a higher prior probability to h than to h', that is:

$$\mathrm{p}(h) > \mathrm{p}(h')$$

If she is open-minded, then $\mathrm{p}(h') \neq 0$.

Two sets of betting quotients are then made for the events $\{e_i\}$ of π on the basis of each assumption respectively. After observing n events in the process, *S* comes to believe that π exhibits dependencies and will therefore accord a higher *posterior* probability to h' given $\{e_n\}$ than to h. That is,

$$\mathrm{p}(h'/\{e_n\}) > \mathrm{p}(h/\{e_n\})$$

with the posterior and prior probabilities *for each hypothesis* being related by Bayes's Theorem as usual.

Given this result, it is then entirely rational for *S* to shift allegiance from h to h', where such a shift involves believing that the latter is more quasi-true. If then asked to bet again at this stage—that is, to bet on event e_{n+1}—it is then also entirely rational for *S* to make this bet on the basis of the betting quotients established according to h'. The fact that these quotients are not obtained by Bayesian conditionalization from the previous ones, based on assumption h, should not come as any surprise, since S is now operating within a different background context and Bayes's Theorem applies only to changes in our probability assignments within the *same* context.

In other words, being "local" in our sense implies the abandonment of the "dynamic assumption," at least as far as changes in the background conditions are concerned (after all, who ever said that Bayes's Theorem had to apply to such changes as well!), and since we have, in this case, no dynamic assumption to justify, Hacking's objection loses its bite. This is not the only implication of "locality," and it is worth noting that it can be invoked to overcome two further fundamental objections to the Bayesian approach in general—namely, the problem of "old evidence" and the Popper-Miller argument. To see this, however, we need to introduce a modification to our notion of quasi truth, through the introduction of the temporal element previously suggested.

Pragmatic Probability and Logical Omniscience

In earlier chapters, we argued that the relationship between theory and evidence, as represented by theoretical models and data models, respectively, might not be perspicuously captured by a straightforward deductive connection but by something rather more complex, such as that of "partial isomorphism." However, a general issue arises in both this more general perspective and the narrower "hypothetico-deductive" conception, and it concerns the way in which quasi truth is understood in the context of our grasp of the connections between theory and evidence. To focus the discussion, let us consider the "hypothetico-deductive" approach.

Thus, let us suppose that the relationship between theory and evidence is captured deductively. A hypothesis, h, is confirmed or disconfirmed by first deducing a consequence of it and then relating this consequence to an observation report through some mechanism or other. How is the quasi truth of h to be understood? One way is to say that it is quasi-true if certain of its *known* empirical consequences have not yet been tested or subjected to experiment. However, this renders the quasi truth of h dependent on the will of the experimenter, the availability of money, the equipment, and other such factors.

A more plausible view would be to say that h is regarded as quasi-true because there always exists the possibility of further experimental evidence being discovered within its domain. This meshes nicely with the formalism of the partial structures approach. However, it conflicts with a fundamental assumption of the standard Bayesian program, which says that all the empirical consequences that *can* be deduced from the hypothesis *have* been deduced (Horwich 1982, p. 26; Garber 1983, p. 104). This is the assumption of "logical omniscience," and it is required by the principle of coherence, which underpins the standard subjectivist account. It can be seen quite straightforwardly as follows (Garber 1983, p. 104): If we were not logically omniscient, then there could exist a possible state of the world in which our degree of belief in a true sentence such as "$h \vdash e$," say, could be less than 1. But in that case we would be allowed to bet that "$h \vdash e$" is false, a bet that we would lose no matter what the state of the world, and thus coherence would be lost.

If logical omniscience is accepted and if we are not waiting for any of the tests of these consequences to come in, then "new" evidence cannot offer either negative or positive support for the hypothesis because, to confirm or disconfirm it, such evidence *e* must bear in some way upon the observational consequences and these have all been taken care of. To put it bluntly, there is simply no opening in the list of consequences through which the evidence can impinge upon the hypothesis. This is not to say that the evidence is irrelevant with respect to the hypothesis, since it may reveal the latter's inadequacy in the sense of a lack of scope; that is, there may exist phenomena within the domain of the hypothesis that are not accounted for by that hypothesis. This, of course, is one way in which theory change is effected, but assuming logical omniscience means that, within the hypothetico-deductive approach at least, a hypothesis can be regarded as quasi-true, not in the sense that there may yet be consequences of it that we are currently unaware of but just that we haven't yet found the time or money or resources to test all of its consequences.

Furthermore, logical omniscience rules out the possibility of evidence that is *already known* being used to support a hypothesis. Unfortunately, this is precisely what happens in science. Thus the assumption of logical omniscience prevents Bayesianism from capturing an important aspect of scientific practice. This has been highlighted by Glymour, for example, as one of the fundamental objections to the Bayesian program in general (Glymour 1980a). If we take Bayes's Theorem

$$p(h/e) = p(e/h)p(h)/p(e)$$

then if $h \vdash e$ and e are already known, then

$$p(e/h) = p(e) = 1$$

and hence

$$p(h/e) = p(h)$$

In other words, evidence that is already known cannot raise the probability of a hypothesis. However,

> Scientists commonly argue for their theories from evidence known long before the theories were introduced. Copernicus argued for his theory using observations made over the course of millennia.... Newton argued for universal gravitation using Kepler's second and third laws established before the *Principia* was published. The argument that Einstein gave in 1915 for his gravitational field equations were that they explained the anomalous advance of Mercury, established more than half a century earlier. (Glymour 1980a, p. 85)

Garber has argued that we can weaken the requirement of logical omniscience and replace this *global* form of Bayesianism, in which the probability function representing the agent's degrees of belief is defined over an ideal, maximally fine-grained scientific language, with a *local* form, in which the probabilities are defined over some problem-relative language only (Garber 1983). The logically possible worlds of the former are then substituted by "epistemically possible" ones, and the application of the Bayesian framework becomes highly contextual, being restricted to the particular sentences and beliefs with which one is concerned in the problem at hand. By treating sentences such as $h \vdash e$ as unanalyzable, atomic sentences, whose structure, as expressed in some richer language, is effectively submerged in the problem-relative language, Garber is then able to deal with the problem of "old evidence": By conditionalizing on $h \vdash e$ in these cases, one can represent the change in belief that occurs when the agent becomes aware of this implication.

Given the nature of partial structures, we're happy to acknowledge the point that the application of the Bayesian approach should be contextual. More significantly, perhaps, with the removal of logical omniscience, the way is open for quasi truth to be understood within the hypothetico-deductive framework, via an appropriate extension (da Costa and French 1988): Given some statement *h* of the language L, representing the hypothesis in question, the set of all empirical statements that are logical consequences of *h*, which we *know* to be consequences, will be denoted by K_h. In other words, $K_h = \{e'$: We know that $h \vdash e$ and e is an empirical statement$\}$. With the introduction of a temporal element into the relevant partial structures, K_h is defined at a certain time, t, just as when we say that P is the set of already verified statements, we mean that these statements were verified at a certain time t. To indicate the introduction of such a temporal element, we can simply add the subscript t to the phrase "quasi truth."

We then say that *h* is "quasi-true$_t$" in the partial structure $\mathcal{A}$ if at time t the following conditions are satisfied:

1. There exists an $\mathcal{A}$-normal model $\mathcal{L}$ such that $\mathcal{L} \models h$.
2. $K_h \subset P$.

We can say, loosely speaking, that *h* is "quasi-true$_t$" in $\mathcal{A}$ when it does not contradict the basic partial relations R_i and all its known consequences are verified (at time t). When *h* is not "quasi-true$_t$" in $\mathcal{A}$, it is said to be "quasi-false$_t$" in $\mathcal{A}$. If the domain under consideration is the entire universe, $\mathcal{A}$ can be called a "cosmic structure." In this case, if *h* is "quasi-true$_t$" in $\mathcal{A}$ for all future t, then we can say that the "quasi truth$_t$" of *h* is "stable," and, following the pragmatists, it seems reasonable to say that we have approached truth. Hence "quasi-true$_t$" can be considered a modification of quasi truth.

We can now modify the usual degree of belief in a statement *r* as the degree of belief in the stable "quasi truth$_t$" of *r*. This allows us to retain the standard subjective probability calculus for this form of pragmatic probability, just as we have indicated. Furthermore, with respect to certain classes of statements, such as the observational and logical in particular, this modified degree of belief coincides with the standard formulation. Finally, adapting our framework, we can now say that our pragmatic probabilities change when we discover that further empirical consequences can be deduced from the hypothesis in question and that the successful testing of such consequences leads to an increase in our pragmatic belief in the hypothesis, in that it is now regarded as "closer to the truth." The kinematics of such modified degree of belief change can then be represented by the instance of Bayes's Theorem, thus:

$$p(h/h \vdash e) = p(h \vdash e/h)p(h)/p(h \vdash e)$$

This not only seems a natural way of understanding how a hypothesis could be partially true in this particular context but also allows us to sidestep the famous Popper-Miller argument against inductive probability in general.

Pragmatic Probability and the Popper-Miller Argument

The argument purports to demonstrate that hypotheses cannot, in fact, receive any inductive support at all. To get the argument off the ground, then, some definition of "support" is needed. One possibility is to take it as the difference between p(h/e) and p(h); that is,

$$s(h/e) = p(h/e) - p(h)$$

On the assumption that $h \vdash e$, we then have:

$$s(h/e) = s(h \vee \sim e/e) + s(h \vee e/e)$$

Now $h \vee e$ follows logically from *e*, so the second term represents what can be described as the purely "deductive" support for the hypothesis. Therefore, any "inductive" support, if it exists, must lie in $s(h \vee \sim e/e)$. However,

$$p(h \vee \sim e/e) < p(h \vee \sim e)^9$$

and $s(h \vee \sim e/e)$ must be negative. Hence there can be no inductive support for *h*.

One very obvious response is to adopt an alternative measure of support, such as:

$$s'(h/e) = p(h/e)/p(h) \text{ (Redhead 1985)}$$

However, this allows *e* to accord $h \wedge k$, where *k* is any hypothesis we like, the same degree of support it accords *h* alone (Rosenkrantz 1983). This objection has been generalized by Gillies, who points out that using this measure, the support for *h* becomes independent of *h* itself. But if s′(h/e) is eliminated, it would seem that s(h/e) is the only alternative, and the Popper-Miller argument goes through.

If s(h/e) *is* adopted, then the corresponding form of the argument for the case of pragmatic probabilities is trivial. If the support accorded *h* by the discovery that $h \vdash e$ is given by

$$s(h/h \vdash e) = p(h/h \vdash e) - p(h)$$

then we obtain,

$$s(h/h \vdash e) = s(h \vee \sim (h \vdash e)/h \vdash e) + s(h \vee (h \vdash e)/h \vdash e)$$

As before, $h \vee (h \vdash e)$ follows logically from $h \vdash e$, and so all the inductive support is again contained in the first term. By analogy with the earlier proof, we have

$$p(h \vee \sim (h \vdash e)) = p(h) + p(\sim (h \vdash e))$$

Conditionalizing on $h \vdash e$, rather than e alone, gives an increase in the first term, according to Garber's rule,

$$p(h/h \vdash e) > p(h)$$

while the second term goes to zero as before.

Again, the conclusion is that there can be no inductive support, *given this definition of support*. But what if we now adopt Redhead's alternative? It turns out that the Popper-Miller result can be avoided.

Thus, we shall now define our degree of support accorded h by $h \vdash e$ as

$$s'(h/h \vdash e) = p(h/h \vdash e)/p(h)$$

If, as before, $h \vdash e$, then $p(h \vdash e/h) \neq 1$, since we do not have that $h \vdash (h \vdash e)$. Thus, from the modified version of Bayes's Theorem given previously, we obtain

$$p(h/h \vdash e) = p(h \vdash e/h) \cdot p(h)$$

giving

$$s'(h/h \vdash e) = p(h \vdash e/h)$$

As in Redhead's attempt, the argument can no longer proceed. However, unlike Redhead's case, this result is immune from Gillies's criticism, since the degree of support in this case is not independent of the hypothesis itself.

At first sight, it might seem strange that the support accorded *h* by $h \vdash c$ is equal to the *likelihood* of $h \vdash e$ given *h*, but it should be remembered that, as in all these proofs, we have assumed from the outset that $h \vdash e$; that is, it is given that $p(h \vdash e) = 1$, and the consequence is known to be deducible from the hypothesis. Now, it might be objected that this assumption should be dropped, since the whole point of our earlier discussion was to try to take account of the fact that $p(h \vdash e)$ may not equal 1. However, the Popper-Miller argument still fails in terms of the prior measure, since the division of the degree of support into deductive and inductive terms is dependent on precisely this assumption. Either way the force of the argument is avoided.

Let us be clear about what is going on here: Since the original definition of support, s—defined as the difference between p(h/e) and p(h)—is in contradiction with (essentially) Bayes's Theorem, according to which hypotheses can be confirmed (in our system), s can not be a "good" measure of support. On the other hand, s′ can be used in our system as an intuitive measure of support. Thus, these considerations show that s′ is a "good" support function within our system—the appropriateness of s′ demonstrates that it is a "good" alternative to s. The Popper-Miller argument can be seen as generating a kind of paradox in the field of inductive logic, more or less similar to the Russell paradox in set theory. As is well known, although the latter is eliminable from the various axiomatic systems of set theory, it nevertheless illuminates certain problematic aspects of intuitive set theory that are difficult to explain informally and philosophically. Likewise, the Popper-Miller argument reveals that a certain, apparently quite intuitive, notion of support is actually deeply problematic, requiring a shift to a better definition.

Quasi Truth and Idealizations

According to our definition, pragmatic or quasi truth in general, and "quasi $truth_t$" in particular, involves, above all, consistency with some knowledge base (and in the case of "quasi $truth_t$," empirical irrefutability). But of course, scientific hypotheses and theories are not accepted *simply* because they are pragmatically true. They typically possess other relevant qualities, such as explanatory and predictive power and simplicity. In other words, a theory has to save the appearances, as if it were true in the correspondence sense of truth. Thus it seems appropriate to use the subjective probability calculus, even if only in qualitative terms, to impose some order on the set of extant theories. In this manner, we can "probabilize" the acceptance of theories and hypotheses in general.

As we have made clear, however, this "variant" of quasi truth is elaborated in a particular context—namely, hypothetico-deductivism. This context may be generalized in two directions. First of all, $p(h \vdash e)$ may be different from 1, not only because we may be ignorant of the fact that $h \vdash e$ but also because of the role of idealizations and approximations.[10] Consider, for example, the quantum mechanical analysis of the hydrogen bond that is so fundamental to chemistry (this case study is taken from French and Ladyman 1998). A recent review has emphasized, "The approximate solution of the appropriate Schrödinger equations generates rather reliable values of the structural parameters and the energies of the hydrogen bonds" (Vanquickenborne 1991, p. 32). As the author goes on to note, one of the methods used to obtain these results is the "Hartree-Fock" or self-consistent field approach. The relevant idealization here is to assume that each electron in a molecule or dimer feels only the average repulsion of the others, rather than the instantaneous repulsion of each one. The necessity for some such idealization in the calculation of atomic structures is emphasized by Hartree himself: "For an atom or ion of q electrons, the equation to be solved is a nonseparable partial differential equation in $3q$ variables (for example, for neutral Fe, $q = 26$, 78 variables). Even for the two-electron system of neutral helium this equation has no exact formal solution in finite terms" (1957, p. 16). One way of representing a solution to the appropriate Schrödinger equation in this case would be by means of a table giving all the numerical values. However, suppose we were to attempt to construct such a table for the solutions of the Schrödinger equation for one stationary state of neutral Fe:

> Tabulation has to be at discrete values of the variables, and 10 values of each variable would provide only a very coarse tabulation, but even this

would require 10^{78} entries to cover the whole field; and even though this might be reduced to, say, $5^{78} \approx 10^{53}$ by use of the symmetry properties of the solution, the whole solar system does not contain enough matter to print such a table. And, even if it could be printed, such a table would be far too bulky to use. And all this is for a single stationary state of a single stage of ionization of a single atom. (Hartree 1957, pp. 16–17)

In such cases, an "exact" solution is simply not an option.[11]

The question now is what particular approximations should be used. Hartree notes that the first is to regard each electron as being in a stationary state in the field of the nucleus and the other electrons. It is only in the context of such an approximation that Pauli's all-important Exclusion Principle makes any sense (Hartree 1957, p. 17).

Formally, this approximation is represented by a product wave function in which each one-electron wave function is separated out:

$$\Phi = \psi_\alpha(1)\psi_\beta(2)\ldots\psi_\pi(\mathrm{p}),$$

where the α, $\beta, \ldots$, π label the one-electron wave functions and $1, 2, \ldots$, p the electrons themselves. Since $|\psi_\alpha(\mathrm{j})|^2$ gives the average charge density resulting from the presence of electron j in wave function ψ_α, each of these separated functions can be determined as a solution of the appropriate Schrödinger equation for one electron in the combined field of the nucleus and the total average charge distribution of the electrons in all the other wave functions. The resultant field is described as "self-consistent" because the field of the average electron distribution derived from the wave functions must be the same as that used in evaluating them (Hartree 1957, p. 18).

To satisfy the Pauli exclusion, principle Φ should be taken as a determinant of one-electron wave functions, which, when expanded out, yields a sum of terms, each with coefficient ± 1 and each representing a permutation of the electrons among the one-electron wave functions, all such permutations being equally probable (Hartree 1957, p. 18). Focusing now on the one-electron wave functions themselves, if a central field is imposed, they can be taken to be products of the radial wave function, the variation of the wave function with direction and the spin wave function (Hartree 1957, pp. 8–13). In such a field, the overall set of wave functions divides up into groups, and the qualitative structure of the atom is then specified by the number of occupied wave functions in each of the groups, giving the configuration of the system (p. 19). This is the origin of the shells and orbitals beloved by the quantum chemists.

The calculation of atomic structure then proceeds in three stages (Hartree 1954, pp. 39ff.). The object of the exercise is to find radial wave functions so that the (approximate) energy value for the whole atomic system is stationary with respect to variations in these wave functions. First of all, an expression for this energy value is derived in terms of the radial wave functions. Then the equations for the radial wave functions themselves are derived from this expression, and finally the latter equations are solved. The system of equations can be greatly simplified if terms resulting from the "exchange" of electrons between different wave functions are ignored (Hartree, 1957, pp. 59–60). Whether exchange is ignored or not, what one obtains is a set of simultaneous second-order differential equations that are typically tackled by the method of successive approximation (pp. 77–100). Crucial to this method is the initial estimate of either the radial wave functions themselves, in the case of equations with exchange, or of the contribution to the field of the average charge distribution of the electrons in the relevant groups, or orbitals, in the "without exchange" approximation. Once calculations of the self-consistent fields for some atoms have been carried out, better initial estimates can be obtained by interpolation between these with respect to atomic number (Hartree 1957, pp. 115–135).

Coming back to the hydrogen bond, the total bond energy can be partitioned into different components corresponding to (1) the purely electrostatic interaction between the two unmodified charge densities, (2) the "exchange" interaction resulting from the action of the exclusion principle, (3) a polarization effect, and (4) a charge transfer contribution resulting from electron transfer between the monomers (see Vanquickenborne 1991, pp. 35–38). A further term in the sum giving the total energy is a "mixing" term that would obviously be zero if the four components were independent of one another. Self-consistent field methods can then be applied, and what one gets are tables of figures giving values for the total bond energy and the various contributions (Vanquickenborne p. 37). Better quantitative agreement can then be achieved by including electron correlation and zero-point vibrational corrections.

Now, what are we to make of all this? First of all, from the "extrinsic" perspective, the structural relations between the topmost theoretical level, where Schrödinger's equation lies, and the bottom-level experimental results, together with all the idealizations and approximations in between, can be nicely captured in terms of the framework of partial isomorphisms holding between partial structures (French and Ladyman 1998). Second, switching to the "intrinsic" view, what is confirmed by the evidence in such a case is not Schrödinger's equation or, more generally, quantum mechanics per se, but rather Schrödinger's equation + assorted idealizations and approximations. This can be seen as further support for the view that what we should be interested in is the *pragmatic* truth of Schrödinger's equation. In this case, when we consider $h \vdash e$, h may be taken to be the relevant hypothesis + associated idealizations and approximations. Conceiving of the relationship between theory and evidence in deductive terms does not adequately match the practice of science, as we have tried to indicate with our structuralist considerations previously, but nevertheless, it is the basis for many, if not all, of the applications of the subjective probability calculus. Our intention is simply to suggest that, in this context, pragmatic probability may offer a more appropriate way forward.

This brings us to the second direction in which one may proceed from these considerations. The strict hypothetico-deductive view represents just one rather narrow approach to the relationship between theory and evidence. As we have indicated before, the complexity of structural relationships between models might lead one to conclude that there is more to the theory-evidence relationship than can be captured purely by entailment. One of the better known alternatives is Glymour's "bootstrapping" strategy (Glymour 1980a; see also the collection of essays in Earman 1983), which, expressed rather crudely (see, for example, Edidin 1983, p. 43), relates a given hypothesis to evidence in the following way: Observation sentences expressing the evidence in question are conjoined with auxiliary hypotheses to entail "intermediate theoretical sentences," which then confirm the given hypothesis, relative to the set of auxiliary hypotheses. The strategy was termed a form of "bootstrapping" because, in the original form in which Glymour presented it, the set of auxiliary hypotheses could include or imply the hypothesis to be tested, with possible trivialization ruled out by the condition that there be possible examples of evidence, consistent with the auxiliary hypotheses, which would disconfirm the hypothesis under test.[12] One of the more significant virtues of this account is that it allows for the *selective* testing of hypotheses and eliminates the kind of holism that "transmits" confirmation, so that confirmation of one hypothesis of a theory confirms all. Glymour's strategy allows for certain hypotheses to be confirmed, relative to the background set, but not others.

The relative aspect of this account is one of its important features. Another is the central idea of "moving up," as it were, from the evidence toward the hypothesis with the help of the auxiliaries; thus the strategy can be viewed as a variant of "deduction from the phenomena," which finds historical exemplars in the work of Newton, for example, and which has been developed and advocated by Dorling (1973).[13] The third aspect to note here concerns the next stage—that is, the relationship between the intermediate sentences and the hypothesis. Glymour originally developed a "Hempelian" version of this strategy whereby the intermediates are understood to be "instances" of the hypothesis being tested. Positive and negative instances give positive and negative confirmation, respectively. However, the strategy could incorporate alternative relationships between the hypothesis and the intermediates.

Before we consider that, there are two further points to note. The first concerns the representation of theories and hypotheses within this account. Glymour originally applied his strategy to theories stated in equational form and then also for theories formulated in a first-order language. Van Fraassen has shown how the central insights of the account were contained in the former and has given a reformulation of it in terms of the model-theoretic approach (van Fraassen 1983, pp. 25–42). Thus from the representational standpoint, at least, there is nothing to prevent us from adopting the strategy. However, there is the further issue of whether it can be adapted to a Bayesian, rather than Hempelian, view of confirmation in general. Despite his concerns about Bayesianism expressed in *Theory and Evidence* (one such being the problem of old evidence discussed previously), Glymour himself has indicated how it might be combined with his bootstrap strategy (Glymour 1980b).[14] What he does is simply to apply subjective probability to the relationship between the intermediate sentences and the hypothesis. However, Horwich has claimed that this collapses the bootstrap strategy into nothing more than a form of probabilistic confirmation theory (Horwich 1980).

A more interesting approach and potentially fruitful approach is suggested by Garber, who extends the approach of defining probability functions over instances of h ⊢ e, to defining it over instances of the relation '*e* confirms *h* (in the bootstrap manner) relative to *h*', where *h'* is some auxiliary hypothesis (or set of such hypotheses; 1983, pp. 123–127). He takes this as doing two things: First, it satisfies the concern that Garber understands Glymour to have—namely, that "what should be of interest to confirmation theory is not degrees of belief and their relations, but the precise nature of the structural or logical or mathematical relations between hypothesis and evidence by virtue of which the evidence confirms the hypothesis" (Garber 1983, p. 124). Second, it fills in a gap that Garber takes to exist in Glymour's account—namely, that as it stands the account doesn't give a comparative evaluation of different bootstrapping confirmations. Yet, Garber insists, this is crucial, since for almost any hypothesis *h* and evidence *e* there exists some auxiliary hypothesis by which *e* confirms *h* in bootstrap fashion. What we want is a way of distinguishing the interesting cases of such confirmation from the less interesting or even trivial.

Garber suggests that his framework is "ready-made" (1983, p. 127) to plug the gap since, he insists, almost everything he said regarding the problem of old evidence holds good for whatever logical relation we like, including the bootstrap form: "Within this framework, we can show how the discovery that a given *e* BS (bootstrap) confirms *h* with respect to *T* may increase our confidence in *h*, given one set of priors, and how, given other priors, the discovery that *e* BS confirms *h* with respect to *T* may have little or no effect on our confidence in *h*" (p. 127). Following along in Garber's wake, we can then further adapt this to our account, where the confidence in *h* is understood not to be that *h* is *true* but that it is "quasi-true_t," of course.

There are two further points to note about this. The first is whether Glymour's approach to confirmation fits better with scientific practice as represented in the brief case study of quantum chemistry given before. Exploring this in appropriate detail would greatly extend the present chapter, but it is worth noting that the bootstrap strategy more naturally accommodates the way in which the confirmation of a particular hypothesis draw upon other hypotheses, such as Pauli's Exclusion Principle. Given that one of the declared advantages of the model-theoretic approach is that it incorporates the way in which a whole nexus of theories and models can feature in an explanation, for example, Glymour's strategy might be the natural one to adopt within the model-theoretic framework in general.

The second point is whether Glymour's confirmational relation appropriately parallels the previously mentioned representational relations of partial isomorphisms. Here we must appeal to the extrinsic-intrinsic distinction again. Partial isomorphisms holding between partial structures concern the former, whereas quasi truth, "quasi truth_t" degree of belief, and the like find their place within the context of the latter. Obviously, it makes no sense to talk of isomorphisms, partial or full, holding between *propositions*. Thus, on the intrinsic side, there may be a variety of ways of capturing the sorts of considerations that led to the introduction of partial structures as representational devices, by means of the familiar relationships that hold between propositions. One has already been indicated: We consider a straightforward deductive relationship between h and e and accommodate idealizations and approximations by shifting to a new hypothesis h'. An alternative is offered by the bootstrap mechanism, which seems to more naturally incorporate the dependence of h's confirmation on other hypotheses. There may, of course, be other confirmatory relationships that better suit those aspects of practice that have led us to introduce partial structures, but we shall leave the exploration of these to future works.

The "Problem" of Induction

What about the so-called problem of induction? This can be conveniently reduced to the following set of three questions (de Oliveira 1985):

1. What is the nature of the attitude of acceptance that we adopt in relation to scientific theories?
2. What are the rules of acceptance for such theories?
3. What is the justification for the adoption of those rules?

We have already given an answer to the first question in terms of the "pragmatic" acceptance of a theory: To pragmatically accept a theory means to believe that it has not been refuted within the domain of knowledge it models and that it is pragmatically or quasi-true. This belief in the theory's pragmatic truth is then reflected in our subjective probability assignment. Thus we claim that one should accept a hypothesis that has an a posteriori pragmatic probability of 1, where this value reflects our degree of belief in the partial or quasi truth only of the hypothesis. It is worthwhile comparing this with Shimony's tempered personalism approach, where he argues that we should accept a hypothesis with an a posteriori probability close to, but not equal to, 1 (Shimony 1970, pp. 120–121 and 130–131), the difference expressing the tentative attitude maintained toward the hypothesis. On our view, however, this tentative attitude is reflected, not in terms of a difference in numerical values, but rather in terms of the difference between quasi truth and truth simpliciter. This, we would insist, is a more intuitive and convenient way of capturing the idea that commitment to a hypothesis is weaker than belief in its literal truth.[15]

Turning now to the second of our questions, in the present context this can be broken down into two parts:

1. What are the rules for the a priori *selection* of theories?
2. What are the rules for the a posteriori *acceptance* of theories?

In the first case, the problem in the context of induction is that of ruling out "absurd" or "ridiculous" hypotheses to begin with, or, put yet another way, how can we guarantee that the hypotheses to which we assign our initial degrees of belief are "serious" (Shimony 1970, pp. 110–114; de Oliveira 1985, pp. 136–137)? We have already indicated an answer in terms of a plausible set of heuristic criteria for theory development and pursuit. Granted that it is difficult to codify a set of rules, per se, that can unambiguously distinguish those hypotheses that are "seriously" proposed from those that are not, nevertheless, such criteria can function as methodologically sensible guidelines for choosing the hypotheses of our initial set. It is, perhaps, worth pointing out that there is usually a general and intuitive agreement within a particular scientific community about which hypotheses are to be taken seriously within the relevant domain. And in particular we would emphasize that scientific hypotheses are not proposed within an intellectual vacuum but are put forward in the context of certain background information that confers an initial plausibility upon them even before "experience" is brought to bear. This "initial plausibility" can then be captured by our prior probability assignments. Thus we may invoke these heuristic criteria to account for the restriction, *in practice*, of the set of alternative hypotheses to be formulated in order to increase the reliability of our inductive conclusions.

In the case of question 2b, our answer is simply that such rules are embodied in the calculus of pragmatic probability. We accept a hypothesis *h* when its pragmatic probability, conditional on evidence *e*, is equal to 1, reflecting our belief in the quasi truth of *h* as we have said. And also, as we have just noted, the conditionalization may not be just on the evidence but, in some cases, on the existence of a connection between the hypothesis and the evidence, as expressed by the statement "*h* implies *e*."

Finally, we come to question 3. This concerns the issue of the justification of our inductive mechanisms, commonly regarded as "the" problem of induction. There has, of course, been an enormous amount written about this, and we do not propose to survey this material here. Our claim is straightforward: If it is rational to accept the axioms and rules of the pragmatic probability calculus, then it is rational to accept inductive inferences. Thus consider Russell's Principle of Induction (Russell 1912):

> (Ia) When a thing of a certain sort A has been found to be associated with a thing of another certain sort B and has never been found not to be associated with a thing of sort B, the greater is the probability that they will be associated in a fresh case in which one of them is known to be present.
> (Ib) Under the same circumstances, a sufficient number of cases of association will make the probability of a fresh association nearly a certainty and will make it approach certainty without limit.
> (IIa) The greater the number of cases in which a thing of the sort A has been found associated with a thing of the sort B, the more probable it is (if no cases of failure of association are known) that A is always associated with B.
> (IIb) Under the same circumstances, a sufficient number of cases of the association of A with B will make it nearly certain that A is always associated with B and will make this general law approach certainty without limit.

Of course, it is difficult to say precisely what Russell really meant by his principle (what kind of probability he had in mind and so forth). However, our intention, again, is not historical exegesis: We simply want to remark here that there is a sense in which it can be envisaged as providing a rationale for induction. It can be shown that Ia to IIb are, with certain restrictions, theorems of the subjectivist probability calculus (da Costa and French 1991b; see appendix).[16] Thus, if the axioms of this calculus are at all rationally grounded, then so is Russell's principle. Since the latter obviously constitutes a good rationale for induction, we thereby obtain a rationale for this kind of inference.

It is important to emphasize that our claim is not that these results show that by means of simple induction we are able to attain "the truth" (which shouldn't be our goal anyway!) but rather that if it is rational to believe in the axioms of the probability calculus, then it is rational to believe in simple induction. Perhaps this is the sole justification of induction that we are able to obtain in this context. And since all forms of inductive argument can be reduced to simple induction, we then have a justification of induction in general. Thus, the gist of our position is not that we are able to show that induction brings us to the truth but only that without it we are not rational.

This brings us up against the "problem of rationality" in general and the question of why it is rational to believe the axioms of the probability calculus in particular. The following points are worth noting: First of all, as Black has emphasized (1970, pp. 57–90), we must step out from under the shadow of deductive reasoning. In particular, "Addicts of deduction should not harbor flattering illusions" (p. 144). If there is a problem of justifying induction, then there is one of justifying deduction also (da Costa 1981, pp. 2–10): Why and how does deduction function in the way that it does? Does the underlying language of a deduction reflect reality? If not, then our deduction is vacuous, but if so, then how do we know that it does? By induction?!

Second, and relatedly, to argue that "anything less than true conclusions following from true premises is not genuinely reasonable" is to use a too restrictive sense of reasonableness (Black 1970, pp. 66–67). Placed in the context of the history of science and our previous discussion, this argument is itself unreasonable! A wider, more pragmatic, more *realistic* sense of reasonableness is clearly required, linked in turn to a clarification of the apparent mystery surrounding the raison d'être of induction (Black 1970, p. 67). The latter, we claim, is to lead us to "partial" or "quasi" truth, rather than truth itself, and it is this which makes it reasonable to believe in inductive arguments.

Appendix: Justifying Induction

In the following H denotes a hypothesis (a possible law, etc.), E the initial evidence, and Q_i, $i = 1, 2, \ldots, n$, n *distinct* consequences of H (and E). We thus have:

$$H \wedge E \vdash Q_i, \ i = 1, 2, \ldots, n.$$

$\bar{\mathbf{Q}}_i$ will denote $Q_1 \wedge Q_2 \wedge \cdots \wedge Q_i$.

In particular, H may be the proposition $\forall x\ (x \in A \rightarrow x \in B)$—that is, "All As are Bs," and Q_i, the statement, "x_i, which belongs to A, also belongs to B." $\bar{\mathbf{Q}}_i$ will then express the fact that i members of A are also members of the class B.

THEOREM 1. If $P(Q_i, \bar{\mathbf{Q}}_{i-1} \wedge E) \neq 1$, $P(H, E) > 0$, and $P(Q_i, \bar{\mathbf{Q}}_{i-1} \wedge E) \neq 0$, then $P(H, \bar{\mathbf{Q}}_i \wedge E) > P(H, \bar{\mathbf{Q}}_{i-1} \wedge E)$.

PROOF. We obviously have:

$$P(H, Q_1 \wedge E) = \frac{P(H, E)}{P(Q_1, E)}$$

$$P(H, Q_2 \wedge Q_1 \wedge E) = \frac{P(H, Q_1 \wedge E)}{P(Q_2, Q_1 \wedge E)}$$

....

$$P(H, \bar{\mathbf{Q}}_i \wedge E) = \frac{P(H, \bar{\mathbf{Q}}_{i-1} \wedge E)}{P(Q_i, \bar{\mathbf{Q}}_{i-1} \wedge E)}$$

Since $P(Q_i, \bar{\mathbf{Q}}_{i-1} \wedge E) \neq 1$, we therefore obtain:

$P(H, \bar{\mathbf{Q}}_i \wedge E) > P(H, \bar{\mathbf{Q}}_{i-1} \wedge E)$.

Thus, when H is $\forall x(x \in A \rightarrow x \in B)$, we obtain part Ia of Russell's principle.

THEOREM 2. (Jeffreys-Wrinch). Suppose that $P(H, E) \neq 0$. Then,

$$\lim_{n \to \infty} P(Q_n, \bar{\mathbf{Q}}_{n-1} \wedge E) = 1$$

PROOF. Clearly,

$$P(H \wedge \bar{\mathbf{Q}}_n, E) = P(H, E) P(\bar{\mathbf{Q}}_n, E \wedge H)$$

Since $E \wedge H \vdash Q_i$, $i = 1, 2, \ldots, n$, we have that:

$$P(H \wedge \bar{\mathbf{Q}}_n, E) = P(H, E)$$

On the other hand, it is easy to see that:

$$P(H \wedge \bar{\mathbf{Q}}_n, E) = P(H, E \wedge \bar{\mathbf{Q}}_n)\ \Pi_{i=1}^{n} P(Q_i, \bar{\mathbf{Q}}_{i-1} \wedge E)$$

Consequently,

$$P(H, E \wedge \bar{\mathbf{Q}}_n) = \frac{P(H, E)}{\prod_{i=1}^{n} P(Q_i, \bar{\mathbf{Q}}_{i-1} \wedge E)}$$

But, $P(H, E \wedge \bar{\mathbf{Q}}_n) \leq 1$. Therefore,

$$\prod_{i=1}^{n} P(Q_i, \bar{\mathbf{Q}}_{i-1} \wedge E) \geq P(H, E)$$

We easily deduce that

$$\prod_{i=1}^{n} P(Q_i, \bar{\mathbf{Q}}_{i-1} \wedge E)$$

converges for $n \rightarrow \infty$ and that:

$$\lim_{n\rightarrow\infty} P(Q_n, \bar{\mathbf{Q}}_{n-1} \wedge E) = 1$$

If H is $\forall x(x \in A \rightarrow x \in B)$, then the theorem just proved shows that when the number of instances of a law increases indefinitely, then the probability of a new positive instance tends to 1. We have thus proved part Ib of Russell's principle.

THEOREM 3. Let A and B be two finite classes such that the proportion of As that are Bs is r. Then the probability that a random $x \in A$ belongs also to B is r.

PROOF. Let us suppose that n is the number of elements of A and that m is the number of elements of $A \cap B$. Then, there are n possible, equally probable cases, of which m are favorable. Therefore, the subjective probability that a random x that belongs to A also belongs to B is m/n—that is, r. (It is convenient to insist that this theorem is not *true* by the definition of probability; it has to be proved.)

THEOREM 4. A and B are two classes, A being finite and very large. Then $P(H, \bar{\mathbf{Q}}_n)$ approaches 1 as n increases, where H is $\forall x(x \in A \rightarrow x \in B)$.

PROOF. Bayes's Theorem gives us:

$$P(H, \bar{\mathbf{Q}}_n \wedge E) = \frac{P(H, E)P(\bar{\mathbf{Q}}_n, H \wedge E)}{\sum_i P(H_i, E)P(\bar{\mathbf{Q}}_n, H_i \wedge E)}$$

where H_i stands for the proposition that the proportion of As that are Bs is i; in particular, H_1 is H.

Since i^n, $i < 1$, approaches 0 as n increases, it follows that:

$$\sum_i P(H_i, E)P(\bar{\mathbf{Q}}_n, H_i \wedge E) = \sum_i P(H_i, E)\ i^n$$

approaches P(H, E) as n increases. Because $P(\bar{\mathbf{Q}}_n, H \wedge E) = 1$, we conclude that $P(H, \bar{\mathbf{Q}}_n \wedge E)$ approaches 1.

We have thus proved IIb for large, finite classes—perhaps the really important case. Two further points regarding this proof should be noted:

i. *Sampling with replacement* is presupposed, although this is, of course, unproblematic.
ii. A need not be assumed to be very large. Suppose that A has k elements. Then the "next" value of i below 1 will be $k-1/k$ and with $n = k$, $(k-1/k)^n$ will be around $1/e$—which is not so small—and we may well still be a very long way from $P(H, \bar{\mathbf{Q}}_n \wedge E)$ being close to 1. (We owe this last point to Jeff Paris.)

There is then no difficulty in proving the next proposition:

THEOREM 5. Given the conditions of the preceding theorem (and presupposing the invariance of probability under "random permutations") we have:

$$P(\bar{\mathbf{Q}}_{n+1}, \bar{\mathbf{Q}}_n) \geq P(\bar{\mathbf{Q}}_n, \bar{\mathbf{Q}}_{n-1})$$

Thus, IIa is also true, for large, finite classes. (The relevance of probabilities with regard to finite sets is discussed in Russell 1948.)

We have therefore essentially proved Russell's principle of induction. In synthesis, these theorems show that the axioms of subjective probability imply Ia to IIb. Thus, if these axioms are rationally grounded, then induction is also. Using second-order probabilities (see Gardenfors and Saklin 1982), we can obtain another version of IIb that does not presuppose that the class A is finite. The same is true with regard to IIa.

A further important result, which may be used to support certain inductive inferences, is the following:

THEOREM 6 (Keynes). If $P(H, E) \neq 0$, then we have:

$$P(H, \bar{\mathbf{Q}}_n \wedge E) = \frac{P(H, E)}{P(H, E) + P(\bar{\mathbf{Q}}_n, E \wedge \sim H)(1 - P(H, E))}$$

PROOF. Obviously:

$$P(\bar{\mathbf{Q}}_n, E) = P(\bar{\mathbf{Q}}_n \wedge H, E) + P(\bar{\mathbf{Q}}_n \wedge \sim H, E)$$

On the other hand,

$$P(\bar{\mathbf{Q}}_n \wedge H, E) = P(H, E) \text{ and } P(\bar{\mathbf{Q}}_n, H \wedge E) = 1$$

But,

$$P(H, \bar{\mathbf{Q}}_n \wedge E)\ P(\bar{\mathbf{Q}}_n, E) = P(\bar{\mathbf{Q}}_n, H \wedge E)\ P(H, E)$$

Therefore,

$$P(H, \bar{\mathbf{Q}}_n \wedge E)[P(\bar{\mathbf{Q}}_n \wedge H, E) + P(\bar{\mathbf{Q}}_n \wedge \sim H, E)] = P(H, E)$$

Consequently,

$$\begin{aligned} P(H, \bar{\mathbf{Q}}_n \wedge E) &= \frac{P(H, E)}{P(\bar{\mathbf{Q}}_n \wedge H, E) + P(\bar{\mathbf{Q}}_n \wedge \sim H, E)} \\ &= \frac{P(H, E)}{P(H, E) + P(\bar{\mathbf{Q}}_n \wedge \sim H, E)} \\ &= \frac{P(H, E)}{P(H, E) + P(\bar{\mathbf{Q}}_n, \sim H \wedge E)P(\sim H, E)} \\ &= \frac{P(H, E)}{P(H, E) + P(\bar{\mathbf{Q}}_n, \sim H \wedge E)(1 - P(H, E))} \end{aligned}$$

COROLLARY. Given the conditions of the theorem, if $P(\bar{\mathbf{Q}}_n, \sim H \wedge E) \rightarrow 0$ as $n \rightarrow \infty$, then it follows that,

$$\lim_{n \rightarrow \infty} P(H, \bar{\mathbf{Q}}_n \wedge E) = 1$$

This corollary gives us a sufficient condition for

$$P(H, \bar{\mathbf{Q}}_n \wedge E) \rightarrow 1 \text{ as } n \rightarrow \infty$$

that is, a condition for the probability of a hypothesis to be near 1, when the number of instances increases indefinitely (IIb in another form).

8

From Pragmatic Realism to Structural Empiricism

Throughout this book we have touched on issues to do with the realism-antirealism debate, without getting too tied up with the details (for an account of the current "state of play," see Psillos 2000). In part, this is because we feel that our framework can be pressed into service by either side in this debate. Indeed, this is precisely what has happened. Ladyman has incorporated partial structures into his version of "structural realism," which he sees as realism's best hope in the face of the ontological perplexities presented by modern physics. On the other side of the divide, Bueno has also used partial structures to elaborate a form of "structural empiricism," which extends van Fraassen's antirealist "constructive empiricism" in significant respects. Before we get to these more recent positions, let us consider what light our account can shed on the debate.

Realism, Antirealism, and Quasi Truth

In what we shall call its "standard" form, realism is encapsulated in the twin theses that (1) the laws of theories that can be described as "mature" are typically approximately *true*[1] and (2) the theoretical and observational terms of such theories typically *refer* (Putnam 1978, pp. 20–21; Boyd 1973). Both theses have come in for severe criticism.

Thesis 1 has had to face the perceived difficulties in articulating a "coherent notion" of approximate truth (Laudan 1996, pp. 118–121). One response has been to appeal to certain accounts of verisimilitude that are not only highly formal but also quantitative in nature (see, for example, Kuipers 2000; for an overview, see Psillos 1999, chapter 11). These, in turn, have run into a series of problems, the most notable of which appear to demonstrate that the "metric" of approximate truth must be highly *contextual*. Such problems have led Giere[2] to dismiss approximate truth as a "bastard semantic relationship" (1988, p. 106), preferring instead his account of "similarity," taken as primitive. Despite this dismissal, Psillos, in his more recent defense of "standard" realism (1999), has suggested that Giere's approach does in fact capture the intuitive appeal of approximate truth (p. 275). However, in building on this skeleton, Psillos not only does not provide a formal account of the latter notion but also insists that such a failing should not be seen as a defect (p. 277). He argues, first, "There is an irreducible *qualitative* element in the notion of approximation, i.e. the respects in and degrees to which one description may be said to approximate another" (p. 278), and second, that we have no need for the equivalent of a Tarskian formal account of approximate truth, since our intuitive understanding of the latter does not generate paradoxes, unlike the case of correspondence truth itself (p. 278).

Now we certainly agree that highly quantitative approaches are about as illuminating as spelling out the change in one's degree of belief to the fifth decimal place. Nevertheless, there are good reasons for going beyond the "intuitive understanding" of approximate truth to a more formal account, reasons similar to those listed by Sklar in the quotation cited in chapter 2.[3] Certainly, one might argue that without such an account, one cannot begin to investigate whether the aforementioned "intuitive understanding" is riddled with problems or even consistent. As Mikenberg, da Costa, and Chuaqui noted in their original paper (1986), the formalism of quasi truth might be employed here as a form of approximate truth. Since quasi truth is a contextual notion, relative to a given partial structure, we would expect to be able to accommodate the contextuality within our approach. Furthermore, we have already indicated how Giere's vague notion of "similarity" can be captured more rigorously through partial isomorphisms holding between partial structures. And finally, the whole point of this book is to defend the claim that quasi truth and the attendant formal mechanisms of partial structures and partial isomorphisms is richer than the Tarskian notion and better suited to capturing scientific practice. However, having said that, one might expect that in applying quasi truth to realism, the standard position will have to be recast.

Recalling the formal expression of a partial structure, let us begin with the following questions, which quite naturally arise: "what is the nature of the domain Δ?" and "What are to be included as members of the set A?" These are, of course, intimately connected.

Taking these questions in order, it is worth noting the distinction between the domain of knowledge that a theory *should* account for and the domain that it actually *does* account for successfully (see, for example, Nickles 1977, p. 583). Relative to the latter, a theory is pragmatically true forever, as we have emphasized: This is why we can still use classical mechanics to build bridges, and, so forth. However, it is by means of the first domain that a theory's deficiencies are illuminated, and thus it is via the mismatch between these two kinds of domain that a theory is shown to be inadequate, leading to the search for a better one. The difference is therefore important because it effectively drives the machinery of theory change and scientific progress in general.

With regard to the exact nature of these domains, and of the former kind in particular, we note that the set of distinguished sentences, or primary statements, P, will in general include both (from the realist's perspective) true decidable sentences, such as observation statements, and certain general propositions encompassing laws already assumed to be true. Thus, Δ is not merely a collection of phenomena but is ordered in some way by these laws, together with symmetry principles and the like. If science is understood as cumulative, in the fashion we have tried to defend here, with at least part of the structure of its predecessor preserved through theory change, then Δ may be taken to include these latter aspects as well (cf. Shapere 1977). In this case, the very concept of "domain" becomes a "pretheoretic" notion, in the sense that it is described at least in part within the language of previous theories. Modeling Δ may then actually generate a *different* domain to be modeled by future theories.[4]

This brings us to the question of what to include in A and our attitude toward the members of this set. As we have said, A will in general include both nontheoretical and theoretical terms. With regard to the latter, further discriminations need to be imposed. Some of the theoretical terms in A will refer to unobservable entities, but some will be "idealization terms" such as "point particle" and "rigid rod." Furthermore, within the subset of terms that are not idealizations, not all will be held to *refer* with the same strength of conviction. The partial structures approach can deal with each of these aspects of the set A.

Idealization Terms

First of all, there is the obvious point that the entities to which idealization terms refer are typically understood not to exist, or at least not in this, the actual, world. And the reasons for such an understanding are, as Shapere has emphasized, primarily scientific (Shapere 1969) in the sense that rigid rods are ruled out by Special Relativity (Shapere 1969, pp. 132–137; Maidens 1998), and likewise it is a consequence of the Lorentz theory of the electron that it cannot be a point particle (Shapere 1969, pp. 138–146). Hence we see immediately that the "understanding" must be amended: These entities are taken not to exist in the world as described by a particular theory. Nevertheless, it is convenient for reasons of computational tractability, for example, to invoke such notions in the treatment of certain problems. Thus they feature in theories or versions of theories or, more generally, models that are correspondingly convenient, computationally tractable, and so on.

How should we regard such idealizations? An extreme view is to take the terms involved as referring to entities that exist in other possible worlds (Nowak 1995). Quite apart from the trans-Lewisian nature of this vision, it has dubious metaphysics at its base: Idealization differs from abstraction in that omitting the dimensions of a particle supposedly no longer yields a physical body at all. Thus whereas abstraction leads to generalizations, idealization takes us out of this world altogether. A metaphysically more conservative account is indicated by Grobler: "idealization consists in specifying in advance the kinds of predicates expected to occur in claims being made in a given context about objects of a given kind, rather than in referring to some fictitious, idealized objects" (1995, p. 42). In particular, describing an electron as a mass-point does not amount to substituting for it by some Platonic object; rather, the description is simply an indication of the relative irrelevance of the particle's dimensions in that theoretical context (Grobler 1995, p. 42).

It is this context-dependence of idealizations that is all important; indeed, it is precisely the context that delineates the idealization, as we indicated previously. Let us be more explicit and take the example of the electron: In describing the electron as pointlike, we are obviously excluding spatial dimension from the list of predicates characterizing it. Nevertheless, the intrinsic properties of the electron are retained—properties like mass, spin, and charge—else we could not refer to what is being described as an electron. And this description features in, and is part of, the construction of an appropriate model. It is not the case that the description stands free of the theoretical context, so that we can talk of this idealized electron as a separate entity, existing in some possible world perhaps; rather, the idealization is bound to the model. Recalling Shapere's point that this model will be acknowledged as deficient precisely in those respects in which the electron is regarded as dimensionless, we can say that idealization involves "as if" description: The electron is treated as if it were pointlike and rods are treated as if they were rigid.

But now this "as if" aspect suggests a way of treating assertions involving idealizations from an epistemic perspective. Shapere notes that entire theories might be conceived of as idealizations in the sense that, although there are good grounds for regarding the theories as false, nevertheless it may be convenient to take them as if they were true. Thus, for example, although geometrical optics is known to be false, in that it cannot account for diffraction phenomena, it is still used as an idealization for treating a wide range of other phenomena (Shapere 1969, pp. 152–154). Likewise, as we have already said, Newtonian mechanics may still be used *as if it were true* for the purposes of building bridges and launching rockets and the like. The theory of quasi truth can then be used to formally underpin the claim that idealizations are regarded *as if* they were true. Furthermore, it can be extended to cover not just the case of theories as idealizations in the sense previously noted but also that of idealization terms, where these are thought of as idealizing descriptions laid down within a theoretical context. To describe an electron as if it were a point particle is to lay down a bundle of properties that have meaning only within a model or, more generally, a structure; thus the "as if" character of such idealization terms gets shifted to that of the embedding context. Our epistemic attitude to the latter can then be grounded in the notion of pragmatic or quasi truth (and as we have indicated in chapter 7, the structural aspects of this approach suggest an account of theoretical idealization in general).

A Differentiated Ontological Attitude

As well as acknowledging the difference between those terms that are "idealizations" and those that are taken to genuinely refer, the realist must also accommodate a "differentiated" attitude with regard to the latter (see Fine 1984). The assertions involving such terms make claims of varying degrees of strength, depending on, among other things, the status of the theories in which they occur. The difference in this degree of strength corresponds to the difference in the scientists' degree of belief in such claims and in the existence of the entities referred to. Thus at one end of the spectrum, they will include claims about entities that were once asserted to exist but are not now regarded as existing, such as phlogiston; at the other, they will include claims about entities that are now definitely considered to exist, such as electrons; and in the middle, there will be claims about entities, such as magnetic monopoles and tachyons, whose existence has not yet been established but that still feature in certain theories. The example of quarks, referred to earlier in this book, provides a nice example of an entity moving through the spectrum, as it were, from something introduced as a kind of bookkeeping device to something elevated to the status of one of the "basic building blocks" of the universe.

What is needed is a way of accommodating all these different attitudes to the "scientific zoo." The most obvious is to introduce a formal characterization of degrees of belief that permits exactly the kind of gradation in our attitude toward theoretical entities that a more sophisticated realism requires. Three issues need to be addressed in setting up such a framework. The first concerns the propositional subject of the belief, the second has to do with the changes in such beliefs and the forces that drive such changes, and the third issue relates to the topic of chapter 7—namely, whether the degree of belief should be taken to be in the *truth* of the relevant proposition or its *quasi truth*. Let us briefly consider each of these in turn.

What we are concerned about here are existence claims, rather than claims to do with the truth or quasi truth of the theories per se, although, as we shall see, the two are related. Typically, of course, the propositions concerned will have the form "electrons/phlogiston/quarks exist," and some of the relevant beliefs will be held with a very high degree of confidence, some with a very low degree of confidence, and others, the majority perhaps, with an intermediate range of such degrees. Such a characterization strongly motivates a probabilistic approach, which Horwich, for example, suggests may cut across the realism-antirealism divide (Horwich 1982, pp. 134–136). Although we are obviously sympathetic to this perspective, the "cut" is not so clean as Horwich appears to think, and his "modulated realism or sophisticated instrumentalism" leaves a number of crucial issues untouched.

Taking realism first, Horwich acknowledges the point that it "comes to grief over the history of science" (1982, p. 135) for reasons to do with what is now known as the "pessimistic meta-induction." This is the argument, usually attributed to Laudan (1996), which has the following form:

Premise 1: Entity *a*, put forward by *x*, in historical period p_1, was subsequently agreed not to exist.
Premise 2: Entity *b*, put forward by *y*, in historical period p_2, was subsequently agreed not to exist.
Premise 3: Entity *c*, put forward by *z*, in historical period p_3, was subsequently agreed not to exist.
⋮
Premise n: Entity *i*, put forward by *w*, in historical period p_n, was subsequently agreed not to exist.
(Inductive) Conclusion: The entities put forward by today's scientists will subsequently be agreed not to exist.

The conclusion is then seen as undermining the traditional realist view that we should accept current scientific theories as the best guide to how the world actually is and that we should take the entities referred to by the relevant theoretical terms as existing (Horwich 1982). Of course, the strength of this conclusion depends on that of the premises and hence on the *kinds* of entities labeled by the *x*, *y*, *z*, and so forth. Thus one might take an eclectic approach to the history of science and include the Aristotelian crystalline spheres, phlogiston, caloric, light as corpuscles, the optical ether, light as waves, the electromagnetic ether, and so on. The realist can then mount a response by rejecting certain of these entities as belonging to "immature" theories, thus cutting down the list of premises. Unfortunately, even with a clear characterization of maturity in terms of predictive success, say (see n. 1), this still leaves a goodly number of premises to take account of. Hence, the conclusion that has often been drawn is that "it would be quite wrong to embrace realism and believe that our present scientific theories are true" (Horwich 1982; see also Laudan 1996).

One alternative is to adopt some form of instrumentalism that takes theories to be no more than convenient and economic ways of systematizing data, whose theoretical terms are not to be taken as referring at all, and therefore the theories themselves are not candidates for being regarded as either true or false; that is, they are not "truth-apt." If realism errs on the side of inclusiveness, then instrumentalism moves too far in the opposite direction. This form of antirealism has now fallen out of favor and the leading contender as an alternative to traditional realism is van Fraassen's constructive empiricism. As we have already indicated in earlier chapters, this retains the truth-aptness of theories and acknowledges that theoretical terms *may* refer but insists that we can never *know* whether theories are true or to what these terms refer. Hence the appropriate epistemic attitude is not belief in truth but acceptance as empirically adequate.[5]

Horwich takes the introduction of degrees of belief as allowing us to appropriate what is good from both sides, without the costs. In particular, we can acknowledge the point that we shouldn't blindly accept the existence of every unobservable entity proposed by science, while allowing theories to posses determinate truth values (and avoiding the imposition of a sharp observable-unobservable distinction). Nevertheless, leaving aside the issue of whether the constructive empiricist would even accept such a move, there are further important issues to be addressed concerning the application of such an approach.

Let us begin by considering the entities involved in the pessimistic meta-induction, focusing in particular on the two extreme ends of the spectrum. First of all, granted that there are entities in whose existence we have a great deal of confidence—such as electrons, atoms, or even light[6]—there are clearly others, such as phlogiston, caloric, and the optical and luminiferous ethers, that have fallen by the wayside, as it were. Even the sophisticated realist with her apparatus of degrees of belief is going to have to show that today's scientific entities are not like these in relevant ways, if the pessimistic meta-induction is not going to bite again. She might, for example, argue that the terms referring to these entities are somehow "idle," in the sense that they do not feature in the relevant deductions of predictions whose empirical success leads to the theory concerned being regarded as true, approximately true, or whatever (Kitcher 1993; Psillos 1999). In other words, they do no useful work within the theory. An example of such a term is *caloric*, and it has been argued, "The laws which scientists considered well supported by the available evidence and the background assumptions they used in their theoretical derivation were independent of the hypothesis that the cause of heat was a material substance: no relevant assumption was essentially used in the derivation-prediction of these laws" (Psillos 1999, p. 113).

The problem with this suggestion is that, in order for it to work, the notion of an "idle" term has to be rigidly delineated, so that any term is idle that does not feature in such deductions. This holds realism hostage to a strongly deductivist view of science. And one might well object that there are terms that are idle in this sense but can still be regarded as useful in various ways, such as the heuristic usefulness of a term which helps to provide the metaphysical framework of a theory, without which the theory could never be proposed in the first place.[7] Consider the example of caloric: The laws and predictions referred to here include the law of conservation of heat, Carnot's theorems, and Laplace's prediction of the speed of sound in air (Psillos 1999, pp. 115–125). Even if we grant that these laws and predictions can be formally detached from the assumption as to the nature of caloric (and insofar as this nature is metaphysical, it should perhaps come as no surprise that physics should be detachable from metaphysics in this way),[8] a strong case can be made that the proponents of these laws and predictions could not have conceived them, much less developed and pursued them, without such an assumption. Given our rejection of the sort of heuristics-justification distinction to which this deductivist view is allied, we are naturally sympathetic to this kind of objection. And here the degrees of belief mechanism might be reintroduced: The attitude of the scientist proposing the caloric theory of heat, say, might be appropriately captured in the differential manner indicated previously. Here a further differentiation might be usefully introduced: Insofar as the theoretical terms, or cluster of terms, corresponding to certain properties of "caloric" (and we shall address the issue of the distinction between the entity itself and its properties shortly) feature in the relevant deductions, and insofar as these accrue empirical success for the theory, the scientist might have a high degree of belief in the proposition that heat has these properties. Insofar as the notion of caloric as a material substance was necessary to conceive of the theory in the first place, the degree of belief in the existence of such a substance will not be zero, although given the problems the notion faced, such as that concerning the weight of caloric, this degree of belief may not be very high.[9] As we have tried to emphasize, straightforward empirical success—as understood by the deductivist—is not the only criterion for evaluating theories. In other words, the degrees of belief proposal ties in quite naturally with a fallibilist view of theories in general, and we see no reason why such a view should not be extended to current examples.

Furthermore, the "idle terms" strategy will clearly not work in all cases. Consider the case of the optical and luminiferous ethers: Here it would appear that one can make an even stronger case that the conception of light as a wave could not have been developed without assuming some such medium. How is the realist to deal with science's abandonment of these? A further strategy is to argue that these terms in fact refer to the same "thing" as certain current terms that fulfill the same causal role; thus, it is claimed, the luminiferous ether performed the same causal role as the electromagnetic field and hence was not actually abandoned after all (Hardin and Rosenberg 1982). Now this is problematic for a number of reasons: First of all, it suggests a historical attitude to the beliefs of proponents of the ether that seems Whiggish in the extreme, in that it holds that when they talked of and expressed belief in the ether, they were "actually" talking of and expressing belief in something they had no idea existed—namely, the electromagnetic field. More acutely, perhaps, the theory of reference underlying this strategy is too liberal since, it is claimed, just about any entity, now abandoned, can be said to have fulfilled the same causal role as some current entity (consider, for example, the following sequence: the "natural place" of Aristotle, the "gravity" of Newton, and the "curved space-time" of Einstein; Laudan 1984). Finally, this strategy effectively detaches reference from its theoretical context and entails that "we can establish what a theory refers to independently of any detailed analysis of what the theory asserts" (Laudan 1984, p. 161).

One might attempt to get around the last two problems, at least, by broadening one's account of reference to include descriptive elements as well as causal roles (Psillos 1999, pp. 293–300). The central idea here is that reference is fixed via a "core causal description" of those properties that underpin the putative entity's causal role with regard to the phenomena in question (Psillos 1999, p. 295). The overall set of properties is significantly open to further developments, so that new properties get added around the core as science progresses. Of course, some of these latter properties may subsequently be deleted, particularly through evolutionary changes, but as long as there is significant overlap via the core set, continuity of reference through scientific change can be maintained.

In the context of this kind of framework, it begins to make more sense to say that the term "luminiferous ether" referred to the electromagnetic field (Psillos 1999, pp. 296–299). In this case, the "core causal description" is provided by two sets of properties—one kinematical, which underpins the finite velocity of light, and one dynamical, which ensured the ether's role as a repository of potential and kinetic energy. Other—typically mechanical—properties to do with the nature of the ether as a medium were associated with particular *models* of the ether, and the attitude of physicists toward these, of course, was epistemically much less robust. The core causal description was then taken up by the electromagnetic field, so that one can say that "the denotations of the terms 'ether' and 'field' were (broadly speaking) entities which shared some fundamental properties by virtue of which they played the causal role they were ascribed" (Psillos 1999, p. 296). It is then a small step to conclude that the terms referred to the same entity. Finally, it is claimed that this avoids the two problems posed by Laudan. First of all, not just any old entity can fulfill the same causal role as the current one since there needs to be a commonality of properties as represented by the core causal description. Second, it is only through a detailed analysis of what a theory asserts that we can pick out the relevant properties in the first place; thus reference is not detached from the theoretical context.

To pursue this issue further would take us too far from the aim of this chapter, but the following points are worth noting. First, there is the emphasis on the openness of the set of properties surrounding the core, and, not surprisingly, we can easily take account of this via the partial structures representation. Those properties that constitute the core description will be included in the R_1, together with others, of course, that may be retained through theory change. As we move through such a change, from ether-based theories of light to Maxwell's theory, for example, some of these "peripheral" R_1 properties will be consigned to R_2. Crucially, however, the openness in the whole process, which parallels that of the corresponding theory, is captured by the R_3, representing those properties and relations that we, or the physicists of the time, simply do not know if they are possessed by either the ether or the electromagnetic field.

Second, the various ether models mentioned played an important heuristic role, which was based on certain positive analogies that held between the ether and elastic solids relating, for example, to the transversal nature of light waves (Psillos 1999, p. 131). An example of a negative analogy is represented by the longitudinal waves that can occur in elastic solids but not in the case of light (p. 132) and, most crucially for the heuristic aspect, there existed certain neutral analogies that powered the development of further models (pp. 135–137). The explicit "Hessian" framework here (Psillos 1999, pp. 140–143) can, of course, be straightforwardly accommodated within the partial structures approach, as indicated in chapter 3. More interestingly for the present discussion is the emphasis on the heuristic role of these models, which relates to our point regarding the example of caloric. Clearly, the exploration of those features of the ether that have subsequently come to be represented within the "core causal description" could not have got off the ground in the absence of the models whose mechanical properties come to be discarded for referential purposes. The blurring of the heuristics-justification distinction may lead one to wonder whether the separation of properties—between those in the core and those outside—can in fact be so cleanly made as this account of reference requires.

This brings us to a related point. Let us recall the first problem with the causal role approach to reference—namely, that it implies that scientists of the time were simply mistaken in their beliefs as to what the relevant terms were referring to. Now let us examine the nature of those beliefs a little more carefully. When scientist x expressed a belief in the existence of the ether, the term is taken to refer to an *entity* possessing certain properties. Of course, the degree of belief, not in the existence of the entity per se but in the claim that it possesses certain properties, may fluctuate over time, so that scientist x may start off with equal degrees of belief in the ether possessing certain mechanical properties and dynamical ones. As time goes on, her degree of belief that the ether possesses mechanical ones may decrease while the degree of belief that it has the relevant dynamical ones may remain constant or even increase.[10] And, as we have indicated, our approach seems admirably suited to capturing this sort of situation, particularly insofar as it allows us to capture the heuristic aspects. But now if the mechanical properties are shunted off to the models, as it were, in what sense can we still say that the scientist is referring to the ether *as an entity*? The question is important because separating off the kinematical and dynamical properties from the mechanical ones in this way may obscure precisely that which was taken to be important in the transition from classical to relativistic physics. As well as these properties, and in virtue of its role as an absolute frame of reference, the ether also possessed certain "positional" properties (Psillos 1999, p. 314 fn. 9). If these are included in the core, then there can be no commonality of reference with the electromagnetic field. However, if they are not included in the core, then the perspective on theory change offered by this approach to reference may seem too conservative.

Consider another scientist (or perhaps the same one) working post-Einstein. She now believes in the existence of the electromagnetic field, and her degree of belief in the possession by this field of certain dynamical properties may be equal to that of her colleague (or other self) who believed in the possession by the ether of the same properties, but the object of the belief, that to which the terms refer, appears very different as an entity. Whereas the ether was conceived of as a kind of substance, possessing certain mechanical qualities and acting as an absolute reference frame, the electromagnetic field was not (at least not post-Einstein). The metaphysical natures of the ether and the electromagnetic field, as *entities*, are clearly very different, and the claim might be pressed that, given this difference, there is no commonality of reference.

Now, the realist might respond that insofar as these metaphysical natures do not feature in the relevant science, she is under no obligation to accommodate them in her theory of reference or her scientific realism as a whole. In other words, she might insist that when she, as a realist, insists that the world is as our best theories say it is, that covers the relevant scientifically grounded properties only and not these metaphysical natures. But then one might legitimately ask, what is it that is being referred to? It cannot be the ether/electromagnetic field as an entity, since this entityhood is cashed out in terms of the metaphysical natures. Thus what is being referred to is only the relevant cluster of properties that are retained through theory change. But now this response to the pessimistic meta-induction looks very different from what we initially took it to be. Instead of claiming that the ether was *not* abandoned—when scientists referred to it they were actually referring to the electromagnetic field—what is actually being claimed is that reference to the ether was secured via a certain cluster of properties that also feature in reference to the electromagnetic field. Now this response to the pessimistic meta-induction amounts to the claim that the ether as an entity was indeed abandoned, but that certain *properties* were preserved and retained in subsequent theories, where they feature in or are the subject of the relevant laws. In other words, what this amounts to is the admission that the proponent of the meta-induction is right in noting that the ontology of theories has changed but that she has failed to see that certain *structural* aspects, as represented by the cited properties, are retained. This suggests a different kind of realism from the traditional form—namely, "structural realism"—which we shall return to later.

So far we have just been addressing the issue of the apparently abandoned terms, but what about the other end of our spectrum, where we have terms such as "electron" that have survived theory change, in which we have a high level of confidence that they will be retained through future such changes, and in whose existence we have a correspondingly high degree of belief? Here, too, there are problems for the realist, and again they arise out of metaphysical concerns. Consider the following question: Does the term *electron* as used by J. J. Thomson, for example, refer to the same entity as the term as used in today's quantum field theory? Clearly, the metaphysical nature of the electron has undergone radical change, from a classical particle, to something that also has "wavelike" properties, to a particular kind of quantum field. Now the realist might respond as before by emphasizing that the core set of properties has remained the same from Thomson's time to today and that in these terms we can claim to have reference to the same thing. Setting aside, for the moment, our concerns relating to entityhood and reference, the danger here is that this may be too conservative a response to the concerns that underlie the pessimistic meta-induction, for if we take the wavelike nature of the quantum electron to be constituted by certain properties, then these are surely central to our current understanding of it. But, if these are included in the "core causal description," then the response collapses into a form of the causal role approach and falls prey to the second problem stated previously: If we are allowed to add to the core, then commonalities can be found between all kinds of apparently disparate entities, including Aristotle's natural place, Newton's gravity, and Einstein's spacetime. (The original core might be rather small and limited, but this is no problem if we are allowed to add to it as we go along.) If these wavelike properties are not included in the core, then we would be shifting to the periphery what many would take to be the essential quantum aspect of the electron. At best, this response would appear extremely conservative in maintaining what is basically a classical core for our understanding of the electron; at worst, it appears to deny the centrality of the very theoretical changes that the proponent of the pessimistic meta-induction presents as a challenge. And again, the nature of the "thing" that this approach to reference suggests is that of a bundle or cluster of invariant properties only, so the ontology of our realism has subtly changed.[11]

We have already alluded, in passing, to our second issue concerning changes in belief with regard to these existence claims. Not only is it the case that unobservable entities come and go, as it were, but the degree of belief in the existence of a certain entity may wax and wane in the course of scientific development. One might start off believing very strongly in the existence of the ether, for example, only for that belief to weaken, perhaps quite dramatically, with the development of Maxwell's theory of electromagnetism and perhaps crash to zero upon reading Einstein's 1905 paper on Special Relativity. Of course, the kinds of factors driving these changes will be precisely those we have touched on in this book: Some will be straightforwardly empirical, and others will have to with well-grounded symmetry principles, for example. These are the factors that drive theory change in general, and, of course, changes in one's degree of belief in the existence of a certain entity do not take place in a vacuum. The confidence one has in the existence of the ether increases as the relevant theory is developed, incorporates certain structural aspects from previous empirically successful theories, and achieves a level of empirical success of its own. Likewise, it decreases with the development of a new theory that does not include the term, has greater explanatory power, achieves greater empirical success, and so on.

This then relates to our third issue, whether the belief is in the relevant proposition as true or as quasi-true only. When our scientist *x* expresses a belief in the existence of the ether, is it the case that she believes, or should believe, the proposition "the ether exists" to be true? Clearly, the kinds of considerations that are expressed by the pessimistic meta-induction push one toward a negative answer to this latter question. However, as we have just seen, the form of fallibilism that should be maintained in the face of this argument appears to be adequately incorporated by adopting a broadly probabilistic approach involving degrees of belief, instead of straight belief per se. Furthermore, the concerns that led us to adopt quasi truth and its modification within this degrees of belief approach—the problem of universal laws, the Hacking problem, the Popper-Miller argument, and so forth—do not appear to come into play in this context. Nevertheless, we would still urge the adoption of our framework for the simple reason that these existence claims are never made in a vacuum; such a claim has the epistemic force that it has only because it is made in the context of a particular theory. Any judgment as to the existence of the ether, for example, occurs in the context of a particular theory in which the properties of the ether are expressed and, of course, related—via an appropriately complex web—to the relevant properties of measuring instruments (or, at least, *some* of the properties of the ether are!). To say, "I believe the ether exists" would be to say that "I believe there is an entity whose properties are the following," where these properties are described within the theory. The realist issue of whether such properties are instantiated "in the world" is entirely bound up with the issue of whether the theory should be regarded as true; indeed, to say that the theory is true, on the usual accounts, is just to say that the world contains or manifests or whatever these properties. Existence beliefs are parasitic upon beliefs about the truth of theories in this way.[12] Hence, a shift to quasi truth for truth claims drags with it a parallel shift for existence claims.[13]

Finally, this discussion should make it quite clear that anyone advocating an epistemic attitude of approximate or partial truth for theories need not be committed to the absurd view that unobservable entities "approximately exist." Of course, from an ontological point of view, the notion of an entity existing only "approximately" does not make any sense, but the fallibilism we are advocating here is wholly epistemic. What the quasi truth of theories motivates, as we have indicated, is the incorporation of this notion into the degrees of belief approach outlined previously.

There are further issues that realism—even one modified by quasi truth—must face, and consideration of these will lead us to the alternative side of the debate.

The Warrant for Realism

Why *should* one be a scientific realist, of whatever stripe? Before turning to the possible answers, let's briefly consider the nature of the question itself. We might view it as a purely philosophical question of the same kind as "Why should one be a realist in general, as opposed to an idealist, say?" or, in the context of the mind-body problem, "Why should one be a dualist?" Any answer to our question would then be presumed to have the same form as the answers to these others: One might, for example, give a kind of "transcendental" response, along the lines of "the necessary conditions for scientific cognition require a realist attitude." Such a view of the issue has fallen out of favor in recent years, to be replaced by a broadly "naturalistic" approach that takes our question, along with all philosophical questions, to be on a par with scientific qucstions in general.[14] What this means is that the question of why one should be a scientific realist is treated in just the same way and as requiring just the same sort of answer as the question of why we should accept theory T. The answer to the latter, it is claimed, will typically be of the form "because T is the best explanation of the relevant phenomena," and hence the answer to our question should have the same form—namely, that of an "inference to the best explanation" (IBE).

What would be the "relevant phenomena" in this case? Most obviously, perhaps, it would be the success of science itself. Thus realism is seen as offering the best explanation of this success. The next question is "best compared to what?" IBE is comparative, and typically theory T will be judged the best explanation of the relevant phenomena compared with theory T′, its competitor. What is the competitor in the case of scientific realism? Recent debate has been dominated by the view that the only competitor isn't much of a competitor at all, since it amounts to the claim that the success of science is a kind of miracle—that is, essentially inexplicable. This then gives us the famous "No Miracles Argument" (also known as "the Ultimate

Argument"), widely regarded as powering modern scientific realism: "The positive argument for realism is that it is the only philosophy that doesn't make the success of science a miracle" (Putnam 1975a, p. 69). Thus the No Miracles Argument is seen as a form of IBE, and the warrant for scientific realism inherits all the respectability of the warrant for scientific theories in general.

This overall strategy is problematic on a number of counts. First of all, one might argue that there are contenders to the realist explanation of the success of science other than "it must be a miracle!" Van Fraassen, for example, has offered a Darwinian alternative: "The success of science is not a miracle. It is not even surprising to the scientific (Darwinist) mind. For any scientific theory is born into a life of fierce competition, a jungle red in tooth and claw. Only the successful theories survive—the ones which in fact have latched on to actual regularities in nature" (1980, p. 40). Now, the export of Darwinian arguments beyond the domain of biology is always problematic. Popper famously argued that the theory of evolution as (crudely) summed up by the slogan "survival of the fittest" was tautologous, since "fitness," he claimed, is defined in terms of survival. Similar worries arise here: To insist that only successful theories survive doesn't explain much if success is ultimately understood in terms of "surviving" experimental refutation and the like. Psillos has pursued these worries further, suggesting that we understand van Fraassen's account as "phenotypical," in the sense that theories that are empirically successful are regarded as having the same phenotype and are selected on this basis (Psillos 1999, pp. 96–97). However, a realist will insist on a "deeper," genotypical explanation in terms of some further, underlying feature that renders the theories successful in the first place: For the realist, this genotype will be approximate truth (Psillos 1999, pp. 96–97). The problem, of course, is that this insistence on a "deeper" explanation begs the question against an antirealist like van Fraassen. In the biological case, where the genotypical explanation is ultimately reduced to a genetic causal account, the antirealist will either argue that the terms of this account—terms referring to the genes and the relevant causal mechanisms—can be reduced to observation terms, if she is an old-fashioned reductionist instrumentalist, or that belief in the existence of such entities and in the truth of the accompanying theoretical mechanism is not appropriate, if she is a new-fashioned constructive empiricist, like van Fraassen. The point of such antirealist positions is precisely to resist accepting as true "deeper" explanations of scientific phenomena, and hence they will see no reason to accept them with regard to the success of science itself.

This accusation of question-begging can then be leveled at the realist strategy as a whole. If the antirealist is not prepared to accept inference to the best explanation at the level of scientific methodology—where the "best" is understood as "most approximately true"—then why should she accept it as an appropriate warrant at the level of the philosophy of science? From this perspective, there appears to be a vicious circularity at the heart of the whole realist strategy, since it employs "the very type of argument whose cogency is the question under discussion" (Fine 1991, p. 82).

One response to such accusations is to admit that the strategy is circular but insist that it is not viciously so. The idea is as follows (Psillos 1999, pp. 81–90). Consider again the No Miracles Argument in the following form:

> Premise: Theories accepted on the basis of IBE are empirically successful.
> Conclusion: The best explanation of this success is that the theories are approximately true.

IBE is used to move from the premise to a conclusion that says something about IBE itself, but it is not the case that this conclusion is assumed in the premise. Thus the No Miracles Argument is only "rule-circular" and not, crucially, "premise-circular." Hence, Psillos concludes, the defender of the No Miracles Argument is, at least, no worse off than the defenders of induction and deduction who employ similar strategies (1999, p. 89).

This is an interesting and well-worked-out response (for criticisms, see Douven 2001, and for a reply, see Psillos 2001), but it is not clear to us that the realist can be placed alongside the defenders of deduction and induction quite so straightforwardly. First of all, suppose we were to accept that a justification of IBE that itself employs a form of IBE is unavoidable. Still, one might argue that this still falls short of a justification of realism: To get there, one has to read "best explanation" as "approximately true explanation." Couldn't one be an antirealist, acknowledge that theories are accepted—not as approximately true, of course—on the basis of IBE, and acknowledge that IBE can be justified only by employing IBE but conclude that the "best explanation" of the success of the theories that are accepted lies not with their approximate truth but with some other virtue? Fine, for example, has suggested that "instrumental reliability" can be regarded as the best explanation of the success of science (1986, pp. 153–154), where this is better than approximate truth on the grounds that the former is less inflationary than the latter and therefore should be preferred. Now, setting aside the question of whether this itself begs the question against realism, Fine himself has a rather peculiar attitude toward truth. He accuses the realist of positing an "extra-theoretical relation between theories and the world" (1996, p. 24)—namely, the relation of approximate truth. And here is how the circularity arises for Fine: "Thus, to address doubts over the reality of relations posited by explanatory hypotheses, the realist proceeds to introduce a further explanatory hypothesis (realism), itself positing such a relation (approximate truth)" (Fine 1996, p. 24).

What's going on here is that Fine thinks the realist who adopts the correspondence theory of truth is committed to truth being a natural kind, in the sense that one would know the "essence of truth." But, as Musgrave points out (1996), the lesson we should learn from Tarski is that there is no such essence, since truths are many and various, and so are the ways they "correspond" to the facts. What Tarski's theory does is capture the correspondence relation as best it can be captured. Fine thinks the realist is committed to going beyond Tarski to the essence of truth, and this is what generates the circularity. No, insists Musgrave, a realist does not have to accept an essentialist view of truth; she can be happy enough with as much correspondence as Tarski gives. Hence it is debatable whether approximate truth is more inflationary than instrumental reliability (see also Psillos 1999, pp. 91–93).

There is the further issue of whether "instrumental reliability" is any explanation at all. To say that a theory is instrumentally reliable is to say that it is good at making predictions, accounting for the phenomena, in short, that it is empirically successful. At this level, instrumental reliability is not put forward by the antirealist as an explanation of the empirical success of the theory, since she doesn't see the latter as requiring any kind of explanation. Likewise, at the metalevel, to explain the success of science in terms of the instrumental reliability of theories is not to offer an explanation at all (Psillos 1999, pp. 92–93). Fine himself seems to have realized this, offering a new interpretation of instrumental reliability in terms of the disposition to produce empirical success (1991), but this smacks of desperation. As Psillos notes, either such dispositions must be taken as "brute facts," as it were, in which case they remain mysterious, or else they must be grounded in some nondispositional property (such as approximate truth; Psillos 1999, p. 93). At this point, the antirealist—and we must recall that Fine himself is not such but is trying to construct a "third way" with his "Natural Ontological Attitude"—might stand up and insist that we just shouldn't be playing this sort of game in the first place! Once we begin to offer explanations of the success of science, we have already bought into the "explanationist strategy" and conceded too much to the realist. An "honest" antirealism would reject the entire strategy to begin with and insist that no explanations are to be had—or at least not of the form the realist apparently wants (the closest the antirealist is willing to get would be van Fraassen's Darwinian line, touched on previously). What begins to emerge here is how difficult it is for either side to get their argument off the ground without begging questions against the other.[15] And this is because the very epistemological bases of the two positions are so very different (see, for example, Wylie 1986 for a nice account of the way the two sides dance around each other without properly engaging).

This discussion also begins to indicate why the realist, in defending an "inference-to-the-best-explanation" justification of IBE *is* worse off than either the inductivist or deductivist. Unlike the latter, she can get the justification going only by begging certain fundamental epistemological questions against the antirealist. Consider what IBE is supposed to lead us to: At the "object" level, the inference leads us to the (approximate) truth of the given theory; at the metalevel it is supposed to lead us to the (approximate?) truth of realism, but at this point all sorts of doubts come crashing in. First of all, on this account, theories are put forward as explanations of the phenomena, and realism is analogously put forward as an explanation of the success of these theories in explaining the phenomena. The question arises: Can the success of science itself be appropriately treated as a phenomenon to be explained in an analogous way to "natural" phenomena? The realist's positive response to this question takes her along a curious path. Consider again Fine's suggestion that the best explanation should be one involving instrumental reliability rather than truth. Consistency would appear to demand that at the metalevel, if his form of instrumentalism is accepted as the "best" explanation of the success of science, "best" should not be equated with truth but rather with "instrumental reliability" or "empirical adequacy" (Psillos 1999, pp. 90–91). But this is bizarre: In what sense can instrumentalism be understood as *empirically adequate* to the "phenomenon" in this case? In the case of a theory, the sense is clear, particularly through van Fraassen's model-theoretic formulation. But are we supposed to think of the success of science being represented by a data model that is then embedded into instrumentalism? If we are, then presumably we should be able to judge the "goodness of fit" but in what terms? Do we move to the meta-metalevel and argue about realist versus instrumentalist views of the relationship between realist and instrumentalist views of science and the phenomena of the success of science? We seem to have avoided circularity but only at the expense of a particularly unattractive regress.[16]

Second, in the case of "natural" phenomena, the use of the plural is significant. In the case of the success of science, that's all there is, so we have only a phenomen*on*. This is important because the relationship in this case can be only vertical and not horizontal, synchronic and not diachronic, so there is no dynamic aspect. What this means is that realism, instrumentalism, or any other putative explanation of the success of science cannot be *independently testable* (Worrall 1996, pp. 141–142; see also Almeder 1987, pp. 67–71). Here is another difference between the explanations offered by theories and the "explanations" offered by philosophical positions: Independent testability is crucial for the former to be acceptable, but this requires there to be phenomen*a*,[17] rather than just a phenomenon. Similarly, novel predictions play an important role in judging the empirical success of theories, but it is difficult to see how there could be any such predictions in the case of either realism or instrumentalism. One could continue: "Natural" phenomena are typically not just observed but are created under certain conditions (Hacking 1983); that obviously does not apply to the success of science. There is, related to this, a whole "epistemology of experiment," but again, that finds no purchase at the metalevel.

The realist will fall back to the view that the relevant relationship here is only meant to be *analogous* to that which holds between theories and the phenomena. However, in that case the analogy seems utterly mysterious and obscure, and the realist owes us an account of how this "analogy" can yet support the same form of inference—IBE—and also a judgment as to the "truth" of realism. This brings us to our final point regarding this strategy: The conclusion of the inference at the object level is that the theory is true, or at least approximately so. What is the inference at the metalevel supposed to show? That scientific realism itself is "true?" But what theory of truth is supposed to apply here? If the realist, still pursuing the strategy, insists that it is truth in the correspondence sense that she means, then to what is realism itself supposed to correspond? In the case of a theory, we can at least grasp the notion that the theory is supposed to be true in the sense of corresponding to "the world," where this includes both observable and unobservable aspects. In the case of realism itself, there is, again, only the success of science; if this is regarded as "observable" (perhaps via the "instrumentation" of the history of science!), then there is no unobservable aspect. But then, turning the tables on the realist, "truth" at this level collapses into "empirical adequacy" (assuming we can make sense of the latter in this context). And if it is not correspondence truth that is meant, then, first of all the realist owes us an account of what "truth" is supposed to mean in this context; second, it is hard to see how any such alternative account could fail to undermine the overall strategy itself (since we would have IBE leading to truth at one level and IBE leading to something else at another).

Perhaps it is time to call a halt to these proceedings. The realist's "explanationist" strategy is beginning to look fatally flawed, and insofar as it is, one might also begin to have doubts about the overall "naturalistic" perspective on which it is based.[18] Does this mean realism itself is fatally flawed? After noting that realism is not independently testable, Worrall concludes, "Scientific realism can surely not be *inferred* in any interesting sense from science's success. The 'no miracles' argument cannot *establish* scientific realism; the claim is only that, other things being equal, a theory's predictive success supplies a prima-facie plausibility argument in favour of its somehow or other having latched on to the truth" (1996, p. 142). But a plausibility argument is not a warrant, and the "psychological force" (Worrall 1996, p. 142) of the argument will not sway the antirealist in the least.

Alternatively, we might try to find some other warrant for realism. Of particular interest to us is Hacking's attempt to bypass IBE entirely by offering essentially "pragmatic" grounds for accepting the existence of certain entities: If we can use these entities as tools to create further phenomena, then we have grounds for believing they exist (Hacking 1983). The example he gives is that of a "spray" of positrons being used in an experiment to alter the charge on a niobium ball; hence his now-famous slogan "if you can spray 'em, they exist." This forms the basis of his "entity realism," in which the grounds for accepting the existence of unobservable entities are not theory-based but rather practice-based. The contexts into which this can be nested are, first, a view of scientific knowledge according to which such knowledge has to do with *intervening* in the world, as well as representing it, and second, Dewey's "transformational" pragmatism. For our purposes, the crucial difference between Dewey's approach and Peirce's is the following: The latter focused on the fixation of belief and can hence be viewed as primarily theory oriented, whereas Dewey "was concerned chiefly with the *instrumental* function of the inquiring intelligence in transforming the problematic situation so as to achieve a satisfactory outcome" (Smith 1978, p. 97; his emphasis). The difference in focus reflects on a different attitude toward truth: Whereas Peirce advocated the view we've outlined in chapter 1, with truth reached as an ideal limit, Dewey "was led to set that ideal aside in favour of a transform theory which requires that the indeterminate or problematic situation which first gives rise to inquiry be replaced by a determinate state of affairs wherein the problem is resolved, not merely by answering a theoretical question but by actual manipulation and control of the relevant objects and conditions involved" (Smith 1978, p. 98).

Situating "entity realism" in this broader pragmatist framework renders it both more interesting than it might at first appear and also more problematic. Let us consider the argument for this view in slightly more detail: Regarding electrons, Hacking maintains that there is no one single theory of the electron that can be viewed as the "best explanation" of the phenomena. Rather, there is a disparate set of models with, he insists, no common core. Various of these models—many of them "low level" or phenomenological, rather than "theoretical"—are then drawn upon in order to construct instruments, such as an electron gun, which are then used to create new phenomena. The claim, then, is that the belief in the existence of electrons arises, not from some theoretical inference, but as the result of having to *use* them, as a tool, to intervene in and change the world. In addition, Hacking makes the historical assertion that it was on just such grounds that physicists came to believe in electrons at the turn of the century.

However, underlying his claim is a form of inference that slides from the observable to the unobservable: The idea seems to be that, just as someone who uses a spanner to "intervene" in her car engine cannot doubt the existence of the spanner in the using of it (or, at least, not in some nonidealist sense), so the physicist cannot doubt the existence of the electrons that she effectively takes down off the shelf in the form of an electron gun, say. Now, the antirealist will obviously respond that the two cases are strikingly different, concerning the observable and the unobservable, respectively. The evidence for the existence of the spanner comes not just—or not at all—from its use but from its observability via different means, such as sight and touch, whereas there is no such directly observable evidence for the existence of electrons. This brings us back to the contentious issue of the observable-unobservable distinction, touched upon in chapter 4, and Hacking does have a response that emphasizes the manner in which what are "observable" and "unobservable" lie at different ends of a spectrum, rather than on different sides of an epistemological divide. The response is twofold: First, Hacking notes that instruments that function on different physical bases—different kinds of microscopes, say—appear to reveal the same phenomena. Although these phenomena are unobservable to human senses, nevertheless, the similarities between what these different kinds of instruments "observe" support the idea that what is being "observed" exists. Second, he appeals explicitly to the way the macroscopic shades over into the microscopic by using as an example the preparation of designs for complex diffraction gratings at the macroscopic level; these designs are then shrunk down and etched onto the appropriate materials at the microscopic level. That these gratings then function in the manner for which they were designed is then supposed to show that there is no sharp division between the macro and the micro.

It should come as no surprise that unpacking these claims and their justifications reveals a number of problematic features (see, for example, van Fraassen 1985a). However, even if we were to grant the point concerning the relationship between the observable and the unobservable, Hacking's position must still face difficulties. In using a spanner, to produce a torque and thereby remove a nut, we rely on certain causal properties it can be said to possess. The same goes for electrons, or so it has been argued: What physicists and engineers are actually relying upon when they build an electron gun is their knowledge of the causal properties of the electrons (Hacking 1983, p. 265). Their existence is actually inferred from these—or, rather, from the low-level theory or model representing these bodies (Reiner and Pearson 1995).

Now Hacking could still resist this low-level representational move by insisting, again, that scientific knowledge is gained not just from representation but also from intervening in the world, as Dewey argued. However, the latter's "transformational" epistemology threatens to undermine the robust response to antirealism that entity realism apparently represented. As indicated in the previous quote, Dewey rejected the underlying assumption of a representational approach—namely, that "situations" are always determinate. Rather, the situations themselves, and not just our representations, are indeterminate.[19] When doubts arise, it is not merely a matter of adjusting our beliefs but of actually intervening in the situation and thereby transforming it into a determinate state of affairs. The worry, of course, is that the notion of situations that are transformed from indeterminacy to determinacy through our intervention hardly seems to sit well with scientific realism. Indeed, Dewey's contention that both the "facts" and our conceptual apparatus play an "operational" role suggest a similarity with Kantian views, according to which "the world," as that which can be known, is not wholly independent of the knowing mind. Hence, if Hacking intends his view of knowledge-via-intervention to be understood in something akin to Dewey's sense, then such interventions, in rendering problematic situations determinate, render "the world" dependent on us.

The program of seeking a non-question-begging, unproblematic warrant for realism looks to be in trouble. Perhaps the source of the difficulty is the very assumption that realism—or antirealism for that matter—is the sort of thing that can be warranted, either in the way (realists claim) theories are warranted, via IBE, or, broadly speaking, pragmatically, the way our belief in the existence in spanners is. This is not to say that *arguments* cannot be given for these philosophical positions, but any such argument will proceed from certain fundamental premises that the opponents of the view concerned will find contentious (Wylie 1986). Perhaps these positions should be stripped of any resemblance to scientific theories by regarding them as no more than "attitudes" that one will adopt according to whether the relevant premises match one's basic philosophical prejudices: If one thinks that one's beliefs should run deep, in the sense of extending beyond what can be observed with the senses, then one will be inclined toward realism; if, on the other hand, one thinks that one should not be profligate with one's beliefs, reserving them for only those cases or situations that represent an appropriately high level of epistemic security, then one will tend toward antirealism. Is there any way of mediating between these two poles? At this point, it might be worth applying a pragmatic approach to the debate itself.

Perhaps talk of realism being "true" or antirealism being empirically adequate should be taken as a mere *façon de parler* at best. But then what is the appropriate epistemic attitude? Let us consider the nature of the inquiry: At the level of theories, inquiry begins with a problem situation, about which doubts arise, and these doubts are resolved, according to Peirce, for example, through the "fixing" of belief (Peirce 1940, p. 10).[20] Doubt is to be understood here as the absence of belief, leading to the contrary of the states of mind that belief produces; that is, doubt gives rise to anxiety and uneasiness. However, doubt itself is not fundamental, arising as it does from practice. Again, the nature of its production is defined in terms of its opposite number, belief. Since beliefs, for Peirce, are defined by "the actions they generate, doubt as the absence of belief, arises when actions are halted, impeded or prevented in some way" (1940, p. 29). Thus doubt is tied to the prevention of activity. In responding to doubt, we therefore attempt to establish those beliefs we need to continue our activity, where such attempts constitute inquiry.

Hence, skeptical doubts for Peirce are not "genuine" doubts, since they do not halt, impede, or otherwise prevent practical activity. And it is precisely because of this that they cannot be resolved. For Peirce, the self-perpetuating nature of the infamous skeptical argument that runs throughout the history of philosophy can be situated in the distancing of its essential core of doubt from practical activity (Alcoff 1986).[21] Skeptical doubts are therefore "paper doubts," because of their very nature in lying at such a remove from practical activity. At the level of the realism-antirealism debate, it is *these* sorts of doubts that arise and cannot be resolved, at least not without begging the very questions at issue.[22]

Instrumentalist and empiricist doubts—about whether theoretical terms genuinely refer and theories themselves can be regarded as true—are essentially skeptical doubts, albeit of a restricted kind (see French 1988). As such, they are not "genuine" because they do not halt or prevent practical activity—in particular, scientific activity—and are hence not resolvable. We recall the empiricist's invocation of the empirical underdetermination of theories (which we shall discuss in detail later): There is no procedure to resolve the empiricist's doubt because any argument we might offer in defense of the claim that a particular theory really does reflect the way the world is would be subject to the empiricist's counterargument that a variety of alternative explanations can be offered that accommodate the same evidence (French 1989a). Thus, by analogy with the skeptic's argument, the empiricist says that all our available evidence is entirely consistent with any number of alternative explanations and therefore there is simply nothing we can offer in terms of the "practical" scientific activity of theory production that can refute this position (cf. Alcoff 1986, p. 9).

However, this does not leave the field clear for the realist. As we have seen, she is unable to offer any non-question-begging warrant for her position and errs precisely in regarding it as true in the same sense as the theories she interprets. Can a broadly pragmatist approach help? Peirce's philosophy was famously closer to realism than that of James or Dewey, but his "proof" of the evidence of "reals"—dropping a piece of chalk—is clearly insufficient. Actually, what this episode demonstrates is how careful one must be both in drawing on these earlier views and in projecting one's current concerns upon them. Peirce was not at all concerned with the realist-antirealist debate in anything like its current incarnation. What he was trying to demonstrate was the existence of something that affects our senses and thus acts as an external constraint on our inquiry (see Rescher 1978, pp. 19–21). Thus all he was trying to "prove" was the existence of a mind-independent world.[23] Of course, this is a central plank of realist belief, but it is one that the constructive empiricist would also accept, with, of course, the crucial proviso that we cannot *know* how this mind-independent world is. Furthermore, circularity arises even in Peirce's account: As is well known, Peirce took the scientific method to be superior to other methods of "fixing" belief, and he also took it to rest on the fundamental hypothesis that there exist "real things" that are governed by laws. However, he also argued that the repeated application of this method would alter people's "commonsensical" beliefs and reveal the nature of these "real things."

Peirce's attempts to evade this circularity are not particularly convincing (Smith 1978, pp. 30–31). Perhaps his best effort lies with the admission that although the scientific method cannot prove there are "reals," at least it does not prove the opposite. Hence the method and the hypothesis of "reals" are harmonious in a way that other methods are not, and so no doubts can arise from employing it.[24] Needless to say, the antirealist will find this utterly unconvincing. In particular, as indicated before, the constructive empiricist would simply acknowledge this hypothesis but offer an empiricist construal of our epistemic situation with regard to both the reals and the laws they "obey."[25]

Our constructive empiricist friend might be more comfortable with James, who situated pragmatism within the history of (British) empiricism. James also differed from Peirce in placing a greater emphasis on pragmatism as a method for resolving metaphysical disputes (1907a, p. 45). The idea is as follows: Given a disagreement between two metaphysical views, we ask, "What difference would it practically make to anyone if this notion rather than that notion were true?" (James 1907a, p. 45). If no such difference can be found, then no genuine differences exist, and the dispute is merely verbal. Used in this way, James claims, pragmatism can resolve a cluster of traditional disputes: between monism and pluralism, determinism and free will, materialism and spiritualism.

There are obvious difficulties with this application of pragmatism, particularly in the present context. First of all, what is the sense of "truth" here? Presumably, the monist will want to argue that her view of the world is true in the correspondence sense, yet James himself held a different theory of truth, of course. If it is the latter that he means, then the monist will respond that rather than taking the disputed positions in their own terms, James is simply imposing a further view via the application of the pragmatist theory of truth. This objection carries particular force in the case of the realist-antirealist debate, where, as we have seen, the realist may insist on the correspondence theory, even for her realism. If, on the other hand, James acknowledges that "truth" should be understood in the disputants' own terms, then the (old style) instrumentalist would reject correspondence truth entirely. However, if the realist's opponent is the (new style) constructive empiricist, then the dispute is no longer metaphysical—in the sense that both sides agree there is a mind-independent world and that scientific theories are truth-apt—but has become epistemological. And in this case, it is not clear how the debate can even be set up in such a way that the pragmatic method is applicable.

Second, supposing that we are able to set up the debate appropriately, how is the notion of difference to be spelled out here? What are the differences that count? (See the discussion in Smith 1978, pp. 35–44.) And *who* are these differences supposed to count for? Does it make a difference to the scientists themselves whether they adopt a realist attitude or not? One might plausibly claim that adopting such an attitude makes sound methodological sense in that it encourages scientists to propose and test hypotheses about unobservables. However, there are numerous representatives of various forms of antirealism among the scientific community, and it is not clear that their philosophical view hindered their scientific work in any way. More important, insisting on realism as a methodological doctrine will not impress the constructive empiricist, who will simply agree to its adoption with the proviso that the appropriate attitude be understood in "as if" terms; that is, scientists should conduct their work "as if" there were unobservable entities. Of course, this does not provide any warrant for realism itself.

What about philosophers of science? If we set aside possible differences at the scientific level, does the adoption of either realism or antirealism make any differences at the philosophical level? Well, James refers to "practical" differences, and here it is hard to see any. Whether one is a realist or an antirealist doesn't seem to make any great difference to one's everyday activities, to "negotiating" one's way around the world! Of course, both the realist and the antirealist will insist that it obviously makes a *philosophical* difference, but it is precisely because they disagree as to what are the important philosophical differences that accusations of question begging arise. Although it might seem that a more "pragmatic" approach could help, we should be cautious. James took the "practical" differences referred to previously to include differences as to how one should conduct one's life, and in this sense, he argued, it does make a difference whether one believes in free will or determinism: The latter will leave you with a sense of hopelessness and oppression, whereas the former generates hope and, of course, a sense of freedom. Now, advocating realism on the grounds that it will make you feel better is not going to impress anyone, least of all the constructive empiricist (who might in turn respond that her position will make you feel better because you won't be carrying around all that unnecessary metaphysical baggage)!

Perhaps, then, we should just quietly leave James's understanding of what constitutes a difference and return to the basic point, that the doubts that seem to have arisen in this debate are not resolvable, precisely because, unlike the case of science, there is no agreement as to the very nature of the framework in which they might be resolved. Let us be clear here: What we are claiming are unresolvable are not doubts about *how* the world is but rather doubts as to whether the latter should even be taken as an appropriate aim for science. Once one has chosen what one takes for this aim, then one can begin the process of doubt resolution at the appropriate level. So, if one adopts a realist perspective, then the question as to *how* the world is will come to be seen as not only meaningful but also answerable (at least to a certain extent; as we shall see, depending on what is taken to constitute an answer to this question, certain aspects may remain beyond the reach of an evidentially based answer). However, the point is that the form of enquiry that may be appropriate for answering this question is not appropriate for dealing with that of why one should be a realist in the first place.[26]

Having adopted the appropriate perspective, there is still the question as to which form of realism or antirealism best fits what one takes—within the chosen framework—to be the relevant aspects of scientific practice. Here again, the epistemological and metaphysical features of "standard" realism let it down.

A Pragmatist Epistemology

Let us be blunt: The realist is hamstrung by the metaphysical aspects of truth as correspondence. These leave her vulnerable to the skeptical challenge: How do you *know* when you have attained the truth in this sense? The realist's response that she knows she has attained "the truth," or something close to it—supposing we can introduce an appropriate metric in this context—when she has the best possible theory available then generates the further question: What reason does she have to believe that the theory she judges to be the "best" is closer to the truth than those theories that fall short? What is going on behind this skeptical demand for such a reason is the following: Either this reason is grounded on the epistemic virtues of the theory concerned, or it is not. If it is so grounded, then the realist falls prey to what has been called "the argument from the bad lot" (also known as the "argument from under-consideration"; see Lipton 1996); if it is not, then the grounds obviously need spelling out. If it is accepted that we do not have independent access to the world to which our theories supposedly correspond, then the prospects for such a spelling out look dim. Hence the realist must face the argument from the bad lot.

The argument runs as follows: Let us accept that theories can be ranked, so that the realist can claim that one is the "best." To go beyond such a comparative ranking, the realist must then equate the best, in this sense, with being closer to the truth. But this requires the prior belief that the truth is already more likely to be found in the set of theories that we are faced with (van Fraassen 1989, pp. 142–143). In other words, to claim that the best of the lot is the closest to the truth of the lot, we have to assume that the lot as a whole is close to the truth to begin with. What reason might we have to doubt this? Well, the set of theories we are faced with is, of course, historically given, and history could have given us, as it were, a different set.

A possible realist response is to claim that *if it is agreed that theories can be ranked*, then it is of the nature of the ranking process that the set of theories under consideration cannot be far from the truth, and hence the gap between the comparative and absolute evaluation can be closed. What is it about the ranking process that can effect this closure? Here the realist appeals to scientific practice and notes that "scientists rank new theories with the help of background theories" (Lipton 1996, p. 100; see also Psillos 1999, pp. 215–222). It is then argued that for this ranking to be "reliable," these background theories must be approximately true; otherwise, the ranking could be skewed and not reliable at all. But then, of course, the background theories themselves were previously accepted as the "best" of a ranked lot of theories, and this earlier ranking was peformed relative to a further set of background theories, and so the iteration goes. The conclusion is that "if scientists are highly reliable rankers . . . the highest ranked theories have to be absolutely probable, not just more probable than the competition" (Lipton 1996, pp. 100–101). Hence, the truth can't be "out there" beyond the set of theories considered, and the argument from the bad lot "self-destructs" (Lipton 1996, pp. 100–101).

There are two points to note about this response. The first is that no details are given regarding the relationship between the theories under consideration and the background theories. Perhaps, it might be said, no further details are required since all that is needed for the purposes of the argument is that the background theories are approximately true. Nevertheless, until it is specified how background theories have an impact on the theories being ranked—presumably in a differential manner—the proposal, while appearing plausible, remains vague. We shall return to this issue in the context of a further antirealist argument—namely, the argument from underdetermination. The second point is more important. The skeptical antirealist may feel uneasy at this apparent proof that the best of a lot of theories is also the closest to the truth of that lot because it seems to have solved the metaphysical issue by appeal to epistemology. The sense of unease may have to do with the way in which the concession that scientists are able to rank theories, in order to determine which is the "best," is rendered as the claim that scientists are *reliable* little rankers. "Reliability," here, is taken to mean always being able to rank a more probable theory ahead of its competitors (Lipton 1996, p. 93). Now suppose a scientist were unreliable; presumably that would mean she would sometimes rank less probable theories ahead of the more probable. But this implies that the judgment of how probable a theory is remains independent of its ranking. Yet the whole point of the argument from the bad lot is that we cannot go beyond the ranking of theories to reach any further, independently based judgment. If the realist were to respond that scientists are, by and large, reliable, the antirealist may ask, how do we know this? We know that they rank their theories and choose the best—that much is granted—but we don't know whether they can do this reliably, in the sense of always ranking as the best the theory most likely to be true.

Let us put the point another way: If it is possible for the scientist not to be a reliable ranker, then it must be possible for a theory to be ranked the best yet not be the most likely to be true. In that case, factors other than those employed in the ranking must come into play in order to determine which is the most likely to be true; that is, these factors must establish the ranking as reliable. But there are no such factors. If it is not possible for the ranking to be unreliable, then the best theory is ipso facto the most likely to be true, according to the realist. But now, likelihood of being true is judged relative to the background theories, since that is how the ranking proceeds. Likewise, the best of these background theories are taken to be most likely to be true, and so we iterate our way backward. But now there is nothing to prevent the background theories from skewing the ranking, and nothing to prevent their background theories from skewing their ranking, and so on. In other words, the whole of science might have got off on the wrong foot at the beginning of the whole process. It seems that the skeptical challenge cannot be resisted so easily.

The skeptic can get her foot in the door, as it were, only because of the gap between metaphysics and epistemology allowed by standard realism. If this gap were to be closed, then the previous argument would not even get off the ground. This is what pragmatism does, of course, by effectively rendering truth epistemological: If what is true is tied to our methods of inquiry, then, of course, those methods will give us the truth! There is then no question as to whether the sorts of epistemic values that come into play when scientists rank theories will lead to the truth or not. As Ellis puts it: "Our epistemic values *must* be adapted to the end of discovering what is true, because *truth* is just the culmination of the process of investigating and reasoning about nature in accordance with these values" (1996, p. 189). Of course, the antirealist may complain that this victory over skepticism is too cheap, and the standard realist may, equally, complain that it is too expensive. It is too cheap because it effectively defines away the problem, and it is too expensive because it ties our understanding of how the world is too tightly to our epistemic values. The response of the pragmatic realist to these complaints is ultimately the same: The antirealist is essentially correct in her counterfactual claim that the history of science *could* have been otherwise but only if *we* had been otherwise, and the realist is essentially correct in that how the world is for the pragmatic realist is how the world is *for us*, but this does not make *the world* dependent upon us. The idea is that if we were beings with a different set of epistemic values, then the theories we would rank as the best and hence as the more approximately true would be very different from the one we currently favor. In this sense, the antirealist is correct in suggesting there may be other theories "out there," as it were. If we had these other epistemic values, then these theories would be pragmatically true for us. And the way the world is, according to these theories, would be the way it is for us, as such beings. This is not to say that there is no mind-independent world (Ellis 1996, pp. 190–192). The point of Peirce's demonstration with the chalk is that we cannot will the chalk to fall upward, any more than we can will heavier bodies to fall faster than lighter ones or will electrons to hit one part of a scintillation screen rather than another. The pragmatic realist accepts that there is a world independent from us—and to that degree, she accepts what is common to both the standard realist and the constructive empiricist—but denies that we can *know* how that world is independently from our epistemic values.[27] According to Ellis, "we can explain anything that can be explained about our knowledge of reality . . . without supposing that there is any way that the world is absolutely. It is enough if there is a way (for us) that the world is *for us*" (1996, p. 192).

Of course, *how* that world is for us is a further issue. Even those who agree with the epistemology underlying standard realism may disagree with the sort of ontological picture it appears to favor. Here, too, partial structures have been put to use.

Structural Realism

Let us recall the problem of ontological change in the history of science, as encapsulated by the pessimistic meta-induction and also the inadequacies of "standard" realist attempts to deal with it. As we saw in the discussion of reference and theory change, the realist seems pushed toward the view that what is retained through such changes are the relevant *structures*.[28] This constitutes the heart of Worrall's attempt to defuse the pessimistic meta-induction by insisting that, with regard to these shifts in ontology, "there was continuity or accumulation in the shift, but the continuity is one of *form* or *structure*, not of content" (Worrall 1996, p. 157).[29] This forms the basis of his "epistemic" form of "structural realism," which maintains that all that we know relates to theoretical structure, whereas the content of our theories remains hidden to us. This raises two fundamental questions:

1. How are we to appropriately characterize this structure?
2. How are we to characterize content?

We shall briefly consider each in turn.

The principal historical example that Worrall draws upon is the theory of light, where we see a succession of shifts from a view of light as corpuscles, to one of waves, then to wave-particle duality, and finally to the current conception in terms of quantum fields. This example is suggestive but also misleading. Worrall argues that although conceptions of the *nature* of light have changed through history, the mathematical equations describing it have not but have been incorporated into successive theories, culminating, classically, in Maxwell's theory of electromagnetism, which is then taken to be subsumed into quantum electrodynamics. The example can be misleading precisely because it has led to the impression that realism is wedded to a consideration of explicitly mathematical theories only and cannot offer much comfort to the realist when it comes to (qualitatively expressed) biological theories, for example. However, this is simply not the case.

Mathematical equations represent one component of theoretical structure, but nonmathematical, qualitative aspects can also be represented through the resources of logic and set theory, giving a broader conception of theory structure, as we have seen.[30] Structural realism has typically been presented within the framework of the Received View of theories, and the relevant structure is expressed via the Ramsey sentence of the theory.[31] Unfortunately, tying this view to the "Ramsification" of theories generates two important problems.

First of all, it means that it cannot, in fact, do what the realist wants, which is to capture the continuity of structure beyond the empirical level. This is because any two Ramsey sentences that are incompatible with one another cannot have all their observable consequences in common (English 1973). Hence theoretical equivalence collapses into empirical equivalence. Since any form of realism must allow for a separation of the former from the latter, this result is disastrous for the attempt to construct a viable form of structural realism (see Ladyman 1998). Second, there is the so-called Newman problem, which we can express as follows: Given any "aggregate" of relata *A*, a system of relations can be found having any assigned structure compatible with the cardinality of *A*; hence, the statement "there exists a system of relations, defined over *A*, which has the assigned structure" yields information only about the cardinality of *A*. Thus, if science tells us only about the *structure* of the world—as the structural realist maintains—then it actually tells us very little. Indeed, it has been argued that structuralism collapses into triviality, since all that the claim that the world has a certain structure amounts to is that the relevant set of elements has a certain cardinality (see Demopolous and Friedman 1985; Psillos 1999, pp. 63–65).[32]

In order to get around these problems, Ladyman (1998) has suggested abandoning Ramsey sentences in favor of the alternative descriptive framework offered by the "semantic" or model-theoretic approach. As we have seen, introducing "partial structures" allows this approach to capture, in a natural fashion, both the relationships that hold between theories, horizontally as it were, and those that hold vertically between a theory and the data models. With regard to intertheory relationships, we have seen (in chapter 6) that partial structures can capture precisely the element of continuity through theory change that is emphasized by the structural realist. In particular, it offers the possibility of accommodating examples of such continuity that have been described as "approximate" or partial. Thus Worrall himself refers to the shift from classical to relativistic mechanics and suggests that "there is approximate continuity of *structure* in this case" (1996, p. 160).[33] He goes on to write that "much clarificatory work needs to be done on this position, especially concerning the notion of one theory's structure approximating another" (p. 161).[34] The partial structures approach can contribute to this clarification by indicating how such intertheoretical relationships can be represented by partial isomorphisms holding between the model-theoretic structures representing the theories concerned.

Concerning intratheoretical relationships, partial structures also allow us to capture these relationships downward from the theoretical models to the data models and also upward, from the physical theory to the mathematical, as we have indicated. We recall the point made earlier that such set-theoretical relationships hold only between the mathematical structures and not between such structures and "the world" itself (French and Ladyman 1999). The realist representation of the relationship between theories and the world must therefore be sought elsewhere, perhaps in a notion of reference appropriate for a broadly structuralist metaphysics. We shall not pursue this here but shall simply note Ladyman's claim that the partial structures form of the semantic approach offers a general account of theoretical structure that extends beyond the mathematical equations and thus represents an appropriate formal framework for structural realism. In particular, it is not clear that the "Newman Problem"—typically formulated within the syntactic approach to theories—has any bite within such a framework.

Let us now consider the second of the questions. How is the notion of "content" to be explicated? Worrall famously drew on a historical precedent in the work of Poincaré, who wrote that theoretical terms "are merely names of the images we substituted for the real objects which Nature will hide forever from our eyes. The true relations between these real objects are the only reality we can ever obtain" (1905, p. 162).[35] It is the nature of these "real objects" that constitutes Worrall's "content." Now this may appear an unattractive claim in the context of modern science, as it puts these "natures" forever beyond our epistemic reach. Indeed, Psillos suggests it is reminiscent of talk of medieval forms and substances, which, he claims, were decisively overthrown by the scientific revolution (1999, pp. 155–157). This is historically incorrect, as such "talk" persisted through the 19th century, with physicists developing their theories within a variety of metaphysical frameworks (Harman 1982). With regard to physical *objects*, in particular, it was typically presupposed that such objects were *individuals*, where the metaphysical sense of this individuality was explicated in terms of either an underlying substance or, more commonly, spatiotemporal location.[36] This kind of talk did become suspect with the advent of quantum mechanics, and the view arose that quantum particles cannot be regarded as individuals in some sense (see French and Krause forthcoming).

However, as is now well known, the physics itself does not determine such a view, and it can be shown that quantum mechanics is compatible with an alternative account that continues to regard the particles as individuals (French 1985 and 1989c; Huggett 1997; van Fraassen 1991). Hence there exists a form of "metaphysical underdetermination" in which we cannot infer the appropriate metaphysics for describing the world from the physics itself. This presents a problem for the realist, who is supposedly committed to the claim that physics can tell us "how the world is." If this "telling how it is" is understood to include telling us whether quantum particles are individuals or not, then the previous results undermine the realist's claim.[37] The constructive empiricist, of course, finds further support in this metaphysical underdetermination and rests content with the weaker claim that physics—appropriately interpreted in metaphysical terms—can tell us how the world *could be* (van Fraassen 1991, pp. 480–482). The challenge for the realist is to accommodate this underdetermination without saying "goodbye to metaphysics" entirely[38] (this phrase forming the title of the final section of van Fraassen's book on quantum mechanics).

It is this challenge that has been taken up by Ladyman's "ontic" or "metaphysical" form of structural realism. The central idea of this view is that we can avoid the epistemologically unattractive aspect of Worrall's account by doing away with the "hidden" objects and presenting an ontology that is entirely structural.[39] On this view, it is not merely that all that we *know*, but all that there *is*, is structure. The gap between the ontology and our epistemic reach is then closed, and furthermore, the challenge set by the metaphysical underdetermination can be met by reconceptualizing the notion of "object" entirely in structural terms (Ladyman 1998; French forthcoming a).[40]

There is still further work to be done, of course, particularly when it comes to explicating the metaphysics and ontology of structure (see French and Ladyman 2002), but at this point we shall consider a further problem for realism and indicate how it might be met through the partial structures approach.

The Underdetermination of Theories by Evidence

We noted toward the beginning of this chapter how the history of science can be drawn upon to generate a serious problem for the realist, in the form of the pessimistic meta-induction. The claim that theories are underdetermined by the evidence represents another piece of weaponry in the antirealist's armory. As we shall see, our framework offers, if not a resolution of this issue, then at least a way of approaching it that demonstrates what is at stake (see Bueno and French forthcoming).

A clear expression of the underdetermination thesis goes as follows: "Suppose that two theories T_1 and T_2 are *empirically equivalent*, in the sense that they make the same observational predictions. Then [according to the UTE thesis] no body of observational evidence will be able to decide conclusively between T_1 and T_2" (Papineau 1996, p. 7). The consequences for standard realism are clear: If UTE is correct then the realist is unable to determine which of T_1 and T_2 is more worthy of belief, where "belief" here is understood as "belief that... is true" and truth is explicated in the usual correspondence sense. The lesson drawn by the constructive empiricist, for example, is that we must abandon the view that theories should be believed to be true in favor of the epistemically more secure position of accepting them as empirically adequate only.

A possible response to UTE is to appeal to factors going beyond the set of observational consequences of the theory to background knowledge or "indirect evidence" of some form. The presence or absence of relationships between this "indirect evidence" and the theories concerned is then taken to provide differential support for one theory over another and thus break the underdetermination. Obviously, the precise form of this indirect evidence is crucial, as is the nature of the relationship it is taken to have with the underdetermined theories.

Thus Laudan and Leplin argue that two empirically equivalent hypotheses H_1 and H_2 may be evidentially distinguished because one of them, H_1, say, receives empirical support from evidence that is not a consequence of H_1 and H_2 does not. The idea is as follows:

> Theoretical hypotheses H_1 and H_2 are empirically equivalent but conceptually distinct. H_1, but not H_2, *is derivable* from a more general theory T, which also entails another hypothesis H. An empirical consequence e of H is obtained. e supports H and thereby T. Thus e provides indirect evidential warrant for H_1, of which it is not a consequence, without affecting the credentials of H_2. Thus one of two empirically equivalent hypotheses or theories can be evidentially supported to the exclusion of the other by being incorporated into an independently supported, more general theory that does not support the other, although it does predict all the empirical consequences of the other. The view that assimilates evidence to consequences cannot, on pain of incoherence, accept the intuitive, uncontroversial principle that evidential support flows "downward" across the entailment relation. (Laudan and Leplin 1991, p. 67; our emphasis)

However, in order for e to provide evidential warrant for H_1, Laudan and Leplin have to claim that T provides such warrant for H_1, since it is through T that the support given by e to H "flows" to H_1. Now, the claim that T provides evidential warrant for H_1 is question-begging, since by hypothesis H_1 is derived from T. In fact, as Mill noticed and Miller has spelled out more recently (see Miller 1994, pp. 51–74), deductively valid arguments are circular. What makes them valid is the fact that the information contained in the conclusion H_1 is already contained in the premise T. Thus, to claim that the premise supplies evidence for the conclusion is, in the limit, to claim that the premise supplies evidence for itself. Thus, although Laudan and Leplin do specify the relationship between hypotheses and background theories, their reliance on entailment leads them into circularity.

Likewise, Ellis also wants to argue that factors beyond the strict empirical consequences of a theory may provide differential support for it, where these factors are themselves ultimately empirical. As we shall now see, by insisting that the relevant relationships can be captured through the notion of embedding, he offers a broader approach that opens the door to partial structures.

The example he gives is of two theories of chemical combination, both of which explain and are supported by the observation that one volume of hydrogen combines with one of chlorine to form two volumes of hydrogen chloride. Here H_1 is Avogadro's theory, which postulates the existence of atoms and molecules and "paramechanical" processes of chemical combination. H_2 postulates the existence of certain "gas numbers" characteristic of elemental and compound gases, together with certain laws relating the gas numbers to combining volumes. The theories are equivalent in terms of their empirically testable consequences but are clearly not equivalent ontologically.

Despite their agreement in terms of consequences, Ellis claims that H_1 and H_2 are not and never were empirically equivalent: "For they offered quite different prospects for development and, as new theories were developed, new facts became relevant to the acceptance of one of them (Avogadro's theory)" (Ellis 1996, p. 183). By new facts Ellis means, for example, observations concerning electrolysis that are not empirical consequences of Avogadro's theory (H_1) but, he claims, certainly supported it, since they could be "readily explained" on the atomic theory that H_1 drew upon. So far this seems very similar to the approach of Laudan and Leplin. Ellis, however, writes of H_1 being *embedded* in T and having "well-established links" with—that is, *having hypotheses in common with*—a wide range of other physical and chemical theories. It is absurd, he claims, to state that Avogadro's theory is empirically equivalent to the "gas number" theory because

> the evidence in favour of Avogadro's theory and against the gas number theory, is now overwhelming. This is so because the original theory of Avogadro has become *embedded* in a very general and powerful theory of chemical combination and has well-established links with (in the sense that it has hypotheses in common with) a wide range of other physical and chemical theories. Avogadro's theory has gained support from these other theories because it has become an integral part of them. The evidence in favour of Avogadro's theory cannot be identified with the confirmation that its observational consequences have received. (Ellis 1996, pp. 183–184; the italics are ours)

Ellis's central point is that theories are open to further development that may lead to the establishment of links with other theories such that the former "pick up" the empirical support accruing to the latter. Since we cannot know in advance what these further developments might be, we cannot say in advance what evidence might distinguish two apparently empirically equivalent theories. And the underlying reason, according to Ellis, is the "openness of the field of evidence," where the "field of evidence" of a theory is the set of possible empirical discoveries relevant to its truth or falsity. This set cannot be identified with the set of empirical consequences of a theory since the latter is a subset of the set of logical consequences of a theory, whereas the former is not. In other words, there can be evidence for a theory not entailed by a theory; the field is *open* in a way that cannot be captured by focusing on the theory's logical consequences.

This seems a plausible response to UTE, and, of course, this "openness" can be appropriately represented by means of partial structures. We recall that the openness of theories is exploited for further developments as the relations effectively move out of the "unknown" R_3 into the R_1 and R_2 components of a partial relation R. Shifting attention to the empirical evidence and its representation, the empirical substructures of a theory may themselves be "open" in interesting ways. Hence they, too, can be represented in terms of partial substructures, allowing for the possibility of further developments (Bueno 1997). In particular, the growth in our information about a certain phenomenon can be represented by a hierarchy of partial structures in which, at each level, previously undefined partial relations (whose elements thus belonged to the R_3) come to be defined, taking their elements as belonging either to R_1 or R_2. In this manner, two theories that effectively "share" an empirical (partial) substructure at a particular level of the hierarchy may come to be evidentially distinguished at higher levels.

At the level of "theoretical" models, new relations may come to be established with other models whose empirical support can then be transmitted, as it were, across these relations. With the models represented by partial structures, the appropriate representation of these relations would be in terms of partial isomorphisms. The idea is that there exists a correspondence between the families of partial relations in the model-theoretic representations of Avogadro's theory and the atomic theory, respectively. Ellis's conception of the "field of evidence" of a theory as the set of *possible* empirical discoveries relevant to its truth or falsity has, as its formal counterpart, the R_3 components in a partial structure $\mathcal{A}$, whose members are not determined beforehand but are relevant, indeed crucial, for the quasi truth of sentences interpreted in $\mathcal{A}$. The openness of the field of evidence is then partially mapped via the partial isomorphisms holding between the structures under consideration.

There is the further issue of the manner in which evidence relates to theory on this view. Here we can draw upon our probabilistic theory of confirmation articulated in terms of quasi truth as outlined in chapter 7. What we have, then, is the following account: The underdetermination between Avogadro's theory and the "gas number" theory is broken through the relationship—represented by partial isomorphisms—between the former and empirically supported atomic theory. This introduces a differentiation in the degree of quasi truth of the two theories such that, from this perspective, they can no longer be regarded as equivalent. It is important to note that the empiricist should have no difficulty with this account, since we are not claiming that the theories in question are true but only quasi-true at best. And quasi truth admits an empiricist reading (see Bueno 1997). Indeed, it turns out that the framework of partial structures and quasi truth can be pressed into service on the side of the constructive empiricist.

Structural Empiricism

As we saw in chapter 4, the central concept of constructive empiricism is that of "empirical adequacy." If one were to accept the development of the model-theoretic approach represented by partial structures, how might this notion be represented within such a framework? The question is interesting precisely because the answer would indicate how constructive empiricism could acquire the advantages of the partial structures framework in terms of capturing the openness of scientific practice and so forth.

One possibility would be to identify "empirical adequacy" with quasi truth by taking a theory's empirical substructures (interpreted as partial) to be embeddable in the theoretical structures (also regarded as partial, of course) and, further, regarding the theoretical elements of A as just convenient fictions that have pragmatic value only (da Costa and French 1990). Since a sentence S being quasi-true means that everything happens in the relevant domain *as if* it were true, S could be said to save the appearances and, it has been claimed, in precisely the sense that the constructive empiricist wants (da Costa and French 1990).

Unfortunately, as Bueno has pointed out, this latter claim is actually false (Bueno 1997). First of all, the constructive empiricist is not a fictionalist with regard to theoretical entities but insists on remaining *agnostic* concerning their existence. Second, it is the *models of data* or appearances, as van Fraassen calls them, that are embeddable within the theoretical structures and that should be appropriately partialized, as it were.

An alternative and improved account can be given by drawing on the notion of *partial isomorphism* to characterize the relationship between the appearances and the empirical substructures (Bueno 1997). In these terms, the former are regarded as partially embeddable within the theoretical structures (as indicated before in our discussion of Ellis's response to the UTE argument), and if this is the case, then the theory can be regarded as partially or quasi-empirically adequate (Bueno 1997, p. 596). As Bueno says, this offers a plausible elaboration of van Fraassen's suggestion that an appropriate notion of approximation needs to be introduced into the consideration of the relationship between theory and evidence.

However, the following criticism must be faced: The cardinality of the domain might not be the same as that of the empirical substructures, in which case a partial isomorphism cannot in fact be defined as holding between them. In response, Bueno has proposed the introduction of a hierarchy of partial models of phenomena of increasing cardinality, which represents the gain in information about the phenomena at each level of the hierarchy (1997, pp. 600–603). He writes, "In this context, we then say that T is *empirically adequate* if it is *pragmatically true* in the (partial) empirical substructure E according to the structure A, where A is the last level of the hierarchy of models of phenomena (being thus a total structure)" (pp. 602–603). Thus the identification of empirically adequacy and pragmatic or quasi truth is achieved by regarding a theory as empirically adequate if it is pragmatically or quasi-true with regard to the last level of the hierarchy of models of the phenomena.

This then yields a characterization of a form of "degree" of empirical adequacy that captures the way, in practice, one theory might be regarded as more empirically adequate than another in that it takes account of more empirical information. The central idea is that the higher the relevant level in the hierarchy with regard to which the theory is quasi-empirically adequate, the more (quasi-) empirically adequate a theory will be (Bueno 1999a). This characterization in turn can be used to model empirical progress in science: Such progress corresponds to an increase in the degree of empirical adequacy, and this, in turn, is represented by an increase in the level of the partial models of the phenomena. Furthermore, theoretical progress can also be accommodated: Where the realist insists that a new theory is accepted over an old one because the former entails new phenomena and these are regarded as true, the empiricist takes such phenomena to be representable in the hierarchy, with the appropriate partial isomorphism holding between them and the empirical substructures of the new theory. Thus new kinds of phenomena can also be accommodated by the empiricist, and the success of science—which, we recall, the realist insisted must be seen as miraculous in the absence of realism—comes to be understood in terms of the structural interconnections between empirical substructures and the hierarchy of partial models of the phenomena (Bueno 1999a).[41]

Of course, such an understanding is far too "thin" for the realist's tastes, and as we have seen, explicating this notion of "success" in non-question-begging terms is a tricky business. However, as we have said, our aim in this chapter is not to defend either realism or antirealism but to indicate how both sides in the debate can benefit from the partial structures approach. We believe that what this debate needs is just the kind of detailed articulation that structural realism and empiricism offer and that these positions represent, if not the best, then at least two of the more interesting hopes for the future.

9

Extensions

As we have already indicated, there are obvious ways in which the partial structures approach can be further extended and applied. Consider, for example, the role of models in explanation, which has received increased attention following the well-known criticisms of the "covering law" approach.[1] We can take as the basis of a model-theoretic account of explanation Hughes's assertion that "We explain some feature *X* of the world by displaying a model *M* of part of the world and demonstrating that there is a feature *Y* of the model that corresponds to *X*, and is not explicit in the definition of *M*" (Hughes 1993, p. 132). Although Hughes himself does not go into details as to how we might characterize the relationships between these features, we can easily see how the mechanism of partial isomorphisms can be put to work here. Earlier, Thompson claimed that the semantic approach can better accommodate "complex explanatory frameworks" than can the Received View (Thompson 1989). Crucially, he argues, it can capture the way in which theories can be conjointly employed in explanation, which he sees as a prominent feature of biological explanations in particular. Again, such features of explanation in science can be straightforwardly represented by using the resources of our approach.

Furthermore, we might consider the question of how it is that an explanation can convey understanding, even when it involves a highly abstract and comparatively unfamiliar model. Hughes's answer is that, first, the resources of the model are typically tucked away, as it were, so that we have to apply it to successive phenomena in order to discover them, and in so doing we increase both our understanding and our familiarity with the model, and second, such models carry with them a "greater volume of associations" than more concrete forms, allowing them to be applied across different domains (1993, p. 147). Again, both these features can be accommodated within our approach: We have emphasized, repeatedly, the openness of partial structures and how they can represent the heuristic fruitfulness of models and theories, and, in particular, the framework of partial isomorphisms holding "horizontally" and "vertically" allows us to represent the way in which various model-based resources can be transported between scientific domains. Finally, Hughes's discussion of the evaluation of explanations, in which he reminds us that models should be regarded as neither true nor false but only as more or less "adequate," can be straightforwardly brought within the framework of quasi truth.[2]

If models are allowed to play a role in explanations in this way, then it is hard to resist the suggestion that they also *represent* (Hughes 1993). The manner in which this suggestion has been followed up has proved contentious, however. Van Fraassen, for example, has argued that an appropriate account of resemblance can be given in terms of isomorphism (1994). This has been strongly criticized by Suárez, who argues that just as isomorphism cannot capture representation in art, so it is inappropriate in the scientific context as well (1999b). Similarly, Hughes himself has drawn on Goodman's rejection of resemblance in favor of denotation, when it comes to representation in art, and, rather confusingly perhaps, favors denotation rather than isomorphism or similarity while also maintaining the model-theoretic view of theories (Hughes 1997).

Casting the debate in terms of partial structures offers the possibility of shedding new light on some of the issues involved and, perhaps, answering some of the objections. Thus Suárez's criticism hinges on certain examples from the history of art that purport to show that not only is isomorphism not *sufficient* for representation—since the latter is taken to involve intention—but also it cannot even be *necessary* (Suárez 1999b). Picasso's *Guernica*, for example, is ambiguous as to what it represents, and the canvas cannot be said to enter into an isomorphic relationship, at least not straightforwardly. An even more extreme example is that of Mondrian, whose work, it is claimed, cannot be said to represent or be isomorphic to *anything*.

But this is too quick. Mondrian himself insisted that what his paintings represented was "pure structure," and this can, of course, be accommodated by the model-theoretic view. Furthermore, if we broaden our notion of representation to include partial isomorphisms, then we can understand *Guernica* as standing in a plurality of such relationships with various objects. Indeed, Budd has shown how an account of representation based on isomorphism can, in fact, account for a wide range of examples, from Holbein's *The Ambassadors*, through various examples of abstract art to Escher's drawings (Budd 1993). Indeed, we would say that what Budd is actually offering is an account based on partial isomorphisms suitably adapted for the artistic domain (see French forthcoming b).

The example of Escher's work is particularly significant because Budd claims his approach is able to accommodate "inconsistent" or contradictory representations. Perhaps the most obvious example on the scientific side is one we've already encountered—namely, Bohr's model of the atom. Hughes argues that this can be accommodated within his "denotational" approach as a "local" model that denotes, and hence represents, a clearly specified type of system (1997, p. 330). Quantum mechanics, on the other hand, can be considered a "global" theory and deals with a heterogeneous collection of physical systems. In this case, we can still say that there is representation insofar as each individual system in this collection can be denoted by a particular model defined in terms of the theory (Hughes 1997). Curiously, however, the inconsistency at the heart of the Bohr model is not discussed at all, and the issue arises as to how Hughes's Goodmanian approach could be extended to accommodate inconsistent denotation. Adapting the account we have given in chapter 5, this aspect can also be captured within a general partial structures approach to representation (French forthcoming b).

Introducing partial isomorphisms into Budd's account suggests a way of extending it to cover other forms of artistic representation. Indeed, this kind of approach might be extended to cover all kinds of representation in general. Consider, for example, the way we represent "everyday" situations. It is on the basis of such representations that we construct our decision making. However, there is now a considerable body of psychological research that suggests that these constructions often violate certain consistency or, more generally, coherence constraints. This work has generated enormous controversy, with reactions ranging from the insistence that what these results reveal is nothing less than widespread irrationality (Stich 1984), to the outright denial that empirical findings of this kind can *ever* show that people are systematically irrational (Cohen 1981). Naturalistically inclined philosophers of science have been quick to seize upon this work, claiming that it shows that the "traditional" model of scientific decision making (and, in particular, the Bayesian variants thereof) must be modified or abandoned altogether (see, for example, Giere 1988). Both laypeople and scientists, it is argued, are simply not the ideal reasoners the traditional view would have them be. Typically, this nonideal character is cashed out in terms of nonepistemic, culturally determined factors that are taken to fill the niche left by empirical success.

Elsewhere we have suggested an alternative way of understanding the nonideal character of "everyday" decision making in terms of the partial nature of the models employed (da Costa and French 1993b). The intuitive judgments that are involved are made under uncertainty, where this uncertainty results from a lack of knowledge of the "true" state of the world. We have argued that this lack of knowledge can be suitably incorporated into the partial structures account, where these structures precisely express our ignorance and fallibility in dealing with "the world." It is through them and the notion of partial truth that we can

1. Accommodate the existence of models that have demonstrable pragmatic worth but that are also formally inconsistent or incoherent, and
2. Accommodate the improvement in and development of these models.

In this manner, the practice of both scientists and "natural" reasoners is unified under a single conceptual framework that stresses the partial, incomplete, and developing nature of our representations of reality. Thus, this work can be seen as an attempt to ground the commonality of reasoning processes within the partial structures approach, emphasizing, in particular, such epistemic factors as empirical success.

We believe we can extend this work even further, to include the "non-scientific" beliefs of other cultures. The "classic" example would be the inconsistent witchcraft beliefs of Azande, which have acquired the status of a cultural datum against which successive philosophers can test their particular views of cross-cultural understanding. They have supplied much of the grist to the mill of those relativists who insist that not only can the *content* of beliefs vary from culture to culture but also so can the underlying formal framework.[3] Hence such beliefs offer the most acute challenge to the rationalist, who is now caught between the devil of logical relativism and the deep blue sea of an imperialistic attribution of irrationality. Both alternatives have tended to trade on an incorrect analogy between the scientific and magico-religious frameworks: the "imperialistic" response views witchcraft in terms of some kind of primitive, personified causes, whereas the relativist has tended to draw a comparison with Kuhn, suggesting that Azande operates within a particular social paradigm (see Barnes 1968).

This has been rejected by Horton and others, who argue that Zande witchcraft beliefs are akin to the theoretical beliefs of Western science (Horton 1967 and 1982). We agree with the sentiment but not with Horton's overly positivistic characterization of science. Instead, we have argued that Zande beliefs can also be accommodated within the partial structures framework (da Costa and French 1995): That is, these beliefs should be regarded, from the intrinsic perspective, as *quasi-true* and the representations of witchcraft understood as semipropositional in Sperber's sense, and therefore, from the extrinsic perspective, these representations are themselves appropriately modeled in terms of *partial structures*. Again, the inconsistency is handled by appropriately modifying our discussion of the Bohr model, and the "openness" of Zande beliefs, the way in which they are only loosely "held together," can be straightforwardly accommodated.

Given the crucial role of this example in debates concerning the appropriate characterization of rationality, it should come as no surprise that we believe the latter should be refashioned. The standard view typically focuses on criteria of *consistency* and *truth*, which are held to be "universal," with belief systems such as those of Azande subject to particular "contextual" criteria (see, for example, Lukes 1970). These *may* violate the laws of logic and be analyzed in terms of the coherence or pragmatist theories of truth (Lukes 1970, p. 211).

What we are suggesting is the elevation of the latter to the level of a universal criterion. The advocacy of universal criteria of truth in terms of correspondence offers too much of a tempting hostage to fortune, as the relativist can point to all the well-known difficulties in maintaining such a view in the face of both scientific revolutions and radically different cultural beliefs. Furthermore, the distinction between universal and contextual criteria—with radically different cultural beliefs regarded as "true" in the latter rather than former sense—runs into problems with the relativists' invocation of scientific beliefs as on a par with cultural ones: For if radically different cultural beliefs are to be taken as true in the contextually based sense only, then so should radically different scientific beliefs. If the rationalist is unwilling to extend parity to such scientific beliefs, then she should explain why these are taken to be privileged. Adopting the realist stance that scientific beliefs are so privileged in that they correspond to reality not only begs the question but also lays her open to the sorts of criticisms we have covered in chapter 8.

We agree with the relativist to this extent: that both scientific and cultural beliefs should be treated as on a par, in the sense that both can be brought under a unitary framework. However, it should also be clear that we agree with the rationalist that universal criteria of truth apply, with the difference being that the appropriate understanding of truth is pragmatic. With both Zande and scientific beliefs taken as "representational" and quasi-true, inconsistency can then be accommodated. Correspondence truth and consistency still come into the picture, of course, but at the level of *factual* beliefs. It is here that the rationalists' "bridgehead" between cultures (or scientific paradigms) is located. What this amounts to is a broadly pragmatist account of rationality, but this is perhaps an extension too far, and we shall leave the details to future works.

NOTES

Chapter 1

1. In a recent paper Fernández Moreno addresses the twin issues of (a) whether Tarski *intended* his theory to be a correspondence theory of truth and (b) whether his theory is "*in fact*" a correspondence theory (2001; see also the other papers in this issue of *Synthese*). With regard to the first issue, if the notion of "correspondence" is understood in a rather weak sense, in terms of the holding of certain relations between linguistic expressions and extralinguistic entities (as Tarski himself appears to have understood it), and not in the stronger sense according to which these relations must be understood as isomorphisms, then, on the basis of a nicely detailed historical analysis, Fernández Moreno concludes that Tarski's intentions were to capture this sense of correspondence (2001, p. 130). With regard to the second issue, he defends Tarski's explanation of truth in terms of satisfaction against accusations that it is either uninteresting or trivial and concludes that it fits his characterization of what a correspondence theory should be (2001, pp. 139–140).

2. As Hodges and Etchemendy have emphasized, the notion of truth in a structure cannot in fact be found in Tarski's 1935 paper and is not presented explicitly until 1957 (Hodges 1986; Etchemendy 1988). This has to do partly with the lack of agreement at the time of what a structure is and partly to do with the treatment of the nonlogical constants of a model-theoretic language (which Hodges views as indexical expressions).

3. According to Suppes, "Tarski himself was skeptical about the application of systematic semantical methods in the analysis of natural language" (Suppes 1988, p. 87), but his student Montague analyzed the semantics of natural language in model-theoretic terms (Suppes 1988 for references).

4. We shall return to this idea of initially different beliefs stabilizing around the same conclusion in our discussion of induction in chapter 7.

5. There is an interesting alteration here, from "cannot" in (1907a) to "do not" in (1932).

6. Thus in a note to his earlier paper (James 1885), which foreshadowed much but by no means all of the account of truth subsequently presented (1907a),

James identifies as a defect in this earlier account "the possibly undue prominence given to resembling, which altho a fundamental function in knowing truly, is so often dispensed with" (1932, p. 41)

7. In his own resituating of James within the concerns of contemporary philosophy, Putnam notes, "To say that truth is 'correspondence to reality' is not false but *empty*, as long as nothing is said about what the 'correspondence' is" (Putnam 1995, p. 10)

8. Thus Smith writes: "Despite significant differences between the views of Peirce and James on this topic, they came together in holding an agreement or correspondence theory of truth. In view of this fact, one cannot say for either of them that there is a special 'pragmatist' conception of truth. If any novel element has been introduced, it is the insistence that 'correspondence' not be conceived as a static or timeless relation having no connection with the process of inquiry and the tracing of consequences. If a label is needed, one could not do better than to call their theory of truth a theory of 'dynamic correspondence'" (Smith 1978, p. 77) Of course, from our perspective, this is just as distorting, in its own way.

9. For the best developed account of the nature and role of "domains of knowledge" in science, for example, see Shapere 1977.

10. It is worth noting that this account is essentially based on classical logic, since every $\mathcal{A}$-normal model is classical; it can, however, be extended to nonclassical logics. For example, instead of invoking classical partial structures, we can consider paraconsistent partial structures and develop a theory of paraconsistent partial (or pragmatic) truth. Moreover, the logic appropriate to handle partial truth, as delineated here, is so-called Jaskowski's logic (see Jaskowski 1969; da Costa 1989; da Costa, Bueno, and French 1998a); analogously, one can show that the logic which is convenient for handling nonclassical partial truth, in particular, paraconsistent partial truth, is a kind of generalized Jaskowski's logic. We shall recall these points when we come to consider inconsistencies in science in chapter 5.

Chapter 2

1. However, as we noted in the previous chapter, Hodges has pointed out that Tarski did not, in fact, define this notion of truth in a structure in his classic paper.

2. An erudite presentation of the origins of this approach is given in Suppe 1989, pp. 3–20.

3. Thus Redhead, for example, has remarked on the heuristic fruitfulness of mathematical reformulations of a theory (Redhead 1975).

4. Here we are thinking in particular of the well-known example of the failure to eliminate by definition the notion of force from axiomatized classical mechanics, which did so much to disabuse us of the notion that force was an anthropomorphic concept that could be eliminated in this manner (see, for example, da Costa 1987a, p. 4).

5. We believe that the denigration of axiomatization that one sometimes comes across in the philosophical literature is a consequence of a failure to adequately comprehend the nature of the axiomatic method. In particular, it is not the case that axiomatization *must* proceed within first-order logic; Carnap, for example, employed higher order logic (type theory). Having said that, in some cases first-order axiomatization is relevant, as in the example of Tarski's axiomatization

of elementary algebra and geometry. In general, it is clear that appropriate formalization is essential for the proof of various metamathematical results. Without formalization, there would be no Gödel's Theorems, for example!

6. It may be useful here to draw on a distinction made by Hellman in his defense of a "modal-structuralist" approach to set theory (Hellman 1988, p. 449). Hellman suggests that axiomatization has two main purposes: The first is to codify proofs, and, given the success in formally capturing first-order logical consequence, axioms are typically first-order and presented "as if being categorically" asserted (1988, p. 449). On the other hand, there is the alternative aim of expressing the kinds of structures that are the object of investigation, and in fulfillment of this aim first-order languages may have to give way to second-order ones. For Hellman these structures are mathematical, of course, but for us they are generally "scientific."

7. Of course, these developments all presuppose classical logic. Another kind of generalization would be obtained by using nonclassical logic as the underlying framework, as we have already indicated.

8. Cf. Laymon 1987, p. 197; and Mac Lane: "All mathematics can indeed be built up within set theory, but the description of many mathematical objects as structures is much more illuminating than some explicit set-theoretic description." (1996, p. 174).

9. There are certain similarities—but also crucial formal differences—between the partial structures approach and the work of Wojcicki. For example, he considers four set-theoretical "pictures" of the relationships between scientific laws and the empirical phenomena within their "scope." The second of these pictures Wojcicki calls "classical-diachronic" and introduces functions that are only partially defined into his set-theoretic structure, giving what he calls a "partial physical system" (1974, p. 338; see also 1975). This allows him to define a sequence of such systems, which, he claims, captures the sense in which our measurements of certain parameters abstracted from the physical phenomena are subject to improvement over time. Clearly what Wojcicki is dealing with here is the issue of intertheory relations, which we shall take up in chapter 6. However, his approach might be seen as more restrictive than ours in focusing on diachronic relations between measurable quantities. Here we might echo Stachel's criticism of similar work by Dalla Chiara: "If we look at physics as the evolution of conceptual and instrumental structures under the impact of the interplay between experimentation with the conceptual and instrumental aspects of the structure, then we are led to focus attention more on these conceptual structures and their changes from theory to theory rather than on the consequent numerical agreements or disagreements between measurements of quantities deemed to correspond to the theories in question" (Stachel 1983, p. 95). It is precisely such a shift in attention that we claim our approach is able to accommodate.

Having said that, from our perspective it is interesting that Wojcicki notes, first of all, that what he calls "phenomena" may be viewed as "arbitrary trees of partial structures" and, second, that since each such tree defines a Kripke model for the relevant (formal) language, this approach allows for the possibility of languages based on nonclassical logics (a possibility we shall also consider, albeit somewhat tangentially) in subsequent chapters.

Wojcicki's third picture, which he calls the "operational picture," attempts to capture the "vertical" relations between theory and phenomena by means of formally defined "idealizations" (1974, pp. 339–342). This allows him to sketch an account of the "approximate truth" of a theory, where a theory is approximately true in an "operational structure" if and only if it is true in an idealization of this structure. This is clearly different from our approach, although in his final "picture" Wojcicki indicates—without going into details—how the "operational" and "diachronic" approaches may be combined through the introduction of "partial operational systems" (1974, p. 342).

10. We are grateful to Otavio Bueno and James Ladyman for bringing our attention to, and pressing us on, this point.

11. Cf. Suppes again: "The separation of the purely set-theoretic characterization of the structure of models for a theory from the axioms proper is significant in defining certain important notions concerning models for a theory" (1957, p. 260)—one of these notions being that of isomorphism, of course.

12. Recall Hellman's distinction stated in n. 6. According to van Fraassen, the axioms of, say, quantum mechanics as presented in a standard physics textbook should not be regarded as an attempt at codification from the perspective of first-order logic but rather as an "expression", to use Hellman's term, of the structures described by the theory.

13. As he acknowledges, van Fraassen expressed a similar view himself earlier in his career (1970) but has since come to renounce it.

14. In response to van Fraassen's claim that the notion of one structure being "embedded" in another cannot be captured syntactically, Turney (1990) has developed a syntactic notion of "implantability," which, he argues, is analogous to the semantic concept of embeddability. This allows him to obtain syntactic equivalents of van Fraassen's all-important definitions of empirical adequacy and empirical equivalence. Nevertheless, Turney agrees that the semantic, or model-theoretic, account is an improvement over the syntactic account, on the grounds that it is better able to represent the relationship between scientific theories and observations.

15. The point was previously made by Suppes:

> Theories of more complicated structure like quantum mechanics, classical thermodynamics, or a modern quantitative version of learning theory, need to use not only general ideas of set theory but also many results concerning the real numbers. Formalization of such theories in first-order logic is utterly impractical. Theories of this sort are similar to the theories mainly studied in pure mathematics in their degree of complexity. In such contexts it is very much simpler to assert things about models of the theory rather than to talk directly and explicitly about the sentences of the theory, perhaps the main reason for this being that the notion of a sentence of the theory is not well defined when the theory is not given in standard formalization. (Suppes 1967, p. 58)

16. A *categorical* theory is one that has all its models isomorphic to each other. The only such first-order theories are the complete theories of finite structures.

17. Bueno has argued that empirical adequacy can be understood in terms of our notion of quasi or pragmatic truth (Bueno 1997). This then leads to a form of "structural empiricism," which we shall consider further in chapter 8.

18. And thus, according to Suppe, this approach should be rejected, for the sorts of reasons already noted. For further—rather intemperate—criticism of the structuralist program, see Truesdell (1984). Truesdell's chief complaint concerns the structuralists' misunderstanding and misuse of the history of science, but he also directs both his fire and ire against McKinsey and Suppes's attempted formalization of classical particle mechanics. His case is undermined somewhat by his bizarre allegation that the "Suppesians" "are terrified of more than 'an

afternoon's work' in formal logic" (1984, p. 568). Given McKinsey's invocation of Tarski in support of his work, one can only speculate as to the existence of cross-purposes with respect to what counts as "logic" and "axiomatization"! A more sympathetic review can be found in da Costa and Bueno (1998).

19. On the structuralist account, the distinction between believing a theory or, more crucially, part of a theory to be true and accepting it as empirically adequate is simply meaningless.

20. This is part and parcel of his general claim that the *language* of science must be taken literally.

21. Of course, we may not *know* whether it is true or false. This will be the case with theoretical, as opposed to empirical, content, and for van Fraassen the appropriate stance with regard to the entities referred to is that of the agnostic. With regard to such content, the best we can do, according to van Fraassen, is to consider how the world *could be* the way the theory says it is (1989, p. 193). This attitude to theoretical content is, of course, the basis of his modal interpretation of quantum mechanics, for example, where his antirealist approach is worked out with regard to a specific theory (van Fraassen 1991).

22. And a description of structure, in this sense, yields also a description of content (1989, p. 226).

23. As noted by French and Ladyman with regard to van Fraassen's account (1999). For a useful and illuminating attempt at such an elaboration, see Bueno 1999b.

24. Today classical mathematics is the study of three basic classes of models and their combinations, as Bourbaki recognized. These are the topological structures, the algebraic structures, and the ordered structures. Any other structure of present-day mathematics is some combination of these basic structures.

25. In more complex structures, there may be various principal sets and various auxiliary sets; both kinds of sets are base sets.

Chapter 3

1. It is interesting, given the motivation behind the development of the formalism of "partial" truth, that the idea of an "iconic" relation is due to Peirce (as noted by Black 1962, p. 221, fn. 1 and Suppe 1997, p. 97): "An Icon is a sign which refers to the Object that it denotes merely by virtue of characters of its own, and which it possesses, just the same, whether any such Object actually exists or not.... Anything whatever... is an Icon of anything, in so far as it is like that thing and used as a sign of it" (Peirce 1931–35, p. 247).

2. Accusations to the effect that meaning and use cannot be separated in this manner are deflected by referring to "well-defined technical contexts" when it comes to meaning, and different kinds of question about the models in mathematical and scientific domains when it comes to use (Suppes 1961, pp. 166–167). With regard to this aspect, Apostel's work can be seen as complementary to Suppes's in focusing on the *function* of models: "The concept of model will be useless if we cannot deduce from its function a determinate structure. The scientist in his comments uses the 'model' concept in all the ways we described and thus discards the simple and clear language of model-theory in formal semantics or syntax. Can we give to his use of the term an adequate rational reconstruction, or are we prevented from doing so? We are convinced that it is possible to derive structural features from the functional characterization" (Apostel 1961, pp. 3–4).

3. However, he then qualifies this by noting that since Suppes allows for a difference in *use* of models, their disagreement may be merely verbal.

4. Cf. Achinstein's distinction between "true," "adequate," and "distorted" models (1968, pp. 209–210).

5. Achinstein distinguishes between "theoretical" and "imaginary" models, where the latter are regarded as devoid of existential commitment (1968, pp. 212–221).

6. Cf. Black, who notes that "nowadays logicians use 'model' to stand for an 'interpretation' or 'realization' of a formal axiom system" (1962, p. 226, fn. 7).

7. And with regard to models in particular, crucial elements of this apparatus can be traced back to Carnap's classic *Introduction to Semantics* (Carnap 1942).

8. These are also referred to as "coordinative definitions," "semantic rules," "epistemic correlations," and so forth.

9. Carnap (1946) uses the phrases "constructing models" and "giving interpretations" synonymously.

10. Nagel can be taken as allied with the "modelists": "For the purposes of analysis it will be useful to distinguish three components in a theory: (1) an abstract calculus that is the logical skeleton of the explanatory system, and that "implicitly defines" the basic notions of the system; (2) a set of [correspondence] rules that in effect assign an empirical content to the abstract calculus by relating it to the concrete materials of observation and experimentation; and (3) an interpretation or model for the abstract calculus, which supplies some flesh for the skeletal structure in terms of more or less familiar conceptual or visualizable materials" (Nagel 1961, p. 90).

11. This is precisely the example used by Psillos in his development of an "analogical" approach to models (Psillos 1995 and 1999).

12. Furthermore, the sorts of examples that Psillos has in mind as giving access to the world—namely, models of the optical ether—were already dismissed by Braithwaite as "imaginary," as we have just noted.

13. "All swans are white" has the same logical form as "All electrons have integral charge," yet from the truth of either we can infer nothing about the plausibility of the other (unless, of course, we were to adopt some sort of structuralist view).

14. Cf. Bunge: "The representation of a concrete object is always partial and more or less conventional. The model object will miss certain traits of its referent, it is apt to include imaginary elements, and will recapture only approximately the relations among the aspects it does incorporate" (1973, p. 92).

15. Cf. Black: "A promising model is one with implications rich enough to suggest novel hypotheses and speculations in the primary field of investigation" (1962, p. 233).

16. Cf. Black, again (1962, pp. 232–233): "To make good use of a model, we usually need an intuitive grasp ... of its capacities, but so long as we can freely *draw inferences* from the model, its picturability is of no importance" (his emphasis). The point is that the properties given in the model must be better known than those in the intended field of application (p. 232).

17. Achinstein lumps Hesse in with the Received View, noting that in her earlier paper (Hesse 1953–54) she failed to distinguish analogies from theoretical models. Although this is corrected in (Hesse 1970) she nevertheless regards her models$_1$ as "total interpretations of a deductive system" (1970, p. 10). The attempt to situate Achinstein and Hesse together in a third "analogical" approach to models, distinct from both the Received View and the model-theoretic approach, is

therefore misplaced (Psillos 1995). Insofar as generic considerations as to the nature of analogies can be abstracted from Hesse's account, they can be accommodated within our version of the model-theoretic approach, as we emphasize.

18. In a later essay, Achinstein emphasizes this distinction in the context of a discussion of Maxwell's notion of "physical analogy" (1991). In particular, he argues that such an analogy based on identity of properties allows for the generation of explanatory hypotheses, as in the case of the kinetic theory of gases, whereas it does not when based on formal identities between the relevant laws, as in the example of the analogy between the electromagnetic field and an imaginary fluid.

19. It is interesting that in his own attempt to accommodate the different functions of models, Apostel introduces the notion of a "partial model," where this is understood as "satisfying" only certain "central" features of a theory, such as well-confirmed laws, and not others (1961, p. 6). In particular, Apostel hints that by introducing such a partial relationship between model and theory we might accommodate the developmental function of models.

20. According to Suppe, iconic models were incorporated into the Semantic Approach from early on (1977, pp. 221–230). From this perspective, the Received View's combination of theorems plus correspondence rules represents a *formulation* of a theory, rather than the theory itself. If this formulation is empirically true, it will describe both the actual world and the theory; hence the formulation will have two mathematical models. And if it is accepted that if the theory is correct then it can be said to stand in an iconic relation to the world, then the relationship between these two models is also ("probably") iconic. Thus, "Campbell, Nagel and Hesse are correct in insisting that the theory cannot just be the partially interpreted formalism, and must include a model—the model being the theory—which, if the theory is true, stands in an iconic relation to its phenomena" (Suppe 1977, p. 101).

21. Here again the issue arises as to whether this model is regarded as a model of the kinetic *theory* of gases or of the gas itself, as a *system*. The characterization of Hesse's position as primarily concerned with models of *systems* rather than theories (Psillos 1995) is perhaps not so accurate, given the fundamental importance of the neutral analogy in rendering the *theory* predictive. That point aside, and more important, we can perhaps say the following: All systems of interest to science, whether observable or unobservable, are represented (or, better, representable), and whether we call this representation a "theory" or "model" is less significant than the nature of the representation itself. The "behavior" of gas atoms, for example, gets represented within the kinetic theory of gases, which, from a formal standpoint, incorporates, or has embedded within it, an empirical substructure isomorphic to some "phenomenological" model of gases. The "billiard ball" model may be regarded as incorporating the same, or a relevantly similar, empirical substructure that effectively "forces" partial similarity at the higher theoretical level.

22. See, for example, Cifuentes 1992, pp. 24–25. Again, Apostel introduces several forms of "approximate" isomorphism (1961, pp. 17–21). In particular, in his "Approximation IV" he introduces the notion of an "indetermination domain" for relations, where this is explicated in terms of "couples for which it is not decidable if they belong to a relation or not" (1961, p. 18; Apostel also introduces indetermination domains for the sets, where the former are understood as elements for which it is not decidable whether they belong to the set or not. This bears a strong resemblance to the more recent notion of a "qua-set" (see Dalla Chiara and Toraldo di Francia 1993). Although Apostel notes the "interesting complexities" that result from these various kinds of approximate isomorphisms,

he fails to develop their possible applications within the philosophy of science (although he does use them to develop notions of "approximate satisfaction" [1961, pp. 24–26]). Czarnocka also introduces the notion of a "quasi-isomorphism" in a similar context, and although the meaning of the term is not entirely clear, its contrast with isomorphism, where the latter ensures "cognitive comfort" and the "lack of sceptical doubts" suggests a close resemblance with the idea expressed here (Czarnocka 1995, p. 31). Needless to say, the idea of "partial isomorphism" as holding between partial structures was developed in complete ignorance of these similar notions.

23. For further discussion of Giere's use of this term, see French and Ladyman 1999.

24. Cf. Bunge's set-theoretic account of the "several degrees of formal analogy" (Bunge 1973, p. 115). "Isomorphism is . . . perfect formal analogy" (p. 115), and what we have called "partial isomorphism" Bunge terms "plain (or some-some) formal analogy" (p. 115).

25. One could go further and give a *quantitative* account. However, the dangers inherent in such accounts are revealed in Gorham's attempt to represent structural similarity as a distance relation in an appropriate n-dimensional space (Gorham 1996). Features of objects are then represented as coordinates in this space, and Gorham imposes a hierarchy of relative importance that proceeds from the "ontological" (as most important), to the "nomological" and then the "mathematical" (as least important). Now, one could obviously propose an alternative hierarchy or question whether the mathematical, say, can be separated quite so cleanly from the "nomological." However, the real problems arise when Gorham applies this metric to the issue of verisimilitude and concludes that the overall "dissimilarity" between the Thomson and Bohr models of the atom is 37%, whereas that between the Rutherford and Bohr models is only 26%; hence the Rutherford model is closer to the truth than Thomson's. First of all, it is explicitly assumed that the Bohr model is true, and this raises the issue as to how an apparently inconsistent model might be regarded as true, an issue we address in chapter 5. Second, the calculation of these particular percentages is crucially dependent on an assignment of a distance of 0.5 with regard to the nomological features of the Rutherford and Bohr models, since although they agree on the law governing the motions of electrons within orbits, the Bohr model also incorporates the quantum condition. But of course, one could insist that the latter precisely represents the "quantum jump" (excuse the pun) between classical models like Thomson's and Rutherford's and quantum models like those of Bohr and Sommerfeld. Assigning a greater distance on account of the quantum condition would yield a correspondingly greater dissimilarity between the Rutherford and Bohr models, putting the Thomson and Rutherford models closer together. Of course, one would still have the Rutherford model closer to Bohr than the Thomson model is, but, first of all, one could construct cases in which the situation was not quite so transparent, and one model might come ahead of another, depending on one assignment of the relative importance given to certain features, and not on another; second, one can reasonably conclude that the Rutherford model is more similar to Bohr's than Thomson's, without having to go through a detailed numerical calculation.

26. "Formal analogy is best analyzed when the objects in question are sets, because set theory does not specify the nature of the elements of a set. This does not mean that formal analogy is clear only when it concerns mathematical objects: since any concrete object can be modeled as (or represented by) a set

endowed with a certain structure, the search for formal analogies among concrete objects can be referred to their set representatives. If two such representatives prove analogous, their respective referents will be pronounced formally analogous" (Bunge 1973, p. 115).

27. A recent and interesting example of such cross-fertilization is given by renormalization group theory, originally developed within condensed matter physics and imported, with enormous success, into quantum field theory; see Cao 1997.

28. Thus Kroes's (1989) attempt to describe structural analogies between physical systems is too narrow, since he regards two such systems as analogous if the relevant mathematical equations have the same form. He goes on to represent this kind of structural analogy set-theoretically, in terms of isomorphisms holding between the relevant structures, and thus our approach can be seen as far more general.

29. "Among the glaring mistakes that can easily be avoided by taking our framework into account, the following may be mentioned: (a) mistaking analogy for the far stronger (transitive) relation of equivalence; (b) speaking of isomorphism (or perfect formal analogy) when a much weaker relation of similarity (usually plain analogy) is involved" (Bunge 1973, p. 130).

30. "By a physical analogy I mean that partial similarity between the laws of one science and those of another which makes each of them illustrate the other" (Maxwell 1965, p. 156), quoted in Achinstein 1991, where the "partial" aspect remains uncommented on. Cf. also Suppes, yet again: "Gibbs was not concerned to appeal directly to physical reality or to establish true physical theories but rather to construct models or theories having partial analogies to the complicated empirical facts of thermodynamics and other branches of physics" (1961, p. 168).

31. A conception that was obtained from Rutherford's work, of course, and which Bohr already had in mind from his earlier work on charged particle absorption.

32. Bohr begins his classic three-part paper "On the Constitution of Atoms and Molecules" by considering thc electron orbits as elliptical, but on the next page he introduces the assumption that they are circular (Bohr 1981, pp. 159–233). He felt justified in this procedure on the basis of a consideration of the general characteristics of the motion produced in Coulomb-type fields (Hoyer 1981, p. 115). These considerations are sketched on two sheets contained in the Neils Bohr Archive and prominently involve Kepler's third law (Bohr 1981, pp. 250–251). It is also interesting to note the simple sketches of elliptical motion given on the first of these sheets, which attest to the minimal picturability encouraged by the analogy.

33. Cf. Bunge's account of hydraulic analogues of electric circuits in terms of relational structures that capture the common structure of the two concrete systems (1973, p. 124).

34. For an assessment of recent developments, see Suppe (2000), who claims that "focusing on inherent advantages of syntactic vs. semantic approaches is fruitless. The Devil is in the details. And the details are that semantic approaches have shown impressive philosophical advances with little distraction by artifacts of infelicitous formalizations" (p. S113–S114).

35. Again, Apostel has noted—albeit briefly—that models may function as "intermediaries" between either theories and experience or theories and "reality" (1961, p. 11; see also Hutten 1953–54, p. 289). In particular, if the model is an intermediary between two formalisms in the sense of being a model of both, then one could use the number and kind of the relations involved to develop a form of "degree of intermediacy."

36. Cartwright et al. (1995) give the example of London and London's 1935 model of superconductivity, which, they claim, was constructed at the phenomenological level, independent of theory in methods and aims. Unfortunately, however, their case study is interestingly inapt: Fritz London himself—one of the more philosophically reflective of the new generation of quantum physicists (see Gavroglu 1995)—insisted that the model was *not* phenomenological! Furthermore, it can be shown that it was not developed in a manner that was "independent of theory in methods and aims" and, further still, that its development can be straightforwardly accommodated within the partial structures version of the semantic approach (French and Ladyman 1997).

37. Morrison (1999) presents an example from the history of fluid dynamics: Prandtl's attempt to model fluid flow around objects via the introduction of a thin boundary layer, which, it is claimed, was based directly on the relevant "phenomenology." It turns out, however, that the sense of "independence" from theory in this case is extremely attenuated and amounts to no more than the point that experimental considerations played an important part in the development of the model. Once again this case study can be easily handled by the semantic approach (da Costa and French 2000).

38. Fisher (forthcoming) refers to such models as "developmental"; for a discussion of the impact of the relevant history on certain versions of the semantic approach (but not the partial structures form, unfortunately!), see Portides (2000).

39. Fermi-Dirac statistics correspond to the antisymmetric representation of the permutation group, whereas Bose-Einstein statistics correspond to the symmetric representation. Other forms of statistics correspond to higher order representations and played an important role in the history of the development of quantum chromodynamics, as we shall see in chapter 6.

40. Frisch has also pointed out that no (classical) models can be used to represent an inconsistent theory, such as Maxwellian electrodynamics (Frisch, in press); that the issue of inconsistency in science poses no special problems for the Semantic Approach is argued in chapter 5.

41. Such models include both "scale" and analogue models, according to Black, "representational" models on Achinstein's list and "remnant" models in Downes's critique. We might also add the biologists' use of mice as "models" in immunology research, for example; cf. Herfel (1995, pp. 75–77), who unfortunately confuses such "concrete" models with the modeling of nonlinear behavior on "concrete" computers.

42. Another interesting example is Smeaton's model waterwheel, which was not a prototype or a scaled-down version of some particular waterwheel but, according to Baird, "serves the more abstract purpose of allowing [Smeaton] to better understand how waterwheels in general extract power from water in motion" (Baird 1995, p. 446). For Baird this material model carries "representational meaning" and he insists on the "central point" that "in quite specific ways it functioned just as theory functions" (p. 446). We shall discuss how this "representational meaning" acts as a form of nonpropositional knowledge in the next chapter.

43. Of course, material models allow for a more immediate appreciation of the spatial relationships between the elements concerned (hence the interest in "virtual" modeling on computers) and are central to what Nersessian calls "imagistic" reasoning (1993). The commonality here, in the sense of that which links diagrams, material models, and iconic models, is the representation of structure.

44. On March 6, 1996, the *Guardian* newspaper reported the surprising discovery that the common pipistrelle bat is in fact two distinct species. As the newspaper reported, "Classification experts face a delay in naming the new bat because the original 'type' specimen of *Pipistrellus pipistrellus* went missing from the Natural History Museum some years ago. A reserve specimen has lost its head in storage, and scientists are not sure which type it is."

45. We could press a form of type-token distinction here in the sense that both wire-and-tinplate models of DNA and their holographic counterparts might be regarded as tokens of the same type, with the latter delineated in terms of its role as a *representation*. Shifting to an appropriately Peircean perspective, it is the *sign* which is important, not the material of which it is composed. Our point is therefore a semiotic one.

46. Conversely, and strictly speaking, a model in the Tarskian sense is not an object, or at least not a physical one. This would seem to be obvious, but it relates to an important issue: whether the relationship between models and "the world" can be straightforwardly understood in terms of satisfaction. Thus it has been argued that, strictly speaking, the relation of satisfaction does not and *cannot* hold between a sentence and a physical object; rather, the sentence is satisfied in a *structure* that represents "the world," or an aspect thereof (see Downes 1992). However, Tarski himself—at least in his early work—took sentences to be true *in the world*. This might be understood as a "concrete" form of truth, corresponding to which is the "abstract" form, with regard to which set theory is deployed. The "abstract" form can then be seen as a mathematization of the "concrete." From this latter perspective, one can say, loosely perhaps, that the relation of satisfaction can hold between a sentence and "the world" but only if the latter is seen as an appropriate set-theoretical structure (constructed in a set theory with Urelemente). Of course, the relation of representation that is presumed to hold between the structure and the world is itself deeply problematic and remains so in the present context, where broadly "theoretical" structures are related via partial isomorphisms to equally broadly "phenomenological" ones and the latter "represent" the phenomena. We shall return to this issue in subsequent chapters, but our point here is that scale models may be represented by such structures just as, say, physical systems like the solar system are.

47. Suppe understands van Fraassen as *identifying* a theory with one of its phase space models and argues that while this is legitimate when quantitative theories are concerned, "the Semantic Conception of Theories must distinguish the theory from its phase space models if its account of theories is to apply to qualitative theories as well" (1989, p. 117, fn. 39). Perhaps this should be more properly regarded as a deficiency of the phase space approach, rather than an indication that the notions of theory and model must be kept separate. Certainly it would seem that the partial structures approach can handle quantitative and qualitative theories on an equal footing without imposing such a separation.

48. Ellis, for example, has argued that we can have no ontic commitments to a model's constituents (1990), whereas Campbell argues, correctly, that this can't be quite right, otherwise the model would be of no use (1994). He then goes on to suggest that what we must be "referential" about are the *relationships* expressed in the model, which obviously suggests a form of "structural" realism, a position we shall explore in chapter 8.

49. Together with the underdetermination that follows from such embedding, this consequent attitude toward theories and models renders the model-theoretic

approach a natural framework for the constructive empiricist. It does not follow, of course, that one cannot adopt this approach and still be a realist.

50. We do not use the word "similarly" here inadvisedly. Van Fraassen has noted how aspects of Cartwright's (early) views mesh quite nicely with his constructive empiricism (van Fraassen 1985a, 1991). However, in her more recent work she has adopted a kind of "patchwork realism" that asserts that "Theory is true only in models" (1995, p. 358), understood as shorthand for "The theory is true only in those situations that resemble its models" (p. 358). Confusingly, because of her emphasis on the role of models in scientific practice, she has also been lumped together with Giere (Psillos 1995; Teller 1995, p. 4, fn. 1), who, of course, is a realist (albeit nonstandard) advocating the model-theoretic approach, which Cartwright explicitly rejects (1995, p. 357; see also 1996 and 1999).

51. For detailed criticisms of Cartwright's position, see French and Ladyman (1997) and Bueno (1997). In particular, any distinction between "theoretical" and "phenomenological" or, not necessarily equivalently, "abstract" and "concrete" models is undermined by the supposedly concrete examples she appeals to: springs, pendula, and, crucially, dipole oscillators (Cartwright 1995, p. 359).

52. Cf. Suppes, again: "The abstract set-theoretical model of a theory will have among its parts a basic set which will consist of the objects ordinarily thought to constitute the physical model" (1961, p. 167).

53. We are grateful to Hartmann for drawing our attention to this work; see Hartmann (1996).

54. It is worth recalling that Bunge represents a "model object" as a relational structure exactly as we have done and furthermore notes that the predicates defined on the set of elements will be satisfied "only approximately, if at all, by the ultimate referents" (1973, p. 95). It is partly because of this latter point, of course, that Bunge feels that "theoretical models," which have such model objects as their "primitive base," cannot be taken in the model-theoretic sense. However, for Bunge a theoretical model is also a *hypothetico-deductive system* concerning the model object (1973, pp. 97 and 100) and furthermore can be regarded as nothing more than a set of statements relating certain variables (p. 102).

55. "In the language physicists use, it is already hard to distinguish between the meaning of 'model' and 'theory.' So, for example, the 'Standard Model' of the strong-electroweak interactions is certainly considered a fundamental theory" (Hartmann 1995, p. 52). Another example comes from Bohr: In the first part of "On the Constitution of Atoms and Molecules," he refers to the Rutherford "atom-model" (1981, pp. 161 and 162), whereas in part 3 it is called "Rutherford's theory of the structure of atoms" (p. 215).

56. Cf. Psillos, who acknowledges the point that "theories, like models, may also be seen as approximate, simplified and restricted descriptions/explanations of the phenomena" (1995, p. 115) and accepts that "there is some leeway here" (p. 115). The difference between a theory and a model is then regarded as one of a degree of belief, that in the former being higher than that in the latter (cf. Hartmann 1995, also). This suggests the sort of "continuum" view with which we are obviously sympathetic; however, our sympathy evaporates when it comes to the contrary claim that "the difference between a model and a theory may be seen as an intentional one, i.e., a difference relating to our having different epistemic attitudes towards them" (Psillos 1995, p. 115, fn. 6) On this point, see also Suppes: "In old and established branches of physics which correspond well with

the empirical phenomena they attempt to explain, there is only a slight tendency ever to use the word 'model.' ... On the other hand, in those branches of physics which gives as yet an inadequate account of detailed physical phenomena with which they are concerned there is a much more frequent use of the word 'model'" (Suppes 1961, pp. 168–169), and, again, "in such uses of the word 'model' it is to be emphasized that there is a constant interplay between the model as a physical or non-linguistic object and the model as a theory" (p. 169; cf. Achinstein 1968, p. 215).

57. Sticking our necks out even farther, perhaps we might suggest that all examples of scientific "representation," broadly construed, might be treated in this way. Consider for example, diagrams of experimental apparatus: At a Leeds seminar on the difficulties in reproducing Joule's famous paddle-wheel experiment, it was pointed out that drawings culled from Joule's papers could not be relied upon for the purposes of accurate reconstruction. However, this is to presume that accurate reconstruction was the ultimate purpose of such diagrams. Rather, we would hazard, their function was to indicate the most pertinent relationships between the various components of the apparatus, these relationships revealing—at least in this case—the lines of force involved. Cf. also Bunge, again: "The representation may be pictorial, as in the case of a drawing, or conceptual, as in the case of a mathematical formula. It may be figurative, like the ball-and-spoke model of a molecule, or semisymbolic, as in the case of the contour map of the same molecule; or finally symbolic like the hamiltonian operator for that same object" (1973, p. 92). Nersessian, in particular, has drawn attention to the role of images and "imagistic reasoning" in theory development and claims that, in the quantitative phase of development, the image "serves primarily to make certain structural relationships visualizable" (p. 24).

Chapter 4

1. According to the former interpretation, "x believes that p" is taken to be true if x, who is supposed to be candid and after due explanation where necessary, overtly professes to accept p as true: we ask x if she does believe p, and the issue is decided by x's answer. The latter construction considers that "x believes that p" obtains if either x gives her assent to p, in accordance with the preceding interpretation, or x is committed to giving assent to p in case p constitutes a logical consequence of propositions believed by x. For further details, see Rescher (1968).

2. This emphasis on commitment is not uncommon in the philosophy of science, at least; see, for example, van Fraassen 1980, p.12; Kaplan 1981; Laudan 1977, p. 108.

3. Cf. Suppes: "I think it is true to say that most philosophers find it easier to talk about theories than about models of theories. The reasons for this are several, but perhaps the most important two are the following: In the first place, philosophers' examples of theories are usually quite simple in character, and therefore are easy to discuss in a straightforward linguistic manner. In the second place, the introduction of models of a theory inevitably introduces a stronger mathematical element into the discussion" (1967, p. 57).

4. As Bogdan puts it, philosophers construe the nature of propositions with "anarchic gusto" (1986, p. 9).

5. We recall Grandy's point: That belief reports themselves are linguistic reflects only the fact that they are composed of sentences and not on the structure of the representation that is the object of the report.

6. That the "appearances" are themselves structures is an important point, as it undermines Downes's confused claim that the relationship between theories, viewed as mathematical structures, and "reality," regarded as nonmathematical, cannot be explicated in terms of isomorphism (Downes 1992). However, as van Fraassen makes clear, the isomorphism holds between one set of structures—the empirical substructures of the theory—and another, the appearances; the question of the relationship between the latter and reality itself (however that is construed) is not one that the model-theoretic approach can be expected to answer!

7. The following discussion is based on French and Ladyman 1999.

8. Van Fraassen gives the example of a model of the data from an Einstein-Podolsky-Rosen experiment in quantum mechanics.

9. "Often the characteristics which data must have to be useful as evidence can only be purchased at the cost of tolerating a great deal of complexity and idiosyncrasy in the causal processes which produce that data" (Bogen and Woodward 1988, p. 319).

10. Kaiser suggests this in an account which he claims runs parallel to that of Bogen and Woodward (Kaiser 1995).

11. Kaiser talks of "inference tickets" in this regard (1995).

12. According to Suárez, a model of the phenomena models a phenomenological "fact" (preprint, p. 5; see also Cartwright, Suárez, and Shomar 1995, p. 139).

13. In expressing the relevant causal relations, such models can be regarded as explanations of the phenomena (Suárez, preprint; Cartwright, Suárez, and Shomar 1995). This fits nicely within Cartwright's Aristotelian account, which stresses that phenomenological models can be regarded as true precisely insofar as they represent the causal properties and "capacities" of the phenomena. Of course, strictly speaking (that is, set-theoretically speaking), to talk of "true models" here is a façon de parler, since it is sentences which should be regarded as true, and not nonlinguistic entities. Cartwright herself, of course, has rejected such a set-theoretic perspective of physical models and thus has no compunction about talking of their truth.

14. What we have is a view of explanation in which phenomena are explained by having their properties represented within models that are then enmeshed within a complex network of other models, in turn represented structurally (we shall briefly return to this view in chapter 9).

15. Suggestions that this implies the "partial existence" of the entities concerned are puerile; claims about the phenomena are to be understood as claims about their descriptions, and what these describe are the *properties* of the relevant entities.

16. Bueno argues that it should be, and as we have already noted, van Fraassen himself has suggested a convergence between the two notions. Again, we wish to maintain our position of relative neutrality within this debate, although we shall return to the issue in chapter 8.

17. At times the argument for such a view seems to consist of nothing more than holding up a recent issue of the *Scientific American*, with all its diversity of topics covered, and insisting, "How could one ever believe that all these subjects, from the structure of a fly's eye to the ability to break bricks with a karate chop, could be explained in terms of theoretical physics?"

18. Noting that the notion of "incomplete" or "nonfundamental" theories has not received the attention it deserves in the philosophy of science, Shapere suggests that arguments to the effect that no theory could ever be complete do not tell the whole story: "Suppose we had a theory which over a period of six thousand years was successful in answering all questions that were posed to it. Despite the correctness of the above-mentioned logical point, I expect that people would begin to suppose that they had a 'complete' and perhaps even a 'fundamental' theory" (Shapere 1969, p. 147 fn. 38).

19. Fuller has criticized Hardwig on the grounds that he establishes only that if expertise is relevant, we should defer to the authority of experts but fails to deal with the crucial normative question of when expertise is relevant (Fuller, 1986). Consideration of the latter question leads to the suggestion that there may in fact be rational grounds for rejecting the authority of experts and thinking for oneself (p. 279). Pierson has recently attempted a reconciliation by distinguishing between "closed-system oriented" expertise and "layperson oriented" expertise: With regard to the former, the layperson should defer to the experts, but when it comes to the latter, where a determination must be made as to whether the benefit of following the experts' advice is cost-effective, she is rationally obliged to think for herself (Pierson 1995). Interestingly, Pierson notes that some of the confusion over this issue may arise because of the use of the notion of "testimony" to describe the cognitive relationships in question and that Quinton has situated this notion within a model of knowledge in which there are distinct atomic facts that speak for themselves (Quinton 1982)—that is, a model in which all our beliefs are "factual."

20. Cf. Schiffer, who proposes "harmlessly" to represent propositions as "ordered *n*-tuples of the things which determine them" (1986, pp. 86–87). Of course, for Schiffer "the real trouble with propositions" has to do less with their ontological status than with the claim that belief involves a relation to them; we shall not pursue this issue here.

21. For a response that argues that possible worlds need not be regarded as "complete" in a problematic sense, see Stalnaker (1986). Of course, possible worlds semantics is also given a model-theoretic formulation, although here the sets are sets of possible worlds rather than individuals.

22. Of course, Cartwright and Baird, although perhaps not Giere, would be unhappy with this attempt to give a unitary account of both "instrumental" and "theoretical" knowledge. However, their efforts to force a distinction between the two are predicated on an incorrect characterization of the latter as entirely propositional.

23. As is well known, Vaihinger also espoused a philosophy of "the as-if" and maintained that although all elements of knowledge except sensations were mere fictions, it was nevertheless useful to act *as if* they were true. The standard objection to this view is that in demanding that one acts as if a proposition were true, it presupposes a (nonfictionalist) notion of truth. Vaihinger's work is often linked with the pragmatists', but we must be candid and admit that we find the latter to be much clearer and better developed. Having said that, Fine has recently attempted to resurrect Vaihinger's approach and render it more attractive in the modern context (Fine 1998).

24. It is here, of course, that, with respect to the attribution of belief states, the behaviorist gains a toehold.

25. Cf. van Fraassen: "In practice, acceptance will always be partial and more or less tentative" (1985a, p. 281).

Chapter 5

1. "Incongruent suppositions" can be thought of as those aspects of scientific and everyday belief structures that, when rendered less vague and more precise, can be represented as inconsistencies.

2. See, for example, Gotesky's clear and comprehensive review of "unavoidable" inconsistencies in moral, "theoretical," "factual," psychological, and "existential" situations (Gotesky 1968).

3. Of course, we recognize that it is debatable whether this example, like the others, is actually inconsistent (see, for example, Rescher 1973); indeed, focusing on the example of Bohr's theory of the atom, it is our intention in this chapter to contribute to the debate.

4. "It [the requirement of consistency] can be regarded as the first of the requirements to be satisfied by *every* theoretical system, be it empirical or non-empirical" (Popper 1972, p. 92). One might speculate here on the influence of Hilbert's attempt to ensure the safety and certitude of mathematical practice by developing appropriate consistency proofs, following the appearance of the set-theoretical paradoxes. If we note Popper's emphasis on the axiomatization of physical theories, in precisely and explicitly Hilbert's sense, with the first "fundamental requirement" being that the axioms must be free from contradiction, whether self- or mutual, then the speculation becomes less idle (Popper 1972, p. 71).

5. There is a common perception that there can be no models of sets of inconsistent sentences, if models are understood as structures in which the relevant sentences are *true*, in the Tarskian sense. However, a paraconsistent theory of models can in fact be constructed; see Alves 1984 and da Costa, Béziau, and Bueno 1997.

6. What the principle of contradiction is taken to be is an interesting issue. We can distinguish the following nonequivalent formulations: Of two contradictory propositions, one is false; $\sim(p \,\&\, \sim p)$; and a predicate cannot simultaneously belong and not belong to the same subject (da Costa and Marconi 1987, p. 2).

7. The claim is based on a reading of the *Second Analytics* (A11, 77a 10–22). Aristotle is better known for arguing that, ontologically, contradictions are impossible and, practically, prevent effective communication and social interaction (for a recent discussion of Aristotle's defense of the principle of noncontradiction on the grounds that the ontological consequences of its negation would be unacceptable, see Cassini 1990). As in the case of many recent works on this issue, the present chapter is an attempt to demonstrate that this latter claim is unfounded (see also Gotesky 1968).

8. Thus Moseley wrote, "I believe that when we really know what an atom is, as we must within a few years, your theory even if wrong in detail will deserve much of the credit" (cited in Bohr 1981, pp. 545–546). Ehrenfest was, characteristically perhaps, much more negative, writing that "Bohr's work . . . has driven me to despair. If this is the way to reach the goal I must give up doing physics" (Klein 1970, p. 278) and referring to the "completely monstrous Bohr model" (p. 286).

9. Einstein said of Bohr: "He utters his opinions like one perpetually groping and never like one who believes he is in possession of definite truth" (quoted on the frontispiece of Pais 1991).

10. Schotch considers the denomination "nonadjunctive" as unfortunate, since this approach represents a *generalization* of the classical rule of adjunction (Schotch 1993, p. 425 fn. 14).

11. Pais refers to Bohr's model of the hydrogen atom as a "Triumph over logic" (1991, p. 146).

12. The remark comes from a 1914 letter to Oseen, reporting Bohr's discussion with Debye.

13. It perhaps needs pointing out that the empirical support for Bohr's theory was extremely strong. Pais, for example, notes that the "capstone" of Bohr's work on the Balmer formula for the spectrum of atomic hydrogen is represented by his theoretical account of the so-called Pickering lines, which Bohr ascribed to singly ionized helium. When it was objected that Bohr's account failed to fit the data, Bohr responded that his theory was based on an approximation in which the nucleus was treated as infinitely heavy compared with the electron and that with more realistic masses included, "exact agreement" with the data was obtained. As Pais says, "Up to that time no one had ever produced anything like it in the realm of spectroscopy, agreement between theory and experiment to five significant figures" (1991, p. 149).

14. Von Laue is recorded as exclaiming, "This is all nonsense! Maxwell's equations are valid under all circumstances. An electron in a circular orbit must emit radiation" (Jammer 1966, p. 86). Einstein, on the other hand, said, at the very same occasion (which was a joint physics colloquium of the University and Institute of Technology in Zurich), "Very remarkable! There must be something behind it. I do not believe that the derivation of the absolute value of the Rydberg constant is purely fortuitous" (Jammer 1966, p. 86).

15. Sommerfeld, of course, greatly extended Bohr's theory, and, indeed, the theory came to be known as the "Bohr-Sommerfeld theory of atomic structure" (Mehra and Rechenberg 1982, Vol. 1, part 1; pp. 1:155–257). Interestingly, Sommerfeld's previous work on the dispersion of light in gases was criticized by Oseen in a letter to Bohr for being inconsistent, since it was based on Bohr's model together with Lorentz's theory of the electron (Bohr 1981, p. 337). In reply, Bohr also rejected Sommerfeld's dispersion theory but on experimental and theoretical grounds (Bohr 1981, p. 337).

16. Hettema seems to miss the point when writing that, "arguably, the Bohr theory is not inconsistent with the Maxwell-Lorentz theory, in that it uses a much lesser portion of the Maxwell-Lorentz theory than Lakatos contends" (1995, p. 323; cf. Bartelborth 1989). Hettema is correct is claiming that Lakatos overstates his case when he asserts that Bohr's program was "grafted" onto Maxwell-Lorentz electrodynamics, since, as Brown notes, Bohr separated the frequency of the emitted radiation from the frequency of the periodic motion of the electrons in the atom, with quantum mechanics applying to the former and classical mechanics to the latter (Hettema 1995, p. 314). Nevertheless, the posited stability of the ground state is in direct conflict with a fundamental component of classical electrodynamics—namely, that accelerating charges radiate.

17. In response to Oseen's demand for a "clear logic which always remembers what the basic assumptions are, and which never combines them with opposite assumptions for the sake of accidental progress" (Bohr 1981, pp. 340 and 571), Bohr wrote, "As you, I feel very much indeed that a logical critic [of quantum theory] is very much needed, but I think I take a more optimistic view on the Quantum Theory than you. In fact, I have written a paper in which I have made an attempt to show that the assumptions of the theory can be put on a logical consistent form, which covers all the different kinds [of] applications (by this I mean all applications which assume a discontinuous distribution of the possible (stationary) states; but I do not think that a theory of the Planck type can

be made logical consistent)" (Bohr 1981, 571). After receiving copies of Sommerfeld's work, Bohr abandoned the publication of the paper he mentions, although it was incorporated in his 1918 account of the quantum theory of line spectra (1981, pp. 342 and 572). In his letter to Sommerfeld, Bohr makes it clear that the "logically consistent form" of which he speaks was based on Ehrenfest's "adiabatic principle" (p. 604); for a consideration of the role of this principle within the nonadjunctive framework, see Brown (1993, pp. 401–402).

18. According to Havas, complementarity is to be understood as "real dialectical contradiction" in the Hegelian sense (1993, p. 33). Perhaps it is because of this sort of understanding that Priest chose Bohr's theory as an example of one of his "dialethias" but if regarding Bohr's atomic model as true is questionable, viewing the whole complementarity interpretation of quantum mechanics similarly is even more so.

19. Presumably Heisenberg is referring to "classical" logic here.

20. Of course, this is not to deny the existence of certain commonalities between the "old" and "new" quantum theories and the heuristic moves that were based on these common elements.

21. Here Norton gives examples of symmetry considerations functioning in this way at the metalevel; for further discussion of the fundamental heuristic role of symmetry and invariance principles, see Post 1971.

22. Thus, insofar as these different formulations of Newtonian cosmology are underdetermined by the evidence, Norton advocates avoidance of inconsistency as a methodological principle for breaking the underdetermination. Such a principle might be more appealing to the realist than the antirealist, of course.

23. Smith calls a set of statements in general a "proposal" and by "theory" means a proposal and its deductive closure (Smith 1988b, p. 429, fn. 1). This reluctance to assign the honorific "theory" to inconsistent representations crops up elsewhere in the philosophy of science. Thus Madan suggests that an objective of economics and the social sciences, but crucially not of science itself, might be to find "maximally inconsistent," factually adequate "arguments" (Madan 1983). Such an objective is rational, given the volatility of the economic environment, for example, where propositions are to be regarded as strictly false of the actual world but "truc enough" to warrant attention.

24. That the scientific community at the time was aware of the inconsistencies involved is revealed in the following passage from a beginning graduate textbook: "It is difficult, in fact it is not too much to say that at present it appears impossible, to reconcile the divergent claims of the photoelectric and the interference groups of phenomena. The energy of the radiation behaves as though it possessed at the same time the opposite properties of extension and localization. At present there seems no obvious escape from the conclusion that the ordinary formulation of the geometrical propagation involves a logical contradiction, and it may be that it is impossible consistently to describe the spacial distribution of radiation in terms of three dimensional geometry" (Richardson 1916, pp. 507–508; Richardson was well known for his work on the thermal emission of electrons, eventually winning the Nobel Prize, and we are grateful to Knudsen's study for drawing our attention to this passage (Knudsen, forthcoming).

25. Smith notes that at the 1911 Solvay conference Poincaré expressed his concern over the use of contradictory principles in such derivations, pointing out the logical anarchy that (classically) results (Smith 1988b, p. 431 fn. 10; Poincaré's remarks are recorded in Langevin and de Broglie 1912, p. 451). Gotesky records that Poincaré also said that contradiction is the "prime stimulus" for

scientific research (Gotesky 1968, p. 490); these two aspects of Poincaré's thought are not necessarily inconsistent, of course!

26. An explicit belief is one which involves an explicit mental representation whose content is the content of the belief; "implicit" beliefs are not held explicitly but are "easily inferrable" from explicit beliefs (Harman 1986, p. 13).

27. Harman is not the only one to suggest doing away with closure: Kyburg, long an opponent of "conjunctivitis" (the view that to believe or accept each of a number of beliefs is to believe or accept their conjunction) has also concluded, "It is not the strict inconsistency of the rational corpus that leads to trouble—it is the imposition of deductive closure" (Kyburg 1987, p. 147). Smith's notion of self-consistent subsets was also prefigured by Kyburg with his notion of "strands" (1987, p. 146 and 1974).

28. Harman also rejects the "Logical Inconsistency Principle" because of the ubiquity of inconsistency and concludes that in the event that we find ourselves holding inconsistent beliefs "it is rational simply to retain the contradictory beliefs, trying not to exploit the inconsistency" (1986, p. 17). How we are to avoid exploiting the inconsistency is not explained, however, and Smith's account does at least have the virtue of indicating how this might be achieved in specific cases.

29. "Acceptance in the sense I am describing (like belief) is closed under combination of contexts" (Brown 1990, p. 287). It is this which, according to Brown, distinguishes a "realist" form of acceptance from the "instrumental" variety, where there is no such closure.

30. Mele, for example, regards exclusivity as a special feature of belief that distinguishes it from, say, desire (Mele 1987, p. 4).

31. In some philosophical accounts, these states are regarded as distinct (Martin 1986; Rorty 1972).

32. Cf. Rescher and Brandom, who claim that even if our "object" theory is inconsistent, our "metatheory" should be consistent (Rescher and Brandom, 1980, pp. 136–141). Also cf. Priest, who denies the distinction between "object" and "meta" theory but accepts that truth should be exclusive on the grounds that contradictions should not be multiplied beyond necessity (1987).

33. The result is self-deception; see da Costa and French 1990b.

34. Laymon captures the sense of idealization at play here through the use of "Scott domains" (1987).

35. Cf. Gotesky on "internal" and "external" contradictions (1968, p. 484).

36. Interestingly, Apostel notes that his "partial models" (see chapter 3, fn. 20) may be inconsistent with each other and suggests (although he does not develop this in detail) that this may accommodate the use of "locally inconsistent" models in science, such as Rutherford's model of the nucleus (1961, p. 6).

37. "If one finds oneself holding two mutually inconsistent ideas and reluctant to give up either, there is a natural fallback position which consists in giving one of them a semi-propositional form" (Sperber 1982, p. 170).

38. "[Semipropositional representations] play a role not only as temporary steps towards full propositionality but also as sources of suggestion in creative thinking" (Sperber 1982, p. 171).

39. Cf. Popper: "The initial stage, the act of conceiving or inventing a theory, seems to me neither to call for logical analysis nor to be susceptible of it" (Popper 1972, p. 31).

40. For a recent response to such claims, see Koertge 1993, which also drew our attention to the Reichenbach quotes here.

41. Thus from the philosophers' perspective the distinction can be regarded as a kind of "semipermeable" membrane (Koertge 1993).

42. That Einstein, for example, saw the heuristic fertility of inconsistency is nicely revealed by the following passage in the diaries of Count Kessler: "I talked for quite a while to Albert Einstein at a banker's jubilee banquet where we both felt rather out of place. In reply to my question what problem he was working on now, he said he was engaged in thinking. Giving thought to any scientific proposition almost invariably brought progress with it. For, without exception, every scientific proposition was wrong. That was due to human inadequacy of thought and inability to comprehend nature, so that every abstract formulation about it was always inconsistent somewhere. Therefore every time he checked a scientific proposition his previous acceptance of it broke down and led to a new, more precise formulation. This was again inconsistent in some respects and consequently resulted in fresh formulations, and so on indefinitely" (quote in Stachel 1983, p. 96).

43. Priest claims that the application of Gödel's first incompleteness theorem to the "naive" notion of proof constitutes a further argument for dialetheism because it shows that our naive proof procedures are inconsistent (1987, pp. 49–63). If we insist that naive proof theory be consistent, then we must accept a certain incompleteness, either expressive or proof-theoretic.

44. Thus Hacking writes of his "Argentine fantasy": "God did not write a Book of Nature of the sort that the old Europeans imagined. He wrote a Borgesian library, each book of which is as brief as possible, yet each book of which is inconsistent with every other. No book is redundant. For every book, there is some humanly accessible bit of Nature such that book, and no other, makes possible the comprehension, prediction and influencing of what is going on. Far from being untidy, this is New World Leibnizianism. Leibniz said that God chose a world which maximized the variety of phenomena while choosing the simplest laws. Exactly so: but the best way to maximize phenomena and have the simplest laws is to have the laws inconsistent with each other, each applying to this or that but none applying to all" (Hacking 1983, p. 219). This "fantasy" can be seen as one aspect of the "disunification" tendency in recent philosophy of science, noted in the previous chapter.

45. "The projected theory serves as a guide for the search for the material content of the proper replacement for the original inconsistent proposal" (Smith 1988b, p. 443). Not surprisingly, Norton suggests a similar strategy; see Norton 1993, p. 417.

46. For an indication of a nonadjunctive account of heuristics, see Brown 1993, p. 408.

47. Immediately before this passage, Post quotes Bradley: "The knowledge of privation like all other knowledge, in the end is positive. You cannot speak of the absent and lacking unless you assume some field and some presence elsewhere" (Bradley 1962, p. 457).

48. Cf. p. 444, where Smith rejects talk of the "outright acceptance" of conflicting representations.

49. In considering the claim (made by Donovan, Laudan, and Laudan 1988) that theory appraisal may be favorable even when scientists do not "fully" believe the theory, Hettema writes: "Since the 'old' quantum theory of the atom was a theory in full and very rapid development, it is difficult to assess whether theorists and experimentalists 'believed' the theory. I would like to distinguish between a pragmatic notion of belief and a principled notion of belief, 'pragmatic belief' designating the belief that a particular solution to a problem works in that it yields

the right solution to that problem, 'principled belief' designating that the believer thinks it is true (for some theory of truth). The answer to the question of 'belief' in the old quantum theory then reads YES for the pragmatic, NO for the principled version of belief" (Hettema 1995, p. 322).

50. As Stalnaker points out, "a person may accept a proposition for the moment without the expectation that he will continue to accept it for very long. If a person expects a particular one of his beliefs to be overturned, he has already begun to lose it, but an assumption he makes may be quite explicitly temporary, and he may presume that something is true even when expecting to be contradicted" (Stalnaker 1987, p. 80; by "beliefs" in this context, we, of course, understand factual beliefs, whereas the "assumptions" Stalnaker talks of are more properly regarded as semipropositional).

51. The death knell came in 1925, when Pauli noted that the application of the "old" quantum theory—in particular, the adiabatic principle—to the hydrogen atom in crossed electric and magnetic fields resulted in the conversion of allowed states into forbidden ones (Brown 1993, p. 401; Mehra and Rechenberg 1982, 1:485–509).

52. Hettema has argued that Lakatos's rational reconstruction, although broadly correct, is deficient in the details, such as the claim that Bohr's initial problem was the radiative instability of the Rutherford atom, whereas it was, in fact, its mechanical instability. Thus, "In his rational reconstruction Lakatos sacrifices a lot of historical and philosophical accuracy on the altar of liveliness" (1995, p. 323). Folse also emphasizes Bohr's concern with mechanical stability but maintains that his theory can still be regarded as inconsistent (1985).

53. "In the heuristic situation of empirical science, truth does not descend deductively from the (highly fallible) most general laws, but ascends inductively from the more modest lower reaches of the deductive structure of the theory held at any one time. Even that infra-structure has to be pruned, and the remaining core is never true for certain, but the modifications forced on us still allow us to refer to that part of the theory as having a likeness to truth" (Post 1971, p. 254).

Chapter 6

1. Although how decisive a break this was has been disputed in recent accounts that highlight the similarities between Carnap's and Kuhn's views of science; see, for example, Reisch 1991 and Irzik and Grünberg 1995.

2. Masterman identified the notion of an exemplar with that of a kind of analogy (1970), and Kuhn himself talked of a "shared model-theoretic element" in his and van Fraassen's work (Kuhn 1993). As he also noted, his philosophy of science was subsequently given a formalized treatment in the work of the structuralists.

3. At the 1991 MIT conference dedicated to his work, Kuhn himself stood up in the audience at one point and expressed how much he lamented this influence, which, he maintained (somewhat disingenuously perhaps, given the content of the *Structure of Scientific Revolutions* [1970]), was the result of a profound misunderstanding of his views.

4. In providing a conception of the relationship between theories and phenomena that, Kuhn believed, was closer to scientific practice, this notion also marks off a crucial distinction from Carnap's view of theories; see Irzik and Grünberg 1995, p. 298. Kuhn himself asserts that "the paradigm as shared

example is the central element of what I now take to be the most novel and least understood aspect of this book" (1993, p. 187). Given Kuhn's own subsequent identification of a model-theoretic element in his work (see n. 2), it is worth noting that a model-based understanding of the paradigm-as-exemplar can be found in Hutten, for example: "Whenever we encounter a new situation, we try to orient ourselves by comparing it to a familiar situation and so to evaluate it. From this attitude arises the *normative* function of the model, or the model as ideal and paradigm, e.g. the model house" (1953–54, p. 286).

5. The influence of Wittgenstein both directly on the sociology of scientific knowledge and indirectly through Kuhn is, of course, central; see, for example, Bloor 1983 and the recent debate between Bloor and Lynch (Bloor 1992; Lynch 1992a and 1992b).

6. The characterization of science in terms of a conceptual network is due to Hesse (1980).

7. "The production of new scientific knowledge entails seeing new situations as being relevantly like old ones" (Pickering 1992, p. 4).

8. Fuller suggests that there is much less emphasis within the sociology of science on the way in which this open-endedness is also the source of error and misunderstanding in science (Fuller 1986, chapter 5). For a very different perspective, see van Fraassen 1991, p. 11, where science is portrayed as an "open text."

9. The term "closure" is also to be found in Kuhn (1970).

10. Gremmen refers to this as the "bucket" view of practice, in the sense that everything is thrown into the bucket (Gremmen 1995). When it comes to this second respect in particular, Hacking's influence cannot be understated and is made explicit in Pickering 1992. It is through Hacking and this antireductionist emphasis on heterogeneity that the connection can be made with Cartwright, of course.

11. The shift over this attitude toward "the social" as one passes from the sociology of scientific knowledge to that of scientific practice, and the vehemence with which it is resisted is nicely documented in the collection presented in Pickering 1992.

12. Again the baleful influence of Wittgenstein can be felt here: "We must do away with all *explanation*, and description alone must take its place" (Wittgenstein 1958, §109; this passage is cited by Lynch 1992a, p. 246).

13. One of the criticisms of the "old" sociology of scientific knowledge is that this supposedly radical shift has taken "us" back toward the more conventional and conservative history and philosophy of science; see Collins and Yearley 1992.

14. "Scientific culture, then, itself appears as a wild kind of machine built from radically heterogeneous parts, a supercyborg, harnessing material and disciplinary agency in material and human performances, some of which lead out into the world of representation, of facts and theories" (Pickering 1995b, p. 446).

15. The "dialectic" of resistance and accommodation is what Pickering refers to as the "mangle of practice" (1995a).

16. Earlier he claims, "At stake here . . . is the idea that the conceptual stratum of scientific culture is itself multiple rather than monolithic, and that many disparate layers of conceptualizations, models, approximation techniques, and so on have typically to be linked together in bringing experiment into relation with the higher levels of theory" (Pickering 1995a, p. 99), and "data and theory have no necessary connection to one another; such connections as exist between them have to be made . . . [and] . . . conceptual practice aims at making associations (translations, alignments) between such diverse elements—here, data and theory" (Pickering 1995a, pp. 117–118).

17. Pickering uses as his case study Hamilton's development of quaternions (Pickering and Stephanides 1992; Pickering 1995a, chapter 4; see also his 1995b, which is a slightly rewritten version of the former); for alternative, albeit far less detailed, perspectives on this history, see Anderson and Joshi 1993 and O'Neill 1986; the former notes the role of analogy in particular.

18. Cf. Post, whose account of heuristics carries an explicit anti-Kuhnian and, in general, cumulativist message: "It is not the case that we rearrange the data within [the old theory] into a new pattern. The pattern is preserved. No insight into why the old theory worked is lost by the new. We have sliced off some very high levels and substituted some new ones, but the low (particularly classificatory) levels remain undisturbed within their confirmed ranges of validity" (Post 1971, p. 229).

19. In the case of Pickering's example of quaternions, the source of these resistances are primarily mathematical in character.

20. Flanagan feels that this talk of "passivity" is "unnecessarily provocative, taking the decentering of the subject too far, and pressing too hard to avoid appearing in any way aligned with the voluntarist self of the humanist" (Flanagan 1995, p. 470). Pickering's reply emphasizes the autonomy of "disciplinary agency" in that the effects of a discipline cannot be understood in terms of the standard voluntarist idea of "self-originating" human agency (Pickering 1995c, p. 478).

21. A critique of this approach from the perspective of the "old" sociology of scientific knowledge is given by Collins and Yearley 1992. An interesting point they make is how this idea of nonhuman agency arises from the methodology of the project, which they characterize as "observation informed by the perspective of the estranged visitor" (1992, p. 311). If you fail to understand what is going on in the laboratory around you and, indeed, if this lack of understanding is built in to your very methodology, then it is not surprising that you will come to regard the machines and apparatus of the laboratory—the "inscription devices"—as autonomous agents. There is, perhaps, a deeper point: Underlying this autonomy—this independence of the machinic—is the heterogeneity of practice. And of course, from the perspective of the "estranged visitor" who makes a point of not understanding what is going on (who in her quest for a metaphysically balanced position refuses even to countenance talk of what is "really" going on), practice will be "patchy," with different "agents" interacting in different ways and at different levels. The heterogeneity and hence the autonomy is a function of the methodology. Someone who adopts an "interpretative approach" (Collins and Yearley 1992) and who attempts to grasp the viewpoint(s) of the practitioners (and here Latour et al. might well accuse us of begging the question) will be much less likely to see heterogeneity and autonomy, as they understand the hidden purposes and interrelationships.

22. In *Constructing Quarks*, Pickering writes that "the most striking feature of the conceptual development of [high-energy physics] is that it proceeded through a process of modelling or analogy" (1984, p. 12) and in a footnote (fn. 12, p. 19) cites Hesse's work, among others.

23. "Every single element or stratum of scientific culture—material, conceptual, social—is revisable in practice: the material contours of machines and instruments and their performances; facts, theories and mathematical formalisms; the scale of social actors and their relations with one another; skills, disciplines, plans and intentions; norms, standards, rules; you name it. *All of these evolve open-endedly into the future*" (Pickering 1995d, p. 415; our emphasis).

24. Cf. 1995b, p. 441, where Pickering talks of human agency "traditionally conceived" and also (1995d), where he talks of human agency "in the traditional voluntarist sense . . . and in other ways besides" (p. 414).

25. Again, we find this view expressed in Kuhn 1970, p. 119, as noted by Nersessian in a paper that addresses many of the issues touched on here from the perspective of a "cognitive" approach (1993, p. 11).

26. Cf. "Scientific choice is in principle irreducible and open, but historically, options are foreclosed according to the opportunities perceived for future practice" (Pickering 1984, p. 405).

27. Cf. McMullin, who notes that the opportunities offered by theory are to be found "primarily in the model associated with the theory, a postulated explanatory structure whose elements are capable of further imaginative development" (1976, p. 427).

28. "There is no algorithm that determines the vectors of cultural extension, which is as much as to say that the goals of scientific practice emerge in the real time of practice" (Pickering 1995a, pp. 19–20) Nevertheless, as previously noted there is at least some "predisciplining," even if it is only partial, and here when the moves are "forced," "the 'disciplined' human is indeed an algorithmic actor or . . . a machinelike actor who performs predictably in a given situation" (1995c, p. 478). This is particularly so in the case of mathematical practice, which, as Pickering notes, is strikingly machinelike and algorithmic.

29. The difference may be denoted, although not, perhaps, clarified by Ryle's distinction between "Procrustean" and "canonical" rules (Ryle 1971, pp. 230–231).

30. The "facts" to be explained would include such examples as Poincaré's famous inspiration while stepping on a bus or, of more relevance to the present discussion, Hamilton's "moment of truth" on Dublin bridge.

31. "We thus see (1) that genius is a *talent* for producing that for which no definite rule can be given; it is not a mere aptitude for what can be learnt by a rule. Hence *originality* must be its first property. (2) But since it also can produce original nonsense, its products must be models, *i.e. exemplary*; and they consequently ought not to spring from imitation, but must serve as a standard or rule of judgement for others. (3) It cannot describe or indicate scientifically how it brings about its products, but it gives the rule just as nature does. Hence the author of a product for which he is indebted to his genius does not know himself how he has come by his Ideas; and he has not the power to devise the like at pleasure or in accordance with a plan, and to communicate it to others in precepts that will enable them to produce similar products" (Kant 1991, p. 168).

32. "Unusually subtle responsiveness and energetic innovativeness is often referred to as 'genius,' 'creativity,' or 'imagination,' but its operations can, I think, be analyzed" (Pickering 1995c, p. 479, fn. 1).

33. This episode actually takes up very little of *Constructing Quarks*, but it can be seen as a fulcrum capable of tipping over the entire edifice. Pickering himself notes that the "traditional reductionist and nonemergent" sciences such as molecular biology and quark theory cannot be reached by his irreductionist and historicizing "mangle," viewed hubristically as a "theory of everything," and that although the domain of these sciences is small—since quark theory, for example, is not only confined to the laboratories and accelerators of particle physics but deals with only the rarest phenomena to be observed in such environments—they cannot be ignored. His response, faced with this small but critical challenge to the

hegemony of the mangle, is to suggest the resurrection of the rival to hierarchical particle theories—namely, the S-matrix bootstrap—but understood as suitably "historized" in the sense that the attempt to discover "eternally stable" solutions to the relevant equations is abandoned (1995a, pp. 246–252). The extremity of this response is indicative of the size of the fly in the ointment!

34. It is worth noting that Greenberg gave two lectures on parastatistics at a NATO Advanced Study Institute summer school in 1962. The school had as its principal theme the application of group-theoretical methods to elementary particle physics and was attended by such notable physicists as Glashow, Salam, and Nambu.

35. In private correspondence, Pickering has acknowledged that, on this point, he relied upon Nambu, who described Greenberg's work as "very formal." Nevertheless, he maintains that the choice to pursue color was made in purely contingent terms regarding the routine techniques that could be applied.

36. Notably, the existence of the photon simply falls out of such a gauge invariant theory of electromagnetism.

37. "Gauge invariance... is used as a guide in constructing new field theories" (Post 1971, p. 227).

38. Here we stand accused—but happily so—of "ontologizing" constraints, in the sense of regarding them as structuring and explaining the "flow of practice" from without (Pickering 1995a, p. 66, fn. 37).

39. Giere incorporates such success into his "satisficing" account of scientific decision making, in which decision makers, instead of choosing the option with the maximum expected utility, select an outcome they deem "satisfactory" (1988, pp. 157–170).

40. This is a factor that also arises within so-called everyday or natural reasoning (see da Costa and French 1993b).

41. For discussions of the nature of symmetry arguments and their role in science, see van Fraassen 1989 and Castellani 1999; for more on the heuristic role of symmetry principles, in particular, see Redhead 1975.

42. "The superlaws of symmetry... are as liable to empirical revision as other laws of physics having a less obviously intuitive character" (Redhead 1975, p. 105). Perhaps these "superlaws" should also be regarded as "quasi-" or partially true. (This suggestion is made but not explored in French 2000.)

43. Cf. van Fraassen: "Symmetry arguments have that lovely air of the a priori, flattering what William James called the sentiment of rationality. And they are a priori and powerful; but they carry us forward from an initial position of empirical risk, to a final point with still exactly the same risk. The degree of empirical fallibility remains invariant" (1989, pp. 260–261).

44. "There is a series of restrictions... which render the activity of the scientist constructing new theories essentially different from that of a clueless rat trying one trapdoor after another (a remark probably also applying to any actual rat)." (Post 1971, p. 218).

45. One can also accommodate the openness of practice within other forms of the semantic approach. Thus Suppe notes that "theories undergoing active development frequently are not put forward as literally or even counterfactually true, but rather as 'promising,' 'worth pursuing,' 'approximately correct,' 'being on to something,' 'being in a certain respect importantly true, but in other respects profoundly wrong as well,' and so on. And in their developmental states, such theories typically are importantly incomplete in their accounts of phenomena" (Suppe 1989, p. 347). This is taken to support Suppe's "quasirealism," articulated

in the context of his "relational systems" version of the semantic approach, according to which theories purport to describe how the world would behave if it were "nice and clean."

46. There is more to be said here, of course. Thus van Fraassen regards models as "incomplete" in two different ways: (1) in the sense that completion can be achieved through empirical extension; (2) in the sense of possessing "meaning gaps," which are plugged by an "interpretation" of the model itself (1991, p. 9).

47. Nersessian makes a similar point and replaces "justification" and "discovery" with the unitary context of "development" (Nersessian 1993, p. 6).

48. As Suppe also notes, following Kuhn's and Feyerabend's claims regarding incommensurability, subsequent discussions of meaning change in science were tacitly concerned with individuating theories linguistically. It may be counted another plus for the distinction between a theory and its linguistic formulation highlighted by the semantic approach that such issues to do with meaning can be seen to be separate from the problem of theory individuation.

49. Sticking our collective necks out even further, perhaps all examples of scientific "representation" broadly construed, might be treated in this way. Consider, for example, diagrams of experimental apparatus: At a recent seminar on the difficulties in reproducing Joule's famous paddlewheel experiment, it was pointed out that drawings culled from Joule's papers could not be relied upon for the purposes of accurate reconstruction. However, this is to presume that accurate reconstruction was the ultimate purpose of such diagrams. Rather, we would hazard, their function was to indicate the most pertinent relationships between the various components of the apparatus, these relationships revealing—at least in this case—the lines of force involved. Nersessian, again, has drawn attention to the role of images and "imagistic reasoning" in theory development and claims that, in the quantitative phase of development, the image "serves primarily to make certain structural relationships visualizable" (1993, p. 24).

50. Cf. Nersessian (1993, p. 13), who credits Sellars with the insight that "analogical reasoning creates a bridge from existing to new conceptual frameworks through the mapping of relational structures from the old to the new." She goes on to claim that "most cognitive theories of analogy agree that the creative heart of analogical reasoning is a modelling process in which relational structures from existing modes of representation and problem solutions are abstracted from a source domain and are fitted to the constraints of the new problem domain" (1993, p. 20). Pertinent features of this process as revealed by psychological studies include the preservation of relational systems and the isomorphic mapping of relationships (1993).

51. Cf. Kuhn again: "The role of acquired similarity relations also shows clearly in the history of science. Scientists solve puzzles by modeling them on previous puzzle-solutions, often with only minimal recourse to symbolic generalizations" (1970, pp. 189–190).

52. The relevant group here is U(1) and, corresponding to the symmetry represented by this group, the property that is conserved is, of course, charge.

53. The status of renormalization as a coherent mathematical technique is problematic; for a comparatively sanguine appraisal, see Teller 1995, chapter 7. Reuger has also argued that it counts as an example of a general heuristic strategy of ensuring that reliable results are preserved through future changes in the theory by confining the unreliable parts (Rueger 1990).

54. For further discussion of gauge invariance and its role as a heuristic criterion, see Redhead 2002 and Earman 2002.

55. "There is no theoretical way of drawing a sharp distinction between a piece of pure mathematics and a piece of theoretical science. The set-theoretical definitions of the theory of mechanics, the theory of thermodynamics, and the theory of learning, to give three rather disparate examples, are on all fours with the definitions of the purely mathematical theories of groups, rings, fields, etc. From the philosophical standpoint, there is no sharp distinction between pure and applied mathematics, in spite of much talk to the contrary" (Suppes 1967, pp. 29–30). Thus there may be no distinction in principle between what Pickering refers to as "conceptual" practice and what we have called scientific practice.

56. Thus "resistance" can be equated with the absence of the appropriate isomorphism holding.

57. Nersessian, again: "According to Maxwell, a physical analogy provides both a set of mathematical relationships and an imagistic representation of the structure of those relationships drawn from a 'source' domain to be applied in analyzing a 'target' domain about which there is only partial knowledge" (1993, p. 17).

58. As Collier notes, the problem is not that of formally establishing isomorphisms between models but rather that of ruling out those that are uninteresting (1992, pp. 294–295). What we are about here, of course, and at least in part, is delineating the sorts of considerations that are typically drawn upon to rule out the uninteresting and unworkable isomorphisms in theory development.

59. Incommensurability is now understood in terms of incompatible primitive similarity sets (Kuhn 1970, pp. 198–207), and this marks the beginning of Kuhn's elaboration of this notion in terms of different systems of natural kinds.

60. This interpretation does not necessarily correspond to Kuhn's, although it is not at all clear exactly what the latter is; for a detailed analysis of Kuhn's evolving views on this issue see Hoyningen-Huene 1993, pp. 90ff.

Chapter 7

1. Of course, this form of inference is part of statistics, which we take to form part of, or an aspect of, scientific practice and hence, as a method, part of inductive logic.

2. This focus on subjective probability is not in any sense intended to diminish the role of objective probability in science; on the contrary, we believe it to be indispensable. The two kinds of probability are connected by what we shall call the Principle of the Coherent Use of Probabilities: If we accept that the objective probability of a certain state of affairs s is p, then our subjective probability in the quasi truth of the proposition that s occurs is also p.

3. One could make a similar remark with respect to logical probability. Of course, we accept the standard statistical methods in certain situations. As far as we are concerned, classical statistics itself constitutes an inductive method and should be regarded as constrained by our views of inductive logic.

4. Strictly speaking, the set of propositions closed by the connectives constitute a Boolean prealgebra; it is when we pass to the quotient, in the standard manner, that we get a Boolean algebra.

5. Cf. the arguments of Salmon 1966 and Shimony 1970, which may be adapted to our account.

6. Salmon (1966) and Shimony (1970) adopt a similar position with regard to scientific inference; cf. also Teller (1975).

7. Again, certain aspects of Salmon's and Shimony's accounts can be adopted here, as can elements of the work of Wrinch and Jeffreys (1921; Jeffreys 1961) and Keynes (1921). However, we want to make it absolutely clear that these results are not being appealed to in order to underpin some sort of justification of induction; rather they simply help us choose certain hypotheses instead of others, in particular situations. We take it that induction is "justified" only as a method that directs our degrees of belief, as it were: We act according to our beliefs. Whether we really arrive at the truth or quasi truth is another question, which we shall touch on toward the end of this chapter.

8. It might be insisted that one of the set of hypotheses be significantly more quasi-true than the alternatives. If we were considering truth simpliciter, instead of quasi truth, this would amount to a realist demand. It is an interesting question whether it remains plausible when it is quasi truth that we are concerned with, but one that takes us too far beyond the present discussion.

9. The proof of this inequality is roughly as follows (further details can be found in Redhead 1985): First of all, $p(h \vee \sim e) = p(h) + p(\sim e)$. If we conditionalize on *e*, p(h) must increase, according to $p(h/e) > p(h)$. However, $p(\sim e)$ then goes to zero, and this decrease more than compensates for the increase in p(h).

10. As Comte remarked, in physics there are no correct deductions.

11. There is a nice similarity here with Cherniak's point that consistency checks are typically not feasible for reasons that are likewise ultimately pragmatic but in a profound sense that edges close to the "in principle" (Cherniak 1984).

12. Edidin has argued that such bootstraps should be cut off, since they generate serious problems for the strategy and actually add nothing to it (1983).

13. For a detailed consideration (in Portuguese) of Newton's argument for the Law of Universal Gravitation within the framework of the bootstrapping approach, see French 1989b.

14. For further discussion of the relationship between the bootstrap and Bayesian approaches, see Horwich 1983, Rosenkrantz 1983, and Garber 1983.

15. This approach obviously meshes nicely with a "cumulative" account of scientific progress. We recall that the General Correspondence Principle—which broadly states that we never lose the best of what we had—is mirrored within our system by the Principle of the Permanent Nature of Pragmatic Truth: Once a theory has been shown to be pragmatically true within a certain domain, it remains pragmatically true, within that domain.

16. See also the work of Hintikka and Niiniluoto 1980 and Kuipers 1978 and 1980; our result is essentially a transformation of Horwich's "pseudoproof" into a "full" or "good" one; see Horwich 1982, especially chapter 3.

Chapter 8

1. The appeal to "maturity" is meant to block certain counterexamples from the history of science. The problem, of course, is to come up with a non-question-begging definition of "mature." "Intuitively," there would seem to be many examples of short-lived or otherwise problematic theories that one would not want to describe as approximately true. However, the Aristotelian "theory" of the crystalline spheres was not so short-lived, appeared to be comparatively unproblematic, and might well be regarded as "mature" on this intuitive understanding, yet, again, many realists would hesitate in describing it as even

approximately true. Hence, maturity has been linked to novel predictive success, in the sense that the theory predicts certain kinds of phenomena where these phenomena have not been "written into" the theory during its construction (see Worrall 1996, pp. 153–154). This understanding of maturity is problematic, since one could argue that Bohr's theory/model enjoyed such success, yet, intuitively at least, cannot be regarded as "mature." Here we simply wish to point out two connections: first, that it is related to a general motivation for realism known as the "no miracles argument" and second, that although it rules out certain counterexamples to the realist's theses, others remain. We shall explore aspects of these connections later.

2. Giere defends a form of "constructive realism" that contains features reminiscent of Hacking's "entity realism" (which we shall mention later), together with certain modal elements; we are grateful to Grant Fisher for pointing out some of the tensions within Giere's account.

3. Psillos offers a framework for a theory of "truthlikeness," according to which a theory is approximately true of a world, if it is strictly true of a world that approximates the actual world under certain conditions (1999, p. 277). Of course, this reduces approximate truth to a relation of approximation holding between worlds, and it is not clear how this might be spelled out.

4. This might be seen as already a big concession to realism, since if the successor theory entails that its predecessor was strictly incorrect, then for an antirealist there can be no reference to such a domain described in "pre-theoretical" terms (see van Fraassen 1977, p. 598). Bueno's form of "structural empiricism"—discussed later—offers a possible way for the antirealist to accommodate this point.

5. As in the case of the older instrumentalism, a distinction must be made between the observable and unobservable, a distinction that is notoriously difficult to draw as the sequence electron, atom, molecule, macromolecule, amoeba demonstrates (Horwich 1982, p. 136).

6. Horwich also includes the planet Pluto, but insofar as this counts as observable, according to some antirealists—such as the constructive empiricist—we shall not regard it as especially problematic.

7. This is a point that has been emphasized by Ladyman in a seminar to the HPS Research Workshop at the University of Leeds.

8. It is interesting that the conservation law also featured in the rival dynamical conception of heat and was thus seen as independent of the hypotheses concerning the nature of heat itself (Psillos 1999, p. 118). In other words, what was detached was a certain *structural* aspect common to both theories and thus independent of the metaphysical assumptions regarding the entity concerned.

9. One can understand the case study presented by Psillos as a study of pursuit, and as Psillos emphasizes, the attitude of the scientists themselves was one of "cautious and differentiated belief" (1999, p. 119).

10. Of course, both may decrease or vanish altogether, in which case the degree of belief in the existence of the ether itself might be said to vanish as well.

11. For a detailed analysis of this case that leads to this conclusion, see Norton and Bain (forthcoming).

12. In this sense we are following Peirce, who argued that the means for justifying our beliefs are the same as those for establishing the truth of our theories. We shall return to the pragmatist approach later.

13. When we are talking about degrees of belief in this context, it is important to bear in mind that normally we are not talking about strictly subjective

probabilities but considering only systems of degrees that are constituted by sets of *qualitative* degrees of belief, ordered by a reflexive, symmetrical, and transitive relation, in general with least and greatest degrees. Sometimes, the ordering is not even linear. This point is important for our entire project, since the scientist (and the philosopher, etc.) usually has only qualitative degrees of belief in her hypotheses, theories, and the like. In very particular situations only do we really need subjective probabilities.

14. Interestingly or bizarrely—depending on your point of view—Cartwright has recently offered a form of transcendental argument for her "patchwork" realism (Cartwright 1996 and 1999).

15. Psillos tries to avoid such accusations by noting that some critics of realism will accept abductive inferences of the form embodied in IBE (1999, p. 89), but the point is, if they're honest antirealists, they shouldn't!

16. Psillos is concerned with a different problem. He notes that "even if instrumentalism were shown to be the best explanation of the instrumental success of science, it could not be more empirically adequate" (1999, p. 91). Hence, by the instrumentalist's own lights, instrumentalism could be no more belief worthy than realism. There is, then, at the metalevel itself a form of underdetermination (see later for discussion of this at the good old object level). Psillos sees this as a problem for the instrumentalist, since she either has to admit that instrumentalism is no better than realism in these terms or admit other explanatory virtues into the framework, in which case she concedes the point to realism. But, of course, the knife cuts both ways: The realist also needs to break the underdetermination, and if she does so by insisting on these other virtues, then she will be accused of question begging. And, of course, there is no possibility of her resorting to the standard realist ploy (see later) of adopting a "wait and see" attitude—the extension of the theories to cover new phenomena may reveal one to be more adequate than the other—since there can be no "new" phenomena at this level! Perhaps one can avoid the underdetermination in the first place with some alternative between realism and instrumentalism. Fine's "Natural Ontological Attitude" represents just such an attempt but is flawed in various respects.

17. Worrall gives the example of Newton's explanation of planetary orbits in terms of the Law of Universal Gravity, which also explained Halley's comet, the oblateness of the Earth, and other matters (1996).

18. Putnam himself seems to have retreated to a weaker position in suggesting that realism is akin to an empirical hypothesis in the sense that it could be false and that it can be supported by "facts," but that doesn't mean that realism is *scientific* (1978). Even this may be too strong: In what sense could realism be regarded as "false"? In what sense could the success of science be regarded as a "fact"? Or, at least, in what sense can it be regarded as a "fact" on a par with the "facts" of science that can falsify theories?

19. Dewey was attacked for proposing a form of idealism, but as James points out, "His account of knowledge is not only absurd, but meaningless, unless independent existences be there of which our ideas take account, and for the transformation of which they work" (James 1932, p. xvii).

20. As we have just seen, for Dewey they are resolved by "fixing" the situation (for a comparison, see Smith 1978, pp. 97–98).

21. Note that there can be no procedure for resolving skeptical doubts since any such procedure would itself be subject to doubt.

22. These are, of course, different from "hypothetical doubts" arising in scientific practice, which Peirce agreed were genuine.

23. It might be viewed as akin to the realist's (somewhat violent) response to the social constructivist, which consists of dropping a piece of lab equipment on the latter's foot!

24. There is a naturalist element in Peirce's thinking in that he sometimes regarded philosophical inquiry as an extension of the scientific method, so that science and philosophy could be regarded as continuous (see Smith 1978, p. 32).

25. Thus the constructive empiricist would reject Peirce's antinominalism.

26. This is, of course, strongly reminiscent of certain moves made by Schlick, Feigl, and Carnap, for example, in responding to earlier concerns in the debate between realists and instrumentalists (see the discussion in Psillos 1999, chapter 3). Of course, it is not at all clear that this debate maps so smoothly onto the current realist-antirealist debate.

27. There is still a problem, however. Should our epistemic values be equated with our cognitive values, as seems implicit in this discussion? If they are, then one could argue that all such values ultimately reduce to those cognitive values that pertain to *observation*. But in that case it is not clear what distinguishes the pragmatic realist from the constructive empiricist, who likewise argues that what is observable is what is observable *for us*, as an epistemic community delineated by our particular cognitive faculties (van Fraassen 1980). On the other hand, if they are not equated, then the notion of an "epistemic value" may be too weak to block a form of relativism. Even if we agree upon a core of transcultural values, what about contentious ones such as simplicity, for example? If different scientific communities adopt different measures of simplicity, then they may end up ranking different theories as "the best" and hence as claiming the world is different *for them*. Pragmatic realism could use some further development on these points.

28. For an illuminating discussion of the history of structuralism, especially as it appears in the views of physicists, as well as a critique of modern forms of structural realism, see van Fraassen (forthcoming).

29. Thus this form of structural realism is tied to the sort of "cumulativist" approach to science we have defended here; see again Post 1971.

30. And indeed, mathematical systems such as topology, ordered systems, game theory, and finite geometry can cope with qualitative ideas. The point is that structural realism need not be tied to highly mathematized theories only.

31. The role of Ramsey sentences in expressing the structural aspects of theories is well known and was famously adopted by Carnap, for example (Psillos 1999, pp. 48–58). It should come as no surprise, therefore, to find elements of structuralism in Carnap's work, as when he refers to the particular structure of the domain of the theoretical language and writes that "the structure can be uniquely specified but the elements of the structure cannot. Not because we are ignorant of their nature; rather because there is no question of their nature" (Carnap 1956, p. 46). Carnap, of course, advocated a form of "neutralism" that regarded the conflict between realism and instrumentalism as essentially linguistic (Psillos 1999, pp. 58–63). The essential idea lies in the claim that from the perspective of the representation of theories in terms of Ramsey sentences, to assert the existence of electrons is just to assert the truth of quantum mechanics (see our previous discussion), where the former is understood as the assertion of the existence of certain kinds of events that are called "electrons" in the theoretical language. This might be viewed as a form of "structural empiricism" (see, for example, Bueno 1999a) rather than structural realism, and we shall consider the former in a little more detail later (Psillos argues that it, too, falls prey to the Newman problem [1999, pp. 63–67]).

32. At a symposium on structural realism at the British Society for the Philosophy of Science annual conference, Worrall sought to defend his Ramsey sentence approach to structural realism against the Newman problem by appeal to the distinction between observational and theoretical terms, articulated syntactically (Worrall 2000).

33. Post refers to this case as an example of what he calls "inconsistent correspondence," since classical mechanics agrees only approximately with the relativistic form, in the sense that the latter asymptotically converges to the former in the limit and the former asserts a proposition that only agrees with the latter in that limit (1971, p. 243). For further discussion, see Pagonis 1996.

34. Bueno has suggested that allowing for approximate correspondence may fatally weaken structural realism, since it apparently grants that there may be structural *losses*, in which case a form of pessimistic meta-induction may be reinstated (private discussion). This is an important point. One way of dealing with it would be to insist that not all structures get carried over, as it were, but only those that are genuinely *explanatory*. We could then avail ourselves of Post's historically based claim that there simply are no "Kuhn-losses," in the sense of successor theories losing all or part of the explanatory structures of their predecessors (1971, p. 229).

35. As Domski has emphasized, Poincaré may not be the most suitable forefather to claim for this form of structural realism, given his Kantian inclinations and rejection of truth as the aim of science (Domski preprint). She offers the early Schlick as a more suitable candidate. If theoretical content is cashed out in terms of the nature of "real objects," understood in a Kantian sense, then the realist side of SR may be radically transformed, to say the least.

36. Boltzmann, for example, wrote such an understanding of objects into Axiom I of his Mechanics.

37. Some realists have insisted that they need not be committed to such an understanding, but then it is not clear what the content of their realism is. They are presumably committed to the existence of electrons, say, and to the claim that these electrons possess certain properties such as charge and spin, but they cannot say what these electrons *are*, fundamentally. This amounts to little more than the insistence that there is something "out there," but we know not what, metaphysically speaking. Ladyman refers to this as "ersatz realism" (Ladyman 1998).

38. If the realist refuses to be drawn on the metaphysics at least at the level of individuality versus nonindividuality, then how are we supposed to make sense of the impact of quantum mechanics?

39. Elements of such a view can be found earlier in the history of structuralism, in the work of Cassirer and Eddington, for example (for further details, see French and Ladyman forthcoming; French forthcoming a).

40. Yet again one can detect something akin to this idea in Apostel's work, where he suggests that we might define the notions of element and relation in terms of that of "system," taken as primitive (1961, pp. 20–21). Thus, "The definition of what is to be called element or relation should . . . be a function of the intersection of systems" (p. 21).

41. In a further program of research, Bueno has extended the "ethos" of constructive empiricism into the philosophy of mathematics by arguing that, in its application to science, mathematics does not need to be true, but only quasi-true (1999b). Shifting—once again—to the extrinsic perspective, the applicability of mathematics can be analyzed in terms of partial homomorphisms holding between

mathematical and empirical (partial) structures (for further details, see French 1999 and 2000; Bueno and French 1999).

Chapter 9

1. We are grateful to Alirio Rosales for recent discussions on this topic.

2. Similarly, unification in science can also be considered from the model-theoretic perspective, and again the partial structures approach might be put to good use here (a preliminary attempt can be found in da Costa 1989).

3. Just as in the case of inconsistency in science, nonclassical logic has been turned to here as well. Barnes and Bloor, for example, have suggested, "It is ironic that logicians, who expose with admirable ruthlessness how problematic, variable and difficult to ground patterns of inference are, and who freely confess how very little is agreed upon by the totality of practitioners in their field, are turned to again and again to provide constraints upon the possibilities of rational thought. Just as there is always a certain demand for iron laws of economics, so there seems always to be a demand for iron laws of logic" (1982, p. 45, fn. 40). For a consideration of the possibility of applying paraconsistent logic to the Zande case, see da Costa, Bueno, and French 1998b.

References

Achinstein, P. (1968), *Concepts of Science: A Philosophical Analysis*, Johns Hopkins Press.

Achinstein, P. (1991), "Maxwell's Analogies and Kinetic Theory," in P. Achinstein, *Particles and Waves: Historical Essays in the Philosophy of Science*, Oxford University Press: 207–232.

Addison, J. W. (1965), "Some Notes on the Theory of Models," in J. W. Addison, L. Henkin, and A. Tarski (eds.), *The Theory of Models*, North-Holland: 438–441.

Alcoff, L. (1986), "Charles Peirce's Alternative to the Skeptical Dilemma," *Auslegung* **13**: 6–19.

Almeder, R. (1987), "Blind Realism," *Erkenntnis* **26**: 57–101.

Alves, E. H. (1984), "Paraconsistent Logic and Model Theory," *Studia Logica* **43**: 17–32.

Anderson, R., and Joshi, G. C. (1993), "Quaternions and the Heuristic Role of Mathematical Structures in Physics," *Physics Essays* **6**: 308–319.

Apostel, L. (1961), "Towards the Formal Study of Models in the Non-Formal Sciences," in H. Freudenthal (ed.), *The Concept and the Role of the Model in Mathematics and Natural and Social Sciences*, D. Reidel: 1–37.

Arruda, A. I. (1980), "A Survey of Paraconsistent Logic," in A. I. Arruda, R. Chuaqui, and N. C. A. da Costa (eds.), *Mathematical Logic in Latin America*, North-Holland: 1–42.

Arruda, A. I. (1989), "Aspects of the Historical Development of Paraconsistent Logic," in G. Priest, R. Routley, and J. Norman (eds.), *Paraconsistent Logic: Essays on the Inconsistent*, Philosophia, Munich: 99–130.

Baird, D. (1995), "Meaning in a Material Medium," in D. Hull, M. Forbes, and R. M. Burian (eds.), *PSA 1994*, Vol. 2, Philosophy of Science Association: 441–451.

Barnes, B. (1977), *Interests and the Growth of Knowledge*, Routledge and Kegan Paul.

Barnes, B., and Bloor, D. (1982), "Relativism, Rationalism and the Sociology of Knowledge," in M. Hollis and S. Lukes (eds.), *Rationality and Relativism*, MIT Press: 21–47.

Barnes, S. B. (1968), "Paradigms, Scientific and Social," *Man* **4**: 94–102.

Bartelborth, T. (1989), "Is Bohr's Model of the Atom Inconsistent?" in P. Weingartner and G. Schurz (eds.), *Philosophy of the Natural Sciences, Proceedings of the 13th International Wittgenstein Symposium*, HPT: 220–223.

Barwise, J., and Perry, J. (1983), *Situations and Attitudes*, MIT Press.

Beth, E. W. (1949), "Towards an Up-to-Date Philosophy of the Natural Sciences," *Methodos* **1**: 178–185.

Black, M. (1962), *Models and Metaphors*, Cornell University Press.

Black, M. (1970), *Margins of Precision*, Cornell University Press.

Bloor, D. (1983), *Wittgenstein: A Social Theory of Knowledge*, Columbia University Press.

Bloor, D. (1992), "Left and Right Wittgensteinians," in A. Pickering (ed.), *Science as Practice and Culture*, University of Chicago Press: 266–282.

Bogdan, R. J. (1986), "The Importance of Belief," in R. J. Bogdan (ed.), *Belief: Form, Content and Function*, Oxford University Press: 1–16.

Bogen, J., and Woodward, J. (1988), "Saving the Phenomena," *Philosophical Review* **12**: 303–352.

Bohr, N. (1981), *Niels Bohr Collected Works*, Vol. 2, ed. U. Hoyer, North-Holland.

Bohr, N., and Wheeler, J. A. (1939), "The Mechanism of Nuclear Fission," *Physical Review* **56**: 426–450.

Bourbaki, N. (1968), *The Theory of Sets*, Addison-Wesley.

Boyd, R. (1973), "Realism, Underdetermination and a Causal Theory of Evidence," *Nous* **7**: 1–12.

Boyd, R. (1984), "The Current Status of Scientific Realism," in J. Leplin (ed.), *Scientific Realism*, University of California Press: 41–82.

Bradley, F. (1962), *Appearance and Reality*, Oxford University Press.

Braithwaite, R. (1953), *Scientific Explanation*, Cambridge University Press.

Braithwaite, R. B. (1962), "Models in the Empirical Science," in E. Nagel, P. Suppes, and A. Tarski (eds.), *Logic, Methodology and Philosophy of Science*, Stanford University Press: 224–231.

Brown, B. (1990), "How to Be Realistic About Inconsistency in Science," *Studies in History and Philosophy of Science* **21**, 281–294.

Brown, B. (1993), "Old Quantum Theory: A Paraconsistent Approach," *PSA 1992*, Vol. 2, Philosophy of Science Association: 397–411.

Budd, M. (1993), "How Pictures Look," in D. Knowles and J. Skorupski (eds.), *Virtue and Taste*, Blackwell: 154–175.

Bueno, O. (1997), "Empirical Adequacy: A Partial Structures Approach," *Studies in History and Philosophy of Science* **28**: 585–610.

Bueno, O. (1999a), "What Is Structural Empiricism? Scientific Change in an Empiricist Setting," *Erkenntnis* **50**: 59–85.

Bueno, O. (1999b), *Philosophy of Mathematics: A Structural Empiricist View*, Ph.D. Thesis, University of Leeds.

Bueno, O. (2000), "Empiricism, Mathematical Change and Scientific Change," *Studies in History and Philosophy of Science* **31**: 269–296.

Bueno, O., and de Souza, E. (1996), "The Concept of Quasi-Truth," *Logique et Analyse* **153–154**: 183–199.

Bueno, O., and French, S. (1999), "Infestation or Pest Control: The Introduction of Group Theory into Quantum Mechanics," *Manuscrito* **22**: 37–86.

Bueno, O. and French, S. (forthcoming), "Underdetermination and the Openness of the Field of Evidence."

Bueno, O., French, S., and Ladyman, J. (2002), "On Representing the Relationship Between the Mathematical and the Physical," *Philosophy of Science* **69**: 497–518.

Bunge, M. (1973), *Method, Model and Matter*, D. Reidel.
Callebaut, W. (1993), *Taking the Naturalistic Turn*, University of Chicago Press.
Callon, M., and Latour, B. (1992), "Don't Throw the Baby Out with the Bath School!" in A. Pickering (ed.), *Science as Practice and Culture*, University of Chicago Press: 343–368.
Campbell, K. (1994), "Selective Realism in the Philosophy of Physics," *The Monist* **77**: 27–46.
Cao, T. Y. (1997), *Conceptual Developments of 20th Century Field Theories*, Cambridge University Press.
Carnap, R. (1939), *Foundations of Logic and Mathematics*, University of Chicago Press.
Carnap, R. (1942), *Introduction to Semantics*, Harvard University Press.
Carnap, R. (1949), *The Logical Syntax of Language*, Routledge and Kegan Paul.
Carnap, R. (1956), "The Methodological Character of Theoretical Concepts," in H. Feigl and M. Scriven (eds.), *Minnesota Studies in the Philosophy of Science Vol. I*, University of Minnesota Press: 38–76.
Carnap, R. (1958), *Introduction to Symbolic Logic and Its Applications*, Dover.
Carnap, R. (1963), *The Logical Foundations of Probability*, University of Chicago Press.
Carnap, R. (1990), "Testability and Meaning," in R. R. Ammerman (ed.), *Classics of Analytic Philosophy*, Hackett: 136–195; *Philosophy of Science* **3** and **4** (1936 and 1937).
Cartwright, N. (1983), *How the Laws of Physics Lie*, Oxford University Press.
Cartwright, N. (1995), "The Metaphysics of the Disunified World," in D. Hull, M. Forbes, and R. M. Burian (eds.), *PSA 1994*, Vol. 2, Philosophy of Science Association: 357–364.
Cartwright, N. (1996), "Fundamentalism vs. the Patchwork of Laws," in D. Papineau (ed.), *The Philosophy of Science*, Oxford University Press: 314–326.
Cartwright, N. (1999), *The Dappled World: A Study of the Boundaries of Science*, Cambridge University Press.
Cartwright, N., Shomar, T., and Suárez, M. (1995), "The Tool Box of Science: Tools for Building of Models with a Superconductivity Example," in W. E. Herfel et al. (eds.), *Theories and Models in Scientific Processes*, Editions Rodopi: 137–149.
Cassini, A. (1990), "Aspectos Semanticos y Ontologicos de la Justificacion Aristotelica del Principio de no Contradiccion," *Elenchos* **11**: 5–28.
Castellani, E. (1998), "Galilean Particles: An Example of Constitution of Objects," in E. Castellani (ed.), *Interpreting Bodies: Classical and Quantum Objects in Modern Physics*, Princeton University Press: 181–194.
Castellani, E. (1999), *Simmetria e Natura: Dalle armonie delle figure alle invarianze delle leggi*, Roma-Bari, Laterza.
Chang, C. C. (1974), "Model Theory 1945–1971," in L. Henkin et al. (eds.), *Proceedings of the Tarski Symposium*, American Mathematical Society: 173–186.
Cherniak, C. (1984), "Computational Complexity and the Universal Acceptance of Logic," *Journal of Philosophy* **81**: 739–758.
Chisholm, R. (1992), "William James's Theory of Truth," *The Monist* **75**: 569–579.
Cifuentes, J. C. (1992), *O Método dos Isomorfismos Parciais*, Coleção CLE (Centro de Logica).
Cohen, C. J. (1981), "Can Human Irrationality Be Experimentally Demonstrated?" *Behavioral and Brain Sciences* **4**: 317–331.

Cohen, L. J. (1989), "Belief and Acceptance," *Mind* **98**: 367–389.

Collier, J. D. (1992), "Critical Notice: Paul Thompson, *The Structure of Biological Theories*," *Canadian Journal of Philosophy* **22**: 287–298.

Collins, H. M., and Yearley, S. (1992), "Epistemological Chicken," in A. Pickering (ed.), *Science as Practice and Culture*, University of Chicago Press: 301–326.

Culp, S. (1995), "Objectivity in Experimental Inquiry: Breaking Data-Technique Circles," *Philosophy of Science* **62**: 430–450.

Czarnocka, M. (1995), "Models and Symbolic Nature of Knowledge," in W. E. Herfel et al. (eds.), *Theories and Models in Scientific Processes*, Editions Rodopi: 27–36.

Da Costa, N. C. A. (1958), "Nota Sobre o Conceito de Contradição," *Anuário da Sociedade Paranaense de Matemática* **1** (NS): 6–8.

Da Costa, N. C. A. (1974a), "On the Theory of Inconsistent Formal Systems," *Notre Dame Journal of Formal Logic* **11**: 497–510.

Da Costa, N. C. A. (1974b), *Sistemas Formais Inconsistentes*, NEPEC.

Da Costa, N. C. A. (1981), *Logica Indutiva e Probabilidade*, Mathematics and Statistics Institute of the University of Sao Paulo.

Da Costa, N. C. A. (1982), "The Philosophical Import of Paraconsistent Logic," *Journal of Non-Classical Logic* **1**: 1–12.

Da Costa, N. C. A. (1986a), "On Paraconsistent Set Theory," *Logique et Analyse* **115**: 361–371.

Da Costa, N. C. A. (1986b), "Pragmatic Probability," *Erkenntnis* **25**: 141–162.

Da Costa, N. C. A. (1987a), "O Conceito de Estrutura em Ciência," *Boletim da Sociedade Paranaense de Matemática* **8**: 1–22.

Da Costa, N. C. A. (1987b), "Outlines of a System of Inductive Logic," *Teoria* **7**: 3–13.

Da Costa, N. C. A. (1989), "Logic and Pragmatic Truth," in J. E. Fenstad et al. (eds.), *Logic, Methodology and Philosophy of Science VIII*, Elsevier: 247–261.

Da Costa, N. C. A., Béziau, J. Y., and Bueno, O. A. B. (1997), *Teorias Paraconsistente de Conjuntos*, Centro de Lógica.

Da Costa, N. C. A., and Bueno, O. A. B. (1998), Review of Balzer, W., and Moulines, C. V., *Structuralist Theory of Science, History and Philosophy of Logic* **19**: 270–272.

Da Costa, N. C. A., Bueno, O. A. B., and French, S. (1998a), "The Logic of Pragmatic Truth," *Journal of Philosophical Logic* **27**: 603–620.

Da Costa, N. C. A., Bueno, O. A. B., and French, S. (1998b), "Is There a Zande Logic?" *History and Philosophy of Logic* **19**: 41–54.

Da Costa, N. C. A., and Chuaqui, R. (1988), "On Suppes' Set-Theoretical Predicates," *Erkenntnis* **29**: 95–112.

Da Costa, N. C. A., and Doria, F. (1992), "Suppes Predicates for Classical Physics," in J. Echeverria, A. Ibarra, and T. Mormann (eds.), *The Space of Mathematics*, Walter de Gruyter: 168–191.

Da Costa, N. C. A., and Doria, F. (1995), "On Jaskowski's Discussive Logics," *Studia Logica* **54**: 33–60.

Da Costa, N. C. A., and Doria, F. (1996), "Structures, Suppes Predicates and Boolean-Valued Models in Physics," in P. I. Bystrov and V. N. Sadovsky (eds.), *Philosophical Logic and Logical Philosophy*, Kluwer: 91–118.

Da Costa, N. C. A., Doria, F., and de Barros, J. A. (1990), "A Suppes Predicate for General Relativity and Set-Theoretically Generic Spacetimes," *International Journal of Theoretical Physics* **29**: 935–961.

Da Costa, N. C. A., and Dubikajtis, L. (1977), "On Jaskowski's Discussive Logic," in A. I. Arruda, N. C. A. da Costa, and R. Chuaqui (eds.), *Non-Classical Logics, Model Theory and Computability*, North-Holland: 37–56.

Da Costa, N. C. A., and French, S. (1988), "Pragmatic Probability, Logical Omniscience and the Popper-Miller Argument," *Fundamenta Scientae* **9**: 43–53.

Da Costa, N. C. A., and French, S. (1989a), "Pragmatic Truth and the Logic of Induction," *The British Journal for the Philosophy of Science* **40**: 333–356.

Da Costa, N. C. A., and French, S. (1989b), "On the Logic of Belief," *Philosophy and Phenomenological Research* **49**: 431–446.

Da Costa, N. C. A., and French, S. (1989c), "Critical Study of 'In Contradiction,'" *Philosophical Quarterly* **39**: 498–501.

Da Costa, N. C. A., and French, S. (1990a), "The Model-Theoretic Approach in the Philosophy of Science," *Philosophy of Science* **57**: 248–265.

Da Costa, N. C. A., and French, S. (1990b), "Belief, Contradiction and the Logic of Self-Deception," *American Philosophical Quarterly* **27**: 179–197.

Da Costa, N. C. A., and French, S. (1991a), "Consistency, Omniscience and Truth," *Philosophical Science* **8**: 51–69 (Russian translation).

Da Costa, N. C. A., and French, S. (1991b), "On Russell's Principle of Induction," *Synthese* **86**: 285–295.

Da Costa, N. C. A., and French, S. (1993a), "Towards an Acceptable Theory of Acceptance," in S. French and H. Kamminga, *Correspondence, Invariance and Heuristics*, D. Reidel: 137–158.

Da Costa, N. C. A., and French, S. (1993b), "A Model Theoretic Approach to 'Natural Reasoning,'" *International Studies in the Philosophy of Science* **7**: 177–190.

Da Costa, N. C. A., and French, S. (1995), "Partial Structures and the Logic of the Azande," *American Philosophical Quarterly* **32**: 325–339.

Da Costa, N. C. A., and French, S. (2000), "Theories, Models and Structures: Thirty Years On," *Philosophy of Science* **67** (Proceedings): S116–S127.

Da Costa, N. C. A., and French, S. (forthcoming), "Inconsistency in Science," in *Proceedings of the World Congress of Paraconsistency*.

Da Costa, N. C. A., and Krause, D. (1994), "Schrödinger Logics," *Studia Logica* **53**: 533–550.

Da Costa, N. C. A., and Marconi, D. (1987), *An Overview of Paraconsistent Logic in the 80s*, Monografias da Sociedade Paranaense de Matemática (reprinted in *Logica Nova*, Akademie-Verlag, 1988).

Dalla Chiara, M. L., and Toraldo di Francia, G. (1993), "Individuals, Kinds and Names in Physics," in G. Corsi et al. (eds.), *Bridging the Gap: Philosophy, Mathematics, Physics*, Kluwer Academic Press: 261–283.

De Finetti, B. (1970), *Teoria delle Probabilita*, Einaudi.

Demopoulos, W., and Friedman, M. (1985), "Critical Notice: Bertrand Russell's *The Analysis of Matter*: Its Historical Context and Contemporary Interest," *Philosophy of Science* **52**: 621–639.

Domski, M. (preprint), "The Epistemological Foundations of Structural Realism: Poincaré and the Structure of Relations," paper given to the Research Workshop of the Division of History and Philosophy of Science, University of Leeds.

Donovan, A., Laudan, L., and Laudan, R. (eds.) (1988), *Scrutinizing Science: Empirical Studies of Scientific Change*, D. Reidel.

Dorling, J. (1972), "Bayesianism and the Rationality of Science," *British Journal for the Philosophy of Science* **23**: 181–190.

Dorling, J. (1973), "Demonstrative Induction: Its Significant Role in the History of Physics," *Philosophy of Science* **40**: 360–372.
Douven, I. (2001), "A Passion for Realism," in review symposium of S. Psillos, *Scientific Realism: How Science Tracks Truth, Metascience* **10**: 354–359.
Downes, S. M. (1992), "The Importance of Models in Theorizing: A Deflationary Semantic View," *PSA 1992*, Volume 1, Philosophy of Science Association: 142–153.
Dupré, J. (1995), "Against Scientific Imperialism," in D. Hull, M. Forbes, and R. M. Burian (eds.), *PSA 1994*, Vol. 2, Philosophy of Science Association: 374–381.
Earman J. (ed.) (1983), *Testing Scientific Theories*, University of Minnesota Press.
Earman, J. (1992), *Bayes or Bust? A Critical Examination of Bayesian Confirmation Theory*, MIT Press.
Earman, J. (2002), "Gauge Matters," *Philosophy of Science* **69**: S209–S220.
Eddington, A. (1939), *The Philosophy of Physical Science*, Cambridge University Press.
Edidin, A. (1983), "Bootstrapping Without Bootstraps," in J. Earman (ed.), *Testing Scientific Theories*, University of Minnesota Press: 43–54.
Ellis, B. (1990), *Truth and Objectivity*, Cambridge University Press.
Ellis, B. (1996), "What Science Aims to Do," in D. Papineau (ed.), *The Philosophy of Science*, Oxford University Press: 166–193; reprinted from P. Churchland and C. Hooker (eds.), *Images of Science: Essays on Realism and Empiricism, with a Reply by Bas C. van Fraassen.*, University of Chicago Press, 1985: 48–74.
Elton, L. R. B. (1961), *Nuclear Sizes*, Oxford University Press.
English, J. (1973), "Underdetermination: Craig and Ramsey," *Journal of Philosophy* **70**: 453–462.
Ereshefsky, M. (1995), "Pluralism, Normative Naturalism and Biological Taxonomy," in D. Hull, M. Forbes, and R. M. Burian (eds.), *PSA 1994*, Vol. 2, Philosophy of Science Association: 382–389.
Etchemendy, J. (1988), "Tarski on Truth and Logical Consequence," *Journal of Symbolic Logic* **53**: 51–79.
Fernández Moreno, F. (2001), "Tarskian Truth and the Correspondence Theory," *Synthese* **126**: 123–147.
Fine, A. (1984), "The Natural Ontological Attitude," in J. Leplin (ed.), *Scientific Realism*, University of California Press: 83–107.
Fine, A. (1986), "Unnatural Attitudes: Realist and Instrumentalist Attachments to Science," *Mind* **95**: 149–179.
Fine, A. (1991), "Piecemeal Realism," *Philosophical Studies* **61**: 79–96.
Fine, A. (1996), "Afterword," in *The Shaky Game*, University of Chicago Press (2nd ed.).
Fine, A. (1998), "The Viewpoint of No-One in Particular," *Proceedings and Addresses of the American Philosophical Association*, **72**: 19.
Fisher, G. (forthcoming), "Developmental Models and Their Lack of Autonomy," paper presented to the annual meeting of the British Society for the Philosophy of Science, Sheffield, July 2000.
Flanagan, O. (1995), "The Moment of Truth on Dublin Bridge: A Response to Andrew Pickering," *South Atlantic Quarterly* **94**: 467–475.
Folse, H. (1985), *The Philosophy of Niels Bohr: The Framework of Complementarity*, North-Holland.
Franklin, A. (1986), *The Neglect of Experiment*, Cambridge University Press.
Franklin, A. (1995), "Commentary," *PSA 1994*, Vol. 2, Philosophy of Science Association: 452–457.

French, S. (1985), *Identity and Individuality in Classical and Quantum Physics*, PhD thesis, University of London.
French, S. (1988), "Models, Pragmatic Virtues and Limited Scepticism: The Three Pillars of Constructive Empiricism," *Manuscrito* **11**: 27–46.
French, S. (1989a), "A Peircean Approach to the Realism—Empiricism Debate," *Transactions of the Charles S. Peirce Society* **25**, 293–307.
French, S. (1989b), "A Estrutura do Argumento de Newton para a Lei da Gravitação Universal," *Cadernos da Historia e Filosofia da Ciencia* Serie 2, **1**: 33–52.
French, S. (1989c), "Identity and Individuality in Classical and Quantum Physics," *Australasian Journal of Philosophy* **67**: 432–446.
French, S. (1990), "Rationality, Consistency and Truth," *Journal of Non-Classical Logic* **7**: 51–71.
French, S. (1995), "The Esperable Uberty of Quantum Chromodynamics," *Studies in History and Philosophy of Physics* **26**: 87–105.
French, S. (1997), "Partiality, Pursuit and Practice," in M. L. Dalla Chiara et al., *Proceedings of the 10th International Congress on Logic, Methodology and Philosophy of Science* (Dordrecht), Kluwer: 35–52.
French, S. (1999), "Models and Mathematics in Physics: The Role of Group Theory," in J. Butterfield and C. Pagonis (eds.), *From Physics to Philosophy* Cambridge University Press: 187–207.
French, S. (2000), "The Reasonable Effectiveness of Mathematics: Partial Structures and the Application of Group Theory to Physics," *Synthese* **125**: 103–120.
French, S. (forthcoming a), "Symmetry, Structure and the Constitution of Objects," paper presented at the workshop *Symmetries in Physics: New Reflections*, University of Oxford, January 2001; available at www.philosophy.ox.ac.uk/physics/contents.html#4.
French, S. (forthcoming b), "A Model-Theoretic Account of Representation," *Philosophy of Science*.
French, S., and Krause, D. (1999), "The Logic of Quanta," in T. Cao (ed.), *Conceptual Foundations of Quantum Field Theory*, Cambridge University Press: 324–342.
French, S., and Krause, D. (forthcoming), *Quantum Individuality*.
French, S., and Ladyman, J. (1997), "Superconductivity and Structures: Revisiting the London Account," *Studies in History and Philosophy of Modern Physics* **28**: 363–393.
French, S., and Ladyman, J. (1998), "A Semantic Perspective on Idealisation in Quantum Mechanics," in N. Shanks (ed.), *Idealization IX: Idealization in Contemporary Physics: Poznan Studies in the Philosophy of the Sciences and the Humanities*, Rodopi: 51–73.
French, S., and Ladyman, J. (1999), "Reinflating the Semantic Approach," *International Studies in the Philosophy of Science* **13**: 99–117.
French, S., and Ladyman, J. (2003), "Remodelling Structural Realism: Quantum Physics and the Metaphysics of Structure," *Synthese*.
Friedman, M. (1982), "Review of *The Scientific Image*," *Journal of Philosophy* **79**: 274–283.
Friedman, M. (1983), *Foundations of Space-Time Theories*, Princeton University Press.
Frisch, M. (preprint), "The World According to Maxwell," paper given to the Research Workshop of the Division of History and Philosophy of Science, University of Leeds.

Fuller, S. (1986), *Social Epistemology*, Indiana University Press.
Garber, D. (1983), "Old Evidence and Logical Omniscience in Bayesian Confirmation Theory," in J. Earman (ed.), *Testing Scientific Theories*, University of Minnesota Press: 99–132.
Gardenfors, D., and Saklin, N. E. (1982), "Unreliable Probabilities: Risk Taking and Decision Making," *Synthese* **53**: 362–386.
Gavroglu, K. (1995), *Fritz London: A Scientific Biography*, Cambridge University Press.
Giere, R. (1988), *Explaining Science: A Cognitive Approach*, University of Chicago Press.
Giere, R. (1995), "Viewing Science," in D. Hull, M. Forbes, and R. M. Burian (eds.), *PSA 1994*, Vol. 2, Philosophy of Science Association: 3–16.
Gillies, D. (1986), "Discussion: In Defense of the Popper-Miller Argument," *Philosophy of Science* **53**: 110–113.
Gillies, D. (1987), "Probability and Induction," in G. M. R. Parkinson (ed.), *The Encyclopaedia of Philosophy*, Croom-Helm.
Glymour, G. (1980a), *Theory and Evidence*, Princeton University Press.
Glymour, C. (1980b), "Bootstraps and Probabilities," *Journal of Philosophy* **77**: 691–699.
Gorham, G. (1996), "Similarity as an Intertheory Relation," *Philosophy of Science* **63** (Proceedings): S220–S229.
Gotesky, R. (1968), "The Uses of Inconsistency," *Philosophy and Phenomenological Research* **28**: 471–500.
Gracia, J. J. (1988), *Individuality*, State University of New York Press.
Grandy, R. (1986), "Some Misconceptions About Belief," in R. Grandy and R. Warner (eds.), *Philosophical Grounds of Rationality*, Clarendon Press: 317–331.
Gremmen, B. (1995), "The Structure of Scientific Practice," paper read at the 10th International Congress of Logic, Methodology and Philosophy of Science, August 19–25, 1995, Florence, Italy.
Griesemer, J. R. (1990), "Modeling in the Museum: On the Role of Remnant Models in the Work of Joseph Grinnell," *Biology and Philosophy* **5**: 3–36.
Grobler, A. (1995), "The Representational and the Non-Representational in Models of Scientific Theories," in W. E. Herfel, W. Krajewski, I. Niiniluoto, and R. Wójcicki (eds.), *Theories and Models in Science*, Poznan Studies in the Philosophy of the Sciences and the Humanities, Rodopi: 37–48.
Hacking, I. (1967), "Slightly More Realistic Personal Probability," *Philosophy of Science* **34**: 310–316.
Hacking, I. (1983), *Representing and Intervening*, Cambridge University Press.
Halmos, P. R. (1960), *Naive Set Theory*, Van Nostrand.
Hardin, C., and Rosenberg, A. (1982), "In Defence of Convergent Realism," *Philosophy of Science* **49**: 604–615.
Hardwig, J. (1985), "Epistemic Dependence," *Journal of Philosophy* **82**: 335–349.
Hardwig, J. (1991), "The Role of Trust in Knowledge," *Journal of Philosophy* **88**: 693–708.
Harman, G. (1986), *Change in View*, MIT Press.
Harman, P. M. (1982), *Metaphysics and Natural Philosophy*, Harvester.
Hartmann, S. (1995), "Models as a Tool for Theory Construction: Some Strategies of Preliminary Physics," in W. E. Herfel et al. (eds.), *Theories and Models in Scientific Processes*, Editions Rodopi: 49–67.

Hartmann, S. (1996), "The World as a Process: Simulations in the Natural and Social Sciences," in R. Hegselmann et al. (eds.), *Simulation and Modelling in the Social Sciences from the Philosophy of Science Point of View*, Kluwer Academic Publishers.

Hartmann, S. (1999), "Models and Stories in Hadron Physics," in M. Morgan and M. Morrison (eds.), *Models as Mediators*, Cambridge University Press: 326–346.

Hartree, D. R. (1957), *The Calculation of Atomic Structures*, Wiley.

Havas, K. G. (1993), "Do We Tolerate Inconsistencies?" *Dialectica* **47**: 27–35.

Hawking, S. (1980), *Is the End in Sight for Theoretical Physics?* Cambridge University Press.

Hellman, G. (1988), "The Many Worlds Interpretation of Set Theory," *PSA 1988* Vol. 2, 1988: 445–455.

Hendry, R. (1997), "Empirical Adequacy and the Semantic Conception of Theories," in T. Childers, P. Kolár, and V. Svoboda (eds.), *Logica '96: Proceedings of the 10th International Symposium, Prague*, Filosofia: 136–50.

Hendry, R. (1998), "Models and Approximations in Quantum Chemistry," in N. Shanks (ed.), *Idealization in Contemporary Physics*, Editions Rodopi: 123–142.

Herfel, W. (1995), "Nonlinear Dynamical Models as Concrete Construction," in W. E. Herfel et al. (eds.), *Theories and Models in Scientific Processes*, Editions Rodopi: 69–84.

Hesse, M. (1953–54), "Models in Physics," *British Journal for the Philosophy of Science* **4**: 198–214.

Hesse, M. (1970), *Models and Analogies in Science*, Oxford University Press.

Hesse, M. (1974), *The Structure of Scientific Inference*, Macmillan.

Hesse, M. (1980), *Revolutions and Reconstructions in the Philosophy of Science*, Harvester.

Hettema, H. (1995), "Bohr's Theory of the Atom 1913–1923: A Case Study in the Progress of Scientific Research Programmes," *Studies in History and Philosophy of Modern Physics* **26**: 307–323.

Heyde, K. (1990), *The Nuclear Shell Model*, Springer-Verlag.

Heyde, K. (1994), *Basic Ideas and Concepts in Nuclear Physics*, Institute of Physics Publishing.

Hintikka, J., and Niiniluoto, I. (1980), "Axiomatic Foundation for a Logic of Inductive Generalizations," in R. C. Jeffrey (ed.), *Studies in Inductive Logic and Probability, Vol. 2*, University of California Press: 157–182.

Hodges, W. (1986), "Truth in a Structure," *Proceedings of the Aristotelian Society* **86**: 135–151.

Hodges, W. (1993), *Model Theory*, Cambridge University Press.

Hollis, M., and Lukes, S. (eds.) (1982), *Rationality and Relativism*, MIT Press.

Horton, R. (1967), "African Traditional Thought and Western Science," *Africa* **38**: 50–71, 155–187; abridged version in B. R. Wilson (ed.) (1970), *Rationality*, Blackwell.

Horton, R. (1982), "Tradition and Modernity Revisited," in M. Hollis and S. Lukes (eds.), *Rationality and Relativism*, MIT Press: 201–260.

Horwich, P. (1980), "The Dispensability of Bootstrap Conditions," *Journal of Philosophy* **77**: 691–699.

Horwich, P. (1982), *Probability and Evidence*, Cambridge University Press.

Horwich, P. (1983), "Explanations of Irrelevance," in J. Earman (ed.), *Testing Scientific Theories*, University of Minnesota Press: 55–65.

Horwich, P. (1991), "On the Nature and Norms of Theoretical Commitment," *Philosophy of Science* **58**: 1–14.
Howson, C. (1985), "Some Recent Objections to the Bayesian Theory of Support," *British Journal for the Philosophy of Science* **36**: 305–309.
Hoyer, U. (1981), "Introduction," in *Neils Bohr Collected Works*, Vol. 2, ed. U. Hoyer, North-Holland: 3–10.
Hoyningen-Huene, P. (1993), *Reconstructing Scientific Revolutions: Thomas S. Kuhn's Philosophy of Science*, University of Chicago Press.
Huggett, N. (1997), "Identity, Quantum Mechanics and Common Sense," *The Monist* **80**: 118–130.
Hughes, R. I. G. (1993), "Theoretical Explanation," in P. A. French, T. E. Uehling Jr., and H. K. Wettstein (eds.), *Midwest Studies in Philosophy, Vol. 18: Philosophy of Science*, University of Notre Dame Press: 132–153.
Hughes, R. I. G. (1997), "Models and Representation," *Philosophy of Science* **64** (Proceedings): S325–S336.
Hutten, E. H. (1953–54), "The Role of Models in Physics," *British Journal for the Philosophy of Science* **4**: 284–301.
Irzik, G., and Grünberg, T. (1995), "Carnap and Kuhn: Arch Enemies or Close Allies?" *British Journal for the Philosophy of Science* **46**: 285–307.
James, W. (1885), "The Function of Cognition," *Mind* **10**: 27–44.
James, W. (1905), "The Sentiment of Rationality," in W. James, *The Will to Believe*, Longmans, Green: 63–110.
James, W. (1907a), *Pragmatism*, Longmans, Green.
James, W. (1907b), "Professor Pratt on Truth," in W. James, *The Meaning of Truth*, Longmans, Green: 162–179.
James, W. (1907c), "A Word More About Truth," in W. James, *The Meaning of Truth*, Longmans, Green: 136–161.
James, W. (1932), *The Meaning of Truth*, Longmans, Green.
Jammer, M. (1966), *The Conceptual Development of Quantum Mechanics*, McGraw-Hill.
Jaskowski, S. (1948), "Rachunek zdán dla systemóv dedukcyjnych sprzecznych," *Studia Societatis Scientiarum Torunensis*, **Sectio A, I**: 171–172.
Jaskowski, S. (1969), "Propositional Calculus for Contradictory Deductive Systems," *Studia Logica* **24**: 143–157.
Jeffrey, R. (1983), "Bayesianism with a Human Face," in J. Earman (ed.), *Testing Scientific Theories*, University of Minnesota Press: 99–132.
Jeffreys, H. (1961), *Theory of Probability*, Oxford University Press.
Kaiser, M. (1991), "From Rocks to Graphs—The Shaping of Phenomena," *Synthese* **89**: 111–133.
Kaiser, M. (1995), "The Independence of Scientific Phenomena," in W. E. Herfel et al. (eds.), *Theories and Models in Scientific Processes*, Editions Rodopi: 180–200.
Kart, I. (1991), *The Critique of Judgement*, Clarendon Press.
Kaplan, M. (1981), "Rational Acceptance," *Philosophical Studies* **40**: 129–145.
Keynes, J. M. (1921), "*A Treatise on Probability*," in *Collected Writings of John Maynard Keynes*, Macmillan (1973).
Kitcher, P. (1993), *The Advancement of Science*, Oxford University Press.
Klein, M. (1970), *Paul Ehrenfest*, Vol. 1, North-Holland.
Knudsen, O. (forthcoming), "O. W. Richardson and the Electron Theory of Matter, 1901–1916."
Koertge, N. (1968), *A Study of Relations Between Scientific Theories: A Test of the General Correspondence Principle*, Ph.D. Thesis, University of London.

Koertge, N. (1987), "Reflections of Empirical, External and Sociological Studies of Science," in A. Fine et al. (eds.), *PSA 1986*, Vol. 2, Philosophy of Science Association: 152–159.

Koertge, N. (1993), "Ideology, Heuristics and Rationality in the Context of Discovery," in S. French and H. Kamminga, *Correspondence, Invariance and Heuristics*, D. Reidel: 125–136.

Koopman, B. O. (1940), "The Axioms and Algebra of Intuitive Probability," *Annals of Mathematics* **41**: 269–292.

Krause, D., and French, S. (1995), "A Formal Framework for Quantum Non-Individuality," *Synthese* **102**: 195–214.

Kroes, P. (1989), "Structural Analogies Between Physical Systems," *British Journal for the Philosophy of Science* **40**: 145–154.

Kuhn, T. S. (1970), *The Structure of Scientific Revolutions*, University of Chicago Press (1964).

Kuhn, T. S. (1993), "Introduction," *PSA 1992*, Vol. 2, Philosophy of Science Association: 3–5.

Kuipers, T. A. F. (1978), *Studies in Inductive Logic and Rational Expectation*, Reidel.

Kuipers, T. A. F. (1980), "A Survey of Inductive Systems," in R. C. Jeffrey (ed.), *Studies in Inductive Logic, Vol. 2*, University of California Press: 183–192.

Kuipers, T. A. F. (2000), *From Instrumentalism to Constructive Realism*, Kluwer.

Kyburg, H. (1974), *The Logical Foundations of Statistical Inference*, D. Reidel.

Kyburg, H. (1987), "The Hobgoblin," *The Monist* **70**: 141–151.

Ladyman, J. (1998), "What Is Structural Realism?" *Studies in History and Philosophy of Science* **29**: 409–424.

Ladyman, J., Douven, I., Horsten, L., and van Fraassen, B. (1997), "A Defence of van Fraassen's Critique of Abductive Reasoning: A Reply to Psillos," *Philosophical Quarterly* **47**: 305–321.

Lakatos, I. (1970), "Falsification and the Methodology of Scientific Research Programs," in I. Lakatos and A. Musgrave (eds.), *Criticism and the Growth of Knowledge*, Cambridge University Press: 91–195.

Langevin, P., and de Broglie, M. (eds.), (1912), *La Théorie du rayonnement et les quanta*, Gauthier-Villars.

Langholm, T. (1988), *Partiality, Truth and Persistence*, Center for the Study of Language and Information, Stanford.

Latour, B. (1987), *Science in Action*, Harvard University Press.

Laudan, L. (1977), *Progress and Its Problems*, University of California Press.

Laudan, L. (1980), "Why Was the Logic of Discovery Abandoned?" in T. Nickels (ed.), *Scientific Discovery, Logic and Rationality*, D. Reidel: 173–183.

Laudan, L. (1984), "Explaining the Success of Science," in J. Cushing et al. (eds.), *Science and Reality*, Notre Dame University Press.

Laudan, L. (1989), "If It Ain't Broke, Don't Fix It," *British Journal for the Philosophy of Science* **40**: 369–375.

Laudan, L. (1996), "A Confutation of Convergent Realism," in D. Papineau (ed.), *The Philosophy of Science*, Oxford University Press: 107–138; originally published in *Philosophy of Science* (1981), **48**: 19–49.

Laudan, L., and Leplin, J. (1991), "Empirical Equivalence and Underdetermination," *Journal of Philosophy* **85**: 1–23.

Laymon, R. (1985), "Idealizations and the Testing of Theories by Experimentation," in P. Achinstein and O. Hannaway (eds.), *Experiment and Observation in Modern Science*, MIT Press: 147–173.

Laymon, R. (1987), "Using Scott Domains to Explicate the Notions of Approximate and Idealized Data," *Philosophy of Science* **54**: 194–221.

Laymon, R. (1988), "The Michelson-Morley Experiment and the Appraisal of Theories," in A. Donovan et al. (eds.), *Scrutinizing Science*, Kluwer: 245–266.

Laymon, R. (1989), "Cartwright and the Lying Laws of Physics," *Journal of Philosophy* **86**: 353–372.

Levi, I. (1987), "The Demons of Decision," *The Monist* **70**: 193–211.

Lindley, D. V. (1965), *Introduction to Probability and Statistics from a Bayesian Viewpoint*, Cambridge University Press.

Lipton, P. (1996), "Is the Best Good Enough?" in D. Papineau (ed.), *The Philosophy of Science*, Oxford University Press: 93–106.

London, F. (1950), *Superfluids*, Vol. 1, Wiley.

London, F. (1954), *Superfluids*, Vol. 2, Wiley.

London, F., and London, H. (1935), "The Electromagnetic Equations of the Supraconductor," *Proceedings of the Royal Society (London)* **A149**: 71–88.

Lukasiewicz, J. (1971), "On the Principle of Contradiction in Aristotle," *Review of Metaphysics* **24**: 485–509.

Lukes, S. (1970), "Some Problems About Rationality," in B. Wilson (ed.), *Rationality*, Harper and Row: 193–213.

Lynch, M. (1992a), "Extending Wittgenstein: The Pivotal Move from Epistemology to the Sociology of Science," in A. Pickering (ed.), *Science as Practice and Culture*, University of Chicago Press: 215–265.

Lynch, M. (1992b), "From the 'Will to Theory' to the Discursive Collage: A Reply to Bloor's 'Left and Right Wittgensteinians,'" in A. Pickering (ed.), *Science as Practice and Culture*, University of Chicago Press: 283–300.

Mac Lane, S. (1996), "Structure in Mathematics," *Philosophia Mathematica* **4**: 174–183.

Madan, D. B. (1983), "Inconsistent Theories as Scientific Objectives," *Philosophy of Science* **50**: 453–470.

Maher, P. (1990), "Acceptance Without Belief," *PSA 1990*, Vol. 1, Philosophy of Science Association: 381–392.

Maidens, A. (1998), "Idealization, Heuristics and the Principle of Equivalence," in N. Shanks (ed.), *Idealization in Contemporary Physics*, Rodopi: 183–200.

Malament, D. (1995), "Is Newtonian Cosmology Really Inconsistent?" *Philosophy of Science* **62**: 489–510.

Martin, M. W. (1986), *Self-Deception and Morality*, University Press of Kansas.

Masterman, M. (1970), "The Nature of a Paradigm," in I. Lakatos and A. Musgrave (eds.), *Criticism and the Growth of Knowledge*, Cambridge University Press: 59–90.

Mates, B. (1974), "Austin, Strawson, Tarski and Truth," in L. Henkin et al. (eds.), *Proceedings of the Tarski Symposium*, American Mathematical Society: 385–396.

Maxwell, J. C. (1965), *The Scientific Papers of James Clerk Maxwell*, ed. W. D. Niven, Cambridge University Press.

McKinsey, J. C. C., Sugar, A. C., and Suppes, P. (1953), "Axiomatic Foundations of Classical Particle Mechanics," *Journal of Rational Mechanics* **2**: 253–272.

McMullin, E. (1968), "What Do Physical Models Tell Us?" in B. van Rootsellar and J. F. Staal (eds.), *Logic, Methodology and Philosophy of Science III*, North-Holland: 385–400.

McMullin, E. (1976), "The Fertility of Theory and the Unit for Appraisal in Science," in R. S. Cohen et al. (eds.), *Essays in Memory of Imre Lakatos*, Reidel: 395–432.

Mehra, J., and Rechenberg, H. (1982), *Historical Development of Quantum Mechanics*, Springer.
Mele, A. (1987), "Recent Work on Self-Deception," *American Philosophical Quarterly* **24**: 2–17.
Mikenberg, I., da Costa, N. C. A., and Chuaqui, R. (1986), "Pragmatic Truth and Approximation to Truth," *Journal of Symbolic Logic* **51**: 201–221.
Miller, D. (1994), *Critical Rationalism: A Restatement and Defence*, Open Court.
Mitchell, S. (1988), "Constructive Empiricism and Anti-Realism," *PSA 1988*, Vol. 1, Philosophy of Science Association: 174–180.
Morrison, M. (1995), "Unified Theories and Disparate Things," in D. Hull, M. Forbes, and R. M. Burian (eds.), *PSA 1994*, Vol. 2, Philosophy of Science Association: 365–373.
Morrison, M. (1999), "Models as Autonomous Agents," in M. Morgan and M. Morrison (eds.), *Models as Mediators*, Cambridge University Press: 38–65.
Moser, P. K. (1983), "William James' Theory of Truth," *Topoi* **2**: 217–222.
Mostowski, A. (1967), "Review of N. Bourbaki's Théorie des Ensembles," *Mathematical Reviews* **34**: 425–426.
Moulines, C. U. (1976), "Approximate Application of Empirical Theories," *Erkenntnis* **10**: 201–227.
Musgrave, A. (1996), "NOA's Ark: Fine for Realism," in D. Papineau (ed.), *The Philosophy of Science*, Oxford University Press: 45–60; *Philosophical Quarterly* **39** (1989): 383–398.
Nagel, E. (1961), *The Structure of Science*, Hackett.
Nagel, E. (1963), "Carnap's Theory of Induction," in P. A. Schilpp (ed.), *The Philosophy of Rudolf Carnap*, Open Court Press.
Nagel, E., Suppes, P., and Tarski, A. (1962), *Logic, Methodology and the Philosophy of Science (Proceedings of the 1960 International Congress)*, Stanford University Press.
Nersessian, N. (1993), "How Do Scientists Think? Capturing the Dynamics of Conceptual Change in Science," in R. N. Giere (ed.), *Cognitive Models in Science*, Minnesota Studies in the Philosophy of Science, Vol. 15, University of Minnesota Press: 3–44.
Nickles, T. (1977), "Heuristics and Justification in Scientific Research," in F. Suppe (ed.), *The Structure of Scientific Theories* (2nd ed.), University of Illinois Press: 571–589.
Nickles, T. (1987), "Twixt Method and Madness," in N. J. Nersessian (ed.), *The Process of Science*, Martinus Nijhoff: 42–67.
Norton, J. D. (1987), "The Logical Inconsistency of the Old Quantum Theory of Black Body Radiation," *Philosophy of Science* **54**: 327–350.
Norton, J. D. (1993), "A Paradox in Newtonian Gravitation Theory," in *PSA 1992*, Vol. 2, Philosophy of Science Association: 412–420.
Norton, J. D. (1995), "The Force of Newtonian Cosmology: Acceleration Is Relative," *Philosophy of Science* **62**: 511–522.
Norton, J. D., and Bain, J. (forthcoming), "What Should Philosophers of Science Learn from the History of the Electron?" in A. Warwick and J. Buchwald (eds.), *The Electron and the Birth of Microphysics*, MIT Press.
Nowak, L. (1995), "Anti-Realism, (Supra-) Realism and Idealization," in W. E. Herfel, W. Krajewski, I. Niiniluoto, and R. Wójcicki (eds.), *Theories and Models in Science*, Poznan Studies in the Philosophy of the Sciences and the Humanities, Rodopi: 225–242.

Olivé, L. (1985), "Realismo y antirealismo en la concepción semántica de las teorías," *Crítica* **17**: 31–38.
Oliveira, M. B. de (1985), "The Problem of Induction: A New Approach," *British Journal for the Philosophy of Science* **36**: 129–145.
O'Neill, J. (1986), "Formalism, Hamilton and Complex Numbers," *Studies in History and Philosophy of Science* **17**: 351–372.
Pagonis, C. (1996), *Quantum Mechanics and Scientific Realism*, Ph.D. Thesis, Cambridge University.
Pais, A. (1991), *Niels Bohr's Times*, Oxford University Press.
Papineau, D. (1996), "Introduction," in D. Papineau (ed.), *The Philosophy of Science*, Oxford University Press: 1–20.
Peirce, C. S. (1931–35), *Collected Papers of Charles Sanders Peirce*, Vols. 1–6. Harvard University Press.
Peirce, C. S. (1940), *The Philosophy of Peirce*, ed. J. Buchler, Routledge and Kegan Paul.
Pérez Ransanz, A. R. (1985), "El concepto de teoría empírica según van Fraassen," *Crítica* **17**: 3–12.
Perry, J. (1986), "From Worlds to Situations," *Journal of Philosophical Logic* **15** (1986): 83–107.
Pickering, A. (1984), *Constructing Quarks*, Edinburgh University Press.
Pickering, A. (1990), "Knowledge, Practice and Mere Construction," *Social Studies of Science* **20**: 682–729.
Pickering, A. (1992), "From Science as Knowledge to Science as Practice," in A. Pickering (ed.), *Science as Practice and Culture*, University of Chicago Press: 1–26.
Pickering, A. (1995a), *The Mangle of Practice: Time, Agency and Chance*, Chicago University Press.
Pickering, A. (1995b), "Concepts and the Mangle of Practice: Constructing Quaternions," *South Atlantic Quarterly* **94**: 417–465.
Pickering, A. (1995c), "Explanation, Agency and Metaphysics: A Reply to Owen Flanagan," *South Atlantic Quarterly* **94**: 475–480.
Pickering, A. (1995d), "After Representation: Science Studies in the Performative Idiom," *PSA 1994*, Vol. 2: 413–419.
Pickering, A., and Stephanides, A. (1992), "Constructing Quaternions: On the Analysis of Conceptual Practice," in A. Pickering (ed.), *Science as Practice and Culture*; University of Chicago Press: 139–167.
Pierson, R. (1995), "The Epistemic Authority of Expertise," *PSA 1994*, Vol. 1, Philosophy of Science Association: 398–405.
Poincaré, H. (1905), *Science and Hypothesis*, Dover (1952).
Popper, K., and Miller, D. (1983), "A Proof of the Impossibility of Inductive Probability," *Nature* **302**: 687–688.
Popper, K. R. (1940), "What Is Dialectic?" *Mind* **49** (1940): 403–436.
Popper, K. R. (1972), *The Logic of Scientific Discovery*, Hutchinson.
Portides, D. (2000), *Representation Models as Devices for Scientific Theory Applications vs. the Semantic View of Scientific Theories*, Ph.D. thesis, London School of Economics.
Post, H. R. (1971), "Correspondence, Invariance and Heuristics," *Studies in History and Philosophy of Science* **2**: 213–255.
Priest, G. (1987), *In Contradiction*, Martinus Nijhoff.
Priest, G., and Routley, R. (1984), "Introduction: Paraconsistent Logics," *Studia Logica* **43**: 3–16.

Priest, G., Routley, R., and Norman, J. (1989), *Paraconsistent Logic: Essays on the Inconsistent*, Philosophia.

Psillos, S. (1995), "The Cognitive Interplay Between Theories and Models: The Case of 19th Century Optics," in W. E. Herfel et al. (eds.), *Theories and Models in Scientific Processes*, Editions Rodopi: 105–133.

Psillos, S. (1999), *Scientific Realism: How Science Tracks Truth*, Routledge.

Psillos, S. (2000), "The Present State of the Scientific Realism Debate," *British Journal for the Philosophy of Science* **51**: 705–728.

Psillos, S. (2001), "Author's response," in "Quests of a Realist," review symposium of S. Psillos, *Scientific Realism: How Science Tracks Truth*, *Metascience* **10**: 366–371.

Putnam, H. (1975a), *Mathematics, Matter and Method*, Vol. 1, Cambridge University Press.

Putnam, H. (1975b), "Degree of Confirmation and Inductive Logic," in *Mathematics, Matter and Method*, Vol. 1, Cambridge University Press.

Putnam, H. (1978), *Meaning and the Moral Sciences*, Routledge and Kegan Paul.

Putnam, H. (1995), *Pragmatism: An Open Question*, Blackwell.

Quinton, A. (1982), *Thoughts and Thinkers*, Duckworth.

Redhead, M. L. G. (1975), "Symmetry in Intertheory Relations," *Synthese* **32**: 77–112.

Redhead, M. L. G. (1980), "Models in Physics," *British Journal for the Philosophy of Science* **31**: 145–163.

Redhead, M. L. G. (1985), "On the Impossibility of Inductive Probability," *British Journal for the Philosophy of Science* **36**: 185–191.

Redhead, M. L. G. (1993), "Is the End of Physics in Sight?" in S. French and H. Kamminga (eds.), *Correspondence, Invariance and Heuristics*, D. Reidel: 327–341.

Redhead, M. L. G. (2002), "The Interpretation of Gauge Symmetry," in M. Kuhlmann, M. Lyre, and A. Wayne (eds.), *Ontological Aspects of Quantum Field Theory: Proceedings of the Bielefeld Conference on Ontological Aspects of Quantum Field Theory*, World Scientific: 281–301.

Reichenbach, H. (1938), *Experience and Prediction*, University of Chicago Press.

Reichenbach, H. (1949), *The Theory of Probability*, University of California Press.

Reiner, R., and Pearson, R. (1995), "Hacking's Experimental Realism: An Untenable Middle Ground," *Philosophy of Science* **62**: 60–69.

Reisch, G. (1991), "Did Kuhn Kill Logical Positivism?" *Philosophy of Science* **58**: 264–277.

Rescher, N. (1968), *Topics in Philosophical Logic*, Reidel.

Rescher, N. (1973), *The Primacy of Practice*, Blackwell.

Rescher, N. (1978), *Peirce's Philosophy of Science*, University of Notre Dame Press.

Rescher, N., and Brandom, R. (1980), *The Logic of Inconsistency*, Blackwell.

Richardson, O. W. (1916), *The Electron Theory of Matter*, Cambridge University Press.

Rorty, A. (1972), "Belief and Self-Deception," *Inquiry* **5**: 387–410.

Rosenkrantz, R. D. (1977), *Inference, Method and Decision: Towards a Bayesian Philosophy of Science*, Reidel.

Rosenkrantz, R. D. (1983), "Why Glymour Is a Bayesian," in J. Earman (ed.), *Testing Scientific Theories*, University of Minnesota Press: 69–98.

Rueger, A. (1990), "Independence from Future Theories: A Research Strategy in Quantum Theory," in A. Fine, M. Forbes, and L. Wessels (eds.), *PSA 1990 Vol. 1*, PSA: 203–211.

Russell, B. (1912), *The Problems of Philosophy*, Oxford University Press.

Russell, B. (1948), *Human Knowledge*, Simon and Schuster.

Russell, B. (1954), *The Analysis of Matter*, Dover.

Ryle, G. (1971), "Why Are the Calculuses of Logic and Arithmetic Applicable to Reality?" in *Collected Papers, Vol. 2*, Barnes and Noble: 226–233.

Salmon, W. (1966), "The Foundations of Scientific Inference," in R. G. Colodny (ed.), *Mind and Cosmos*, University of Pittsburgh Press: 135–275.

Salmon, W. (1978), "Unfinished Business: The Problem of Induction," *Philosophical Studies* **33**: 1–79.

Sapir, E. (1929), "The Status of Linguistics as a Science," *Language* **5**: 207–214.

Schiffer, S. (1986), "The Real Trouble with Propositions," in R. J. Bogdan (ed.), *Belief: Form, Content and Function*, Oxford University Press: 83–117.

Schotch, P. K. (1993), "Paraconsistent Logic: The View from the Right," *PSA 1992*, Vol. 2, Philosophy of Science Association: 421–429.

Schotch, P. K., and Jennings, R. E. (1980), "Inference and Necessity," *Journal of Philosophical Logic* **9**: 329–340.

Shapere, D. (1969), "Notes Towards a Post-Positivistic Interpretation of Science," in P. Achinstein and S. F. Barker (eds.), *The Legacy of Logical Positivism*, Johns Hopkins University Press: 115–160.

Shapere, D. (1977), "Scientific Theories and Their Domains," in F. Suppe (ed.), *The Structure of Scientific Theories*, University of Illinois Press 1977: 518–565.

Shapere, D. (1982), "The Concept of Observation in Science and Philosophy," *Philosophy of Science* **49**: 485–525.

Shimony, A. (1970), "Scientific Inference," in R. G. Colodny (ed.), *The Nature and Function of Scientific Theories*, University of Pittsburgh Press: 79–172.

Sklar, L. (1977), "Facts, Conventions and Assumptions in the Theory of Space-Time," in J. Earman, C. Glymour, and J. Stachel (eds.), *Foundations of Space-Time Theories*, University of Minnesota Press: 206–274.

Smith, J. (1988a), "Scientific Reasoning or Damage Control: Alternative Proposals for Reasoning with Inconsistent Representations of the World," *PSA 1988*, Vol. 1, PSA: 241–248.

Smith, J. (1988b), "Inconsistency and Scientific Reasoning," *Studies in History and Philosophy of Science* **19**: 429–445.

Smith, J. E. (1978), *Purpose and Thought: The Meaning of Pragmatism*, University of Chicago Press.

Sperber, D. (1975), *Rethinking Symbolism*, trans. A. Morton, Cambridge University Press.

Sperber, D. (1982), "Apparently Irrational Beliefs," in M. Hollis and S. Lukes (eds.), *Rationality and Relativism*, MIT Press: 149–180.

Stachel, J. (1983), "Comments on 'Some Logical Problems Suggested by Empirical Theories' by Professor Dalla Chiara," in R. S. Cohen and M. W. Wartofsky (eds.), *Language, Logic and Method*, D. Reidel: 91–102.

Stalnaker, R. (1986), "Possible Worlds and Situations," *Journal of Philosophical Logic* **15**: 109–123.

Stalnaker, R. C. (1987), *Inquiry*, MIT Press.

Stegmüller, W. (1979), *The Structuralist View of Theories*, Berlin.

Stich, S. (1984), "Could Man Be an Irrational Animal?" *Synthese* **64**: 115–135.

Suárez, M. (1999a), "Theories, Models and Representations," in M. Morgan and M. Morrison (eds.), *Models as Mediators*, Cambridge University Press: 168–196.

Suárez, M. (1999b), "Theories, Models and Representations," in L. Magnani, N. J. Nersessian, and P. Thagard (eds.), *Model-Based Reasoning in Scientific Discovery*, Kluwer: 75–83.

Suárez, M. (preprint), "Idealization and the Semantic View of Scientific Theories."

Suppe, F. (1977), *The Structure of Scientific Theories*, University of Illinois Press.

Suppe, F. (1989), *Scientific Realism and Semantic Conception of Theories*, University of Illinois Press.

Suppe, F. (2000), "Understanding Scientific Theories: An Assessment of Developments 1969–1998," *Philosophy of Science* **67** (Proceedings): S102–S115.

Suppes, P. (1957), *Introduction to Logic*, Van Nostrand.

Suppes, P. (1961), "A Comparison of the Meaning and Uses of Models in Mathematics and the Empirical Sciences," in H. Freudenthal (ed.), *The Concept and the Role of the Model in Mathematics and Natural and Social Sciences*, D. Reidel: 163–177.

Suppes, P. (1962), "Models of Data," in E. Nagel, P. Suppes, and A. Tarski (eds.), *Logic, Methodology and the Philosophy of Science: Proceedings of the 1960 International Congress*. Stanford University Press: 252–267.

Suppes, P. (1967), "What Is a Scientific Theory?" in S. Morgenbesser (ed.), *Philosophy of Science Today*, Basic Books: 55–67.

Suppes, P. (1968), "The Desirability of Formalization in Science," *Journal of Philosophy* **65**: 651–664.

Suppes, P. (1969), *Studies in the Methodology and Foundations of Science*, D. Reidel.

Suppes, P. (1970), *Set-Theoretical Structures in Science* (mimeograph), Stanford University.

Suppes, P. (1988), "Philosophical Implications of Tarski's Work," *Journal of Symbolic Logic* **53**: 80–91.

Tarski, A. (1935), "The Concept of Truth in Formalized Languages," in A. Tarski, *Logic, Semantics, Metamathematics: Papers from 1923 to 1938*, Clarendon Press (1956): 152–278.

Tarski, A. (1936), "The Establishment of Scientific Semantics," in A. Tarski, *Logic, Semantics, Metamathematics: Papers from 1923 to 1938*, Clarendon Press (1956): 401–408.

Tarski, A. (1953), "A General Method in Proofs of Undecidability," in A. Tarski, A. Mostowski, and R. M. Robinson (eds.), *Undecidable Theories*, North-Holland.

Teller, P. (1975), "Shimony's A Priori Arguments for Tempered Personalism," in G. Maxwell and R. M. Anderson (eds.), *Induction, Probability and Confirmation*, University of Minnesota Press: 166–203.

Teller, P. (1995), *An Interpretive Introduction to Quantum Field Theory*, Princeton University Press.

Thayer, H. S. (1983), "James' Theory of Truth: A Reply," *Topoi* **2**: 223–226.

Thompson, P. (1989), "Explanation in the Semantic Conception of Theory Structure," *PSA 1988* Vol. 2: 286–296.

Toulmin, S. (1953), *The Philosophy of Science*, Hutchinson.

Truesdell, C. (1984), "Suppesian Stews," in *An Idiot's Fugitive Essays on Science: Methods, Criticism, Training, Circumstances*, Springer Verlag: 503–579.

Turney, P. (1990), "Embeddability, Syntax, and Semantics in Accounts of Scientific Theories," *Journal of Philosophical Logic* **19**: 429–451.

Ullian, J. S. (1990), "Learning and Meaning," in R. B. Barrett and R. F. Gibson (eds.), *Perspectives on Quine*, Blackwell: 336–346.

Van Fraassen, B. (1970), "On the Extension of Beth's Semantics of Theories," *Philosophy of Science* **37**: 325–334.

Van Fraassen, B. (1977), "Discussion," in F. Suppe (ed.), *The Structure of Scientific Theories* (2nd ed.), University of Illinois Press: 598–599.
Van Fraassen, B. (1980), *The Scientific Image*, Oxford University Press.
Van Fraassen, B. (1983), "Glymour on Evidence and Explanation," in J. Earman (ed.), *Testing Scientific Theories*, University of Minnesota Press: 25–42.
Van Fraassen, B. (1985a), "Empiricism in the Philosophy of Science," in P. Churchland and C. Hooker (eds.), *Images of Science*, University of Chicago Press: 245–308.
Van Fraassen, B. (1985b), "On the Question of Identification of a Scientific Theory," *Crítica* **17**: 21–25.
Van Fraassen, B. (1989), *Laws and Symmetry*, Oxford University Press.
Van Fraassen, B. (1991), *Quantum Mechanics: An Empiricist View*, Oxford University Press.
Van Fraassen, B. (1994), "Interpretation of Science: Science as Interpretation," in J. Hilgevoord (ed.), *Physics and Our View of the World*, Cambridge University Press: 169–187.
Van Fraassen, B. (1997), "Structure and Perspective: Philosophical Perplexity and Paradox," in M. L. Dalla Chiara et al. (eds.), *Logic and Scientific Methods*, Kluwer Academic Publishers: 511–520.
Vanquickenborne, L. G. (1991), "Quantum Chemistry of the Hydrogen Bond," in P. L. Huyskens, W. A. P. Luck, and T. Zeegers-Huyskens (eds.), *Intermolecular Forces: An Introduction to Modern Methods and Results*, Springer-Verlag: 31–53.
Vaught, R. L. (1974), "Model Theory Before 1945," in L. Henkin et al. (eds.), *Proceedings of the Tarski Symposium*, American Mathematical Society: 153–172.
Von Neumann, J. (1955), *Mathematical Foundations of Quantum Mechanics*, Princeton University Press (1932).
Von Weizsäcker, C. F. (1935), "Zur Theorie der Kernmassen," *Zeitschrift für Physik* **96** (1935): 431–458.
Von Wright, G. H. (1951), *A Treatise on Induction and Probability*, Harcourt.
Whitt, L. A. (1990), Theory Pursuit: Between Discovery and Acceptance," in A. Fine, M. Forbes, and L. Wessels (eds.), *PSA 1990, Vol. 1*, PSA: 467–483.
Wittgenstein, L. (1958), *Philosophical Investigations*, trans. G. E. M. Anscombe, Blackwell.
Wojcicki, R. (1974), "Set Theoretic Representations of Empirical Phenomena," *Journal of Philosophical Logic* **3**: 337–343.
Wojcicki, R. (1975), "The Factual Content of Empirical Theories," in J. Hintikka (ed.), *Rudolf Carnap, Logical Empiricist*, Reidel: 95–122.
Woodward, J. (1989), "Data and Phenomena," *Synthese* **79**: 393–472.
Worrall, J. (1984), "The Background to the Forefront," *PSA 1984*, Vol. 2, Philosophy of Science Association.
Worrall, J. (1996), "Structural Realism: The Best of Both Worlds?" in D. Papineau (ed.). *The Philosophy of Science*, Oxford University Press: 139–165 (originally published in *Dialectica* [1989] **43**: 99–124).
Worrall, J. (2000), "Miracles and Models: Saving Structural Realism?" paper given to the Annual Meeting of the British Society for the Philosophy of Science, Sheffield.
Wrinch, D., and Jeffreys, H. (1921), "On Certain Fundamental Principles of Scientific Inquiry," *Philosophical Magazine* **42**: 369–390.
Wylie, A. (1986), "Arguments for Scientific Realism: The Ascending Spiral," *American Philosophical Quarterly* **23**: 287–297.

Index